# Climate Change Coaching

# Climate Change Coaching

## The Power of Connection to Create Climate Action

*Charly Cox*
*Sarah Flynn*

Open University Press

Open University Press
McGraw Hill
8th Floor, 338 Euston Road
London
England
NW1 3BH

email: enquiries@openup.co.uk
world wide web: www.openup.co.uk

Executive Editor: Eleanor Christie
Editorial Assistant: Zoe Osman
Content Product Manager: Ali Davis

A catalogue record of this book is available from the British Library

ISBN-13: 9780335250059
ISBN-10: 033525005X
eISBN: 9780335250066

Library of Congress Cataloging-in-Publication Data
CIP data applied for

Typeset by Transforma Pvt. Ltd., Chennai, India

# Praise page

*"Climate change and coaching are made for each other. While activists can raise general awareness, coaches are key to enabling leaders to focus on making climate change happen. Future generations will judge the coaches and coaching organisations of today on how we meet that challenge."*

*David Clutterbuck, Special Ambassador,*
*European Mentoring and Coaching Council*

*"Have you ever had a conversation that inspired and energised you to play your part well in responding to an emergency? If so, perhaps the person you were talking with has read this book or has drawn on the understanding it shares. The human change dimension is so often a missing piece of the jigsaw in responding to our climate crisis. This practical guide fills many of the gaps, with tools that can help us become better allies to each other in supporting the personal and systemic shifts needed in our time."*

*Chris Johnstone, co-author of Active Hope and*
*trainer at ActiveHope.Training*

*"Climate Change Coaching is an invaluable resource for anyone working in sustainability today. It provides practical techniques that get us to the heart of what is holding back progress, and tools to enrol others in our vision of change. For professionals engaged in tackling the biggest challenge ever faced by humanity, it also teaches us how to coach ourselves to build our resilience and prevent burnout. Climate Change Coaching gives us powerful relational tools to complement our technical abilities, to help us move faster and farther, together."*

*Patrick Burgi, Co-Founder of South Pole*

For Dimitri, Nina and Phoebe, and all of our team at the
Climate Change Coaches – CC

For my husband and children … and for those working
to change the world – SF

# Contents

## PART A: What on earth does coaching have to do with climate change?

## PART B: Transform the way you communicate

## PART C: A coaching approach in systems change

# Preface

In 2007, when I worked in Sierra Leone, West Africa, I was sent to interview farmers who had resolved a local conflict. A group of cattle herders had wandered their cows into arable fields, where the cows had eaten the farmers' crops. Machetes were brandished, and people were injured. This was seen as a farming issue and the conflict resolution charity that had commissioned me had helped both sides to coexist more peacefully. However, behind this story was an untold tale of climate change, because the reason that the herders had moved from their usual spot was because it was parched and no longer fertile. This is the reason that I wanted to write this book.

As the climate crisis worsens, we may see far more examples of livelihoods colliding, people suffering and homes vanishing. Many of us did not know to call it climate change when I sat with those farmers in that village, but we do now, and because we know, we need to do something about it. I became a coach because I saw that the difference between success and failure for my team in Freetown was their confidence in themselves and their ability to make good decisions. From this flowed the choices they made around work, education, healthcare and their rights. As a leadership coach in the UK, I discovered that confidence is at the heart of all change and that our powerlessness holds us back from taking responsibility. When I connected the dots between our feelings of powerlessness, the skills of coaching and the issue of climate change, it was like realizing the real reason that the farmers were migrating all over again. I felt compelled to do something with that knowledge.

In 2019, I found myself at the International Coach Federation UK Conference, addressing colleagues on coaching's role in climate action. I was not a formal speaker, rather I had been allowed a 15-minute open mic slot during the lunch break. People were restless, the other speakers ran over and I had just 7 minutes to stake my claim. To say that the room was underwhelmed would be kind; they looked bemused. But several people found me afterwards and reassured me that this was something we could all care about. One of those people was Laura Pacey from Open University Press, who asked me there and then to write this book. I am not too proud to admit that I thought she would change her mind when she realized I did not have anything to write about. Now 150,000 words later, it turns out that Laura was right, and I was wrong.

This book was no solo effort, however. It is the result of the work of a whole cast of characters, whom we have tried to thank in our acknowledgements at the end. The most important person in bringing this book to life is my colleague in the Climate Change Coaches, and early collaborator, Sarah Flynn. Sarah brought her background in psychology and her passion for coaching to this book, holding nothing back and seeking out new ideas and research with an enthusiasm that kept us both excited. Sarah's insights underpin so many of the best ideas in this book, and her fingerprint is on many of the pages. Sarah's hard work over the 2 years that it took us to produce this book meant a lot of sacrifice, for which I am deeply grateful. One sacrifice that neither of us wanted to make, however, was to our health, and so just before we started recruiting contributors and editing our raw copy into something more polished, Sarah very sadly needed to stop work in order to look after herself. At that point, I was very fortunate to be joined by another of our Climate Change Coaches team, Emily Buchanan, who as well as being an exceptional coach, happily has a professional background in book editing. Emily's red pen and her understanding of the subject matter not only significantly improved the following text, but her support as an informed sounding board was invaluable.

I wanted to write this book to put the skills that we use as coaches into the hands of as many people as possible. We know that not everyone can train to be a coach, or even afford their own, and yet we also know that these skills are both easy to grasp and powerful when used. Fundamental to this book is the idea of being in better relationships, with each other, with our communities and organizations, with nature and even with the issue of climate change itself. Relationships are not just nice to have, in big systems change they are key to creativity, motivation and commitment. We talk a lot in this book about the power of connection and the danger of isolation when it comes to taking climate action. In much the same way that the dynamism, ideas and encouragement of our team at the Climate Change Coaches took that idea from a sticky on my wall to a fully fledged business, so too the writing of this book was only made possible by the energy that comes from being part of a team, of writers, editors and contributors. Everyone in this book can take pride in the fact that you are reading it today.

While in 2019, it felt like there were a smattering of coaches openly talking about climate change, just a year later I was headlining the final day of the same ICF UK conference, to an audience of over 500 people. The Climate Coaching Alliance also formed as an umbrella body. Meanwhile, many in the climate movement are recognizing the importance of our emotional responses to the crisis and the need to build relationships to galvanize action. Katharine Hayhoe, Per Espen Stoknes and Kimberly Nicholas, who kindly wrote the foreword to our book, and all of our contributors, who are at the forefront of climate action, are borrowing from social psychology to be even more effective.

We founded the Climate Change Coaches because we had a hunch that you could coach climate change, just as you could any other topic, and we were amazed at the early results we got from testing that hypothesis on the general public. We knew that we were sitting on something powerful and that we needed to teach others to do it too. Now, every day we train sustainability professionals, environmental scientists and community activists who tell us how useful a coaching approach is in their work, and how it also benefits their resilience. In this book we bring together coaches and practitioners from across the climate world, to show you practically how you can use these skills, in the same way that they are. Please do get in touch to let us know how you got on.

Thank you for choosing this book. We hope that it will not only give you a helpful new perspective on climate action, but that you may dip in and out of it, as a useful reference whenever you need it, long into the future.

Charly Cox
Oxford, UK, 2021

# Foreword

This book is an empowering guide to helping all of us to use our strengths to bring about a world where people and nature can thrive. Coaching can meet a huge need to make this dream a reality. The personal, ongoing relationship that a climate change coach, or someone using climate coaching skills, can have, is useful to hold people accountable, to help them identify action for themselves and to work out where their values are in relation to climate change. Coaching can also help people to identify their strengths and how they can apply them to best be of service. And serve we must because climate change threatens the future of life on earth. The data say that if we do not act, in our lifetimes we will not recognize our planet. We are the last possible people who can avert really catastrophic outcomes for ourselves, our children, everybody else's children, indeed all of humanity and the 8 million species with whom we share the planet, for countless generations to come.

Just as the diamonds that contain it, carbon is forever. Some of the carbon we emit today will last twice as long as the Great Pyramids or Stonehenge have been in existence. With our carbon pollution, we are leaving an incredibly long-lasting, important legacy in the atmosphere, and that legacy sets the thermostat for life on Earth. That thermostat determines the basis of our lives and determines what crops can grow, and where, who can live, and where, and how they can make a living. It determines the way we live, both for our human societies and for all the species and ecosystems with which we share the planet. Everything we love and care about depends on a stable climate. In many cases we are already feeling the harm. We know now that limiting global warming to 1.5 degrees Celsius is really the difference between life and death for many people and places around the world. There are limits to what systems can tolerate, and we will exceed them if we do not stop warming really fast. To stop warming, we have to completely stop adding carbon to the atmosphere.

That is the hard news. The better news is that we can do something about this if we choose to act and act now. And the really great news is that many of us do want to act. After publishing my own book, *Under the Sky We Make*, the main feedback I received from people was 'I want to make change, but I don't know how to make change happen' and 'I don't know where to start'. Most people know climate change is happening, but when the scale of the challenge is so vast, it is easy to feel insignificant. Coaching exists to help people who say, 'I want to, but I don't know how'. I believe coaching skills have a really powerful role to play in helping people regain a sense of agency, and take action for the climate.

As you'll read here, when facing the need for big systems change, it can be all too easy to feel that your individual action will not make any difference. Businesses and governments have huge responsibility for the bad situation we are in (and some should be held to account for the fact that we are decades behind where we could have been or should have been, according to the science). Governments must set the course to stop burning fossil fuels; companies must make fossil-free business plans. At the same time, the data show us that to meet the Paris Agreement, about two-thirds of the changes we need to make involve individuals making societal and behavioural changes. In other words, stopping climate change is not about either collective or personal action; it is about both collective and personal action.

The changes we need depend on individuals being aware, engaged and involved, both for legitimacy (so that these social changes are seen as democratic and fair) and also for effectiveness, so that they actually happen. It does not do any good if engineers and scientists do their jobs, and then the technology just sits around waiting to be adopted. If it does not get built and used, it cannot replace fossil fuels and reduce emissions. Connection is essential to engaging people with each other and with the issues they care about, and here you will learn what makes and breaks those connections. As human beings, what gives our life meaning and purpose is primarily the relationships that we have with others. Tapping into that sense of meaning is powerful, both to motivate how we engage with being part of climate solutions, and also to connect to what we love and care about and are willing to work to protect and save. Here you will learn not only how to help someone identify their own sense of purpose but also to develop shared purpose. Being in a community with others is what makes it fun, joyful and sustainable (physically and mentally) to do what is sometimes really tough work.

This book will help you to be more effective in galvanizing action, by relating to others more skilfully, and better identifying and working with the different perspectives blocking climate action. Whether you are working with people expressing the scarcity of 'not enough' or others who feel overwhelmed and defeated, this book offers practical coaching skills that can help you shift mind-sets, while building deeper relationships that can support your own resilience and energize those you coach.

I believe that the biggest mindset shift we need is from one of exploitation to one of regeneration. The regeneration mindset is about shifting to an ethos of care and respect for both people and nature, and really putting both at the centre of what we do. I define regeneration in terms of three principles: centring both people and nature, reducing harm at its source and increasing resilience to thrive through change. We can connect to nature both to care for it, but also to allow it to care for us, and here you will learn how. Emphasized throughout this book is the need to fundamentally relate differently to ourselves, to others and to the natural world. I believe that doing so will lead us to a better way of life that will both make us happier, and start to uproot exploitation and thereby tackle the root cause of climate change.

Kimberly Nicholas PhD
(author of *Under the Sky We Make* and Associate Professor of Sustainability Science at Lund University)

# Introduction

In the course of writing we came upon this quote by Arundhati Roy: 'Another world is not only possible, she is on her way. On a quiet day, I can hear her breathing.' For us, this sums up several important principles that you will see repeated here: believing in the possibility, an implicit capability of our world and all of its species, including humans; recognizing that we are already turning towards great change; and most importantly, connecting powerfully and listening deeply to each other, to nature and to ourselves.

When we talk about climate change with others, we do not always manage to do those things. Sometimes we assert our knowledge and lecture, and sometimes we censor ourselves and say little. Very regularly we presume to know what others are thinking and feeling. All of these things are expressions of our love for the world, our mindfulness of our relationships, our fear of what is happening and our desire to turn it around. If you come to this book looking for another way to engage and motivate people to change, then this book is for you.

## Is this book for you?

We aim to put these easy-to-use, quick-to-grasp tools into the hands of people driving climate action and to equip professional coaches to step out of individualized coaching into a more systemic form. Perhaps you are someone who yearns to do more, knows that you need some new tools and suspects the way you engage people is part of the solution. If you want to understand how behaviour change happens, how you can deeply influence others by empowering rather than telling them and specifically how to handle conversations about human change when human life itself is under threat… then you are in the right place. Specifically if you are:

- A sustainability professional – gain the skills to build motivation and set compelling goals, while managing the overwhelm of information.
- A community organizer or activist – learn how to talk more easily about climate with others and how to overcome defeatism.
- A team leading organizational change – understand how to overcome resistance and bring people with you on the change journey.
- A professional coach – gain confidence in your existing skills and broaden them to include new, climate specific coaching methods.
- A psychologist, counsellor or therapist – learn practical ways to apply your skills outside of the consulting room, and to incorporate them in an action-focused approach.

– An environmental professional – learn new ways to engage others by marrying your technical knowledge with tried and tested relational tools.

## What is it all about?

In our work as climate change coaches, both in private practice and in the training room, we have met many people who feel deeply powerless about the climate crisis. Sometimes these are also powerful people whose authority feels insignificant in comparison to the challenge. Drawing directly on our years of experience as coaches and on the work that we have done on systems change, here we share a new way of framing the human change challenge that the climate crisis poses. This all begins with the belief that powerlessness is a root cause that is holding us back. Powerlessness may look like avoidance, doubt and even obfuscation; all are examples of fear of something. We have seen, however, that if people are first helped to regain their sense of agency, they can next relate to the climate crisis with more logic and less threat. Feeling agency helps people to take action that is motivated by their values and interests, rather than their guilt or shame. Crucially for systems change, agency helps us to relate better to others. Rather than blaming the system or avoiding the subject, we feel more able to voice our worries or share the meaning we find in the action we are taking. This (and not fear-mongering) is how we form deep connections that enrol others to take part too. It is not, however, the case that we all have to be cheerful when we are feeling anxious or sad. Rather it is to recognize that having the courage to be honest and vulnerable in those moments is what builds connection. As you will read in countless case studies here, when we connect well to others, we can mobilize movements and shift systems. Individual action is the missing ingredient in systems change, less for what it does and more for how it makes people feel and for how that mood becomes trend data.

Connection helps us take action and boosts our resilience for what is ahead. This book is about connection, curiosity and courage. It takes all three to be up-close in emotional conversations and in a relationship with truly big, systemic problems. Yet it is when we connect emotionally to the problem and to each other that we stand the greatest chance of making change happen. This book is about relational, not technical solutions. It is also not about proselytizing to convert people to the cause, and it does not extensively explain why we need to act. Our foreword hopefully framed that need, and there are many other great sources of climate science and the change we should make collectively and individually. We trust that if you are here, it is because you are all too aware of why change is needed and fast. The people you want to influence may not be, however, and this book aims to help you bridge that gap, by building sturdier relationships and managing your own response to climate change. We will show you how to address the missing emotional components of change, to help people better connect to the issue and take action.

As the world turns more fully towards the need to address climate change, coaching skills have never been more relevant and useful. With coaching's dual focus of transforming powerlessness into possibility while creating accountability for getting things done, coaching can play a key role in reviving agency, helping people to connect better and generating action. We learned early on that using coaching skills in the context of climate change required a slightly different mindset and range of skills, however. Traditional coaching largely stops at the door of the individual, marriage, team or organization and does not often tackle the societal or existential levels of threat that climate change poses. At the Climate Change Coaches, we have developed methods for detriggering that fear of survival and the often sophisticated phrases that show up with it, and we share them in this book.

In climate change coaching we are working with two big fears: the first is the self-doubt of 'can I do it?' and the second is a systemic doubt of 'will WE do it?' While all of us, coaches or otherwise, are good at championing each other when we doubt ourselves, very few of us know what to say when someone doubts the whole world's ability. This can also trip us up because 'will we do it?' is often a shared doubt, from which it is hard to step back. In this book we will show you how to disentangle yourself from someone else's responses so that you can work with the 'who' they are, not the 'what' they say. When we help people with *their* lived experience of the climate crisis – whether lived in the sense of the felt effects now or in the future, or the organizational changes they are experiencing as a result of a new carbon target – we are able to help them engage more easily with climate change and build a trusting relationship with us that can help them stay the course.

This is fundamentally a book about action. Coaching is a fabulous combination of inner reflection and outer world change, and climate change coaching amplifies this to impact the biggest issue of our time. We will teach you tried and tested climate change coaching skills and show you how to apply a coaching approach with individuals and in organizations to catalyse and sustain change. We will also teach you how to coach yourself as you do it. If you have realized that tackling climate change requires a shift in human behaviour but know that you do not have the right tools, this book is for you. If you recognize that changing individual behaviour and collective culture is as much about psychology as it is about knowledge or process, but you do not know where to start, then read on.

## What is in this book?

Each chapter contains a mix of theory and practical steps. We have also created fictional characters, based on people we have coached, trained or spoken to as part of our research. These are composites, not real people, and are there to ground the ideas into reality. We also include real stories from environmental and sustainability practitioners and professional coaches, to show you how these ideas are being applied in the real world. This book is divided into four main parts.

In **Part A** we explain what a climate change coaching approach is and what role it can play in changing behaviours and systems. We look at why people fail to see the climate crisis as a human change issue; at the range of emotions people experience in relation to it; and how the principles of human change help to move people into sustained action.

In **Part B** we give you tools to use in one-to-one conversations to influence and empower others to act. We provide tools for common problems like scarcity, overwhelm and resistance to change, and that motivate people by reviving purpose. We also look at big emotions that we find uncomfortable, specifically covering anger and grief. We give you practical skills alongside examples of real conversations, showing not just what happened but how it was done.

In **Part C** we zoom out to the organizational and social level, to look at how you can use a climate change coaching approach in your company or community. Bringing together theories of systems change with practical tools, we look at how to create public commitment, how to bring teams and groups with you and how to connect a big-scale vision to transformative operational plans. We conclude with social movements and the glue that holds them together.

In **Part D**, we attend to your own emotional needs and teach you how to coach yourself. Here you get the tools to proactively build your resilience and define your contribution, to steady the ground beneath you. We explore how to recover from stress and burnout and the triggers that cause them. We end the book considering how our grief can be a source of love and hope.

## A note on terminology

Climate change, climate crisis, eco-action, biodiversity loss, deforestation, ocean acidification, soil degradation, plastic pollution … There are so many different ways in which the human race has had an impact upon the planet and life upon it. We do not use the terms climate change or climate crisis in exclusion of the inter-related issues that have been created by human consumption and our use or exploitation of the planet. In writing this book we are looking at the nature of human beings and how to work with each other, so for reasons of simplicity, we use the umbrella terms 'climate change' and 'climate crisis' for all the intercon-necting challenges that currently exist, rather than breaking them down further. We do not do it to oversimplify the challenges, but rather to acknowledge that this is not our area of specialism.

## About our cover design

We hope that you like our front cover. We wanted to situate the human challenge of supporting each other on a steep journey, inside the environmental challenge of the climate crisis itself. To do this, we took a piece of real coast from Flores near Bali and plotted both the actual coastline and the area threatened by sea-level rise. This is the blue, bottom area of the cover. Meanwhile the two green mountain ranges above are cross-sections of the Himalayas, themselves threatened by melting. For all of this inventiveness and skill, we are indebted to the cartogra-pher David Cox (who is also Charly's dad) for creating this for us. We hope that you see the poignancy in it that we do.

## Quick reference guide

This is a big book. Our intention is for you to use it as a reference as much as reading it all. Here is a quick guide to find a solution when you need one, together with a list of the contributors bringing it to life:

**Quick reference guide**

| If you are looking to ... | Including an 'in practice' example from | Chapter |
|---|---|---|
| Understand why people find it hard to see climate change as a human change issue and what enabled that | Irina Feygina, behavioural scientist<br><br>Angela Natividad, advisor, Creatives for Climate | 1 |
| Learn about human change and how that applies in a climate context | Asher Minns, executive director, Tyndall Centre for Climate Change Research | 2 |
| Understand the role of empowerment in taking action and the relationship of powerlessness to taking responsibility | Elizabeth Bechard, health and climate coach and author | 3 |
| Learn what climate change coaching is, how it differs from other methods of coaching and how it can help you to achieve more impact | Sarah Taylor, senior specialist climate change adaptation, Natural England<br><br>Sarah Higginson, Research Knowledge Exchange Manager, Centre for Research into Energy Demand Solutions | 4 |
| Grasp the basics of simple, effective coaching skills that will transform your conversations | Rakel Baldursdottir, climate change coach | 5 |
| Understand the crucial components of influencing others to change | Jill Bruce, Lead Climate Ambassador, Federation of Essex Women's Institutes<br><br>Charlie Williams, climate scientist, University of Bristol | 6 |
| Learn to start climate conversations and rescue them when they go wrong | Adam Lerner, coach and founder, Solvable | 7 |
| Shift the mindset of someone who believes there 'is not enough' to overcome their defeatism | Kelly Isabelle De Marco, climate action coach<br><br>Ruma Biswas, coach, Climate Change Coaches<br><br>Anna Cura, senior researcher, Food, Farming and Countryside Commission | 8 |

| If you are looking to ... | Including an 'in practice' example from | Chapter |
| --- | --- | --- |
| Help someone who feels overwhelmed to take decisions and make a start | Maia C. Rossi, sustainability consultant<br><br>Liz Hall, editor, *Coaching at Work* magazine | 9 |
| Overcome someone's resistance to change and find common ground by connecting to values | Tom Crompton, co-founder, Common Cause Foundation<br><br>May Bartlett, regenerative futures and coaching lead at Solvable | 10 |
| Help someone who feels rudderless to reconnect to meaning and motivation | Alison Maitland, coach<br><br>Jess Scott, volunteer campaigner, London Citizens | 11 |
| Help someone who is raring to go to plan in a way that takes account of uncertainty in the wider world | Charlotte Lin, climate change coach<br><br>Ellie Austin, sustainability consultant and co-founder, Twelve | 12 |
| De-escalate anger and unskilful behaviour in others, and convert it into forward-moving energy instead of damaging relationships | Megan Fraser, coach and founding member of Climate Change Coaches<br><br>Scott Johnson, founder, Kung Fu Accounting and Climate-Conscious Accountants Network | 13 |
| Learn how to best support someone who is experiencing deep sadness and climate grief | Antonia Godber, climate activist<br><br>Megan Kennedy-Woodard, author and co-director of Climate Psychologists | 14 |
| Gain the step you need to help an organization commit to a climate goal | Linda Freiner, group head of sustainability, Zurich Insurance<br><br>Fiona Ellis, professional coach and director, Business Declares | 15 |
| Learn how to take teams on a change journey, and make it safe for people to voice dissent that improves decisions | Nicolas Ceasar, lead coach for NatWest Leadership & Coaching Facility | 16 |
| Gain tools to translate big-scale vision to organizational operations | Sheila O'Hara, sustainability lead, IBM France<br><br>Frank Nigriello, director of corporate affairs, Unipart | 17 |

| If you are looking to ... | Including an 'in practice' example from | Chapter |
|---|---|---|
| Understand how social movements create change at scale and how to overcome polarization | Rowan Ryrie, founder member and mother, Parents for Future UK | 18 |
| | Alex Evans, founder, Larger Us | |
| | Elizabeth Lloyd, district councillor, coach and writer | |
| Put in place proactive measures to protect and boost your resilience, and determine your unique contribution | David Loy, Buddhist scholar and author | 19 |
| | Jenny Ekelund, director of engagement and green transition, The Partnering Initiative | |
| Gain tools to recover from chronic stress and burnout, and return to better balance | Jessica Serrante, co-founder, Radical Support Collective | 20 |
| | Gillian Benjamin, founder, Make Tomorrow | |
| Connect to your own love, grief and feelings of hope for the future | Sophie Tait, climate campaigner, activist and founder of Trash Plastic | 21 |
| | Hamish Mackay-Lewis, leadership coach and nature-based facilitator | |

## Part A

# What on earth does coaching have to do with climate change?

# 1 | A human problem with a human solution

It is not easy to change the world. Well, isn't that an obvious and inconvenient truth! Yet when our own survival is at stake, it is curious, to say the least, that we have not done more to change. So why, in the context of climate change, has the change we need not happened at the pace or scale at which it should?

In this chapter we look beyond the science and the environmental issues we face (and past the strategic obfuscation of the story by those with the most to lose), to consider this crisis as a human behavioural problem with a human solution. We will explore how we have failed to identify with the climate crisis and framed the debate in a way that keeps many of us out of it. We will look at how our economies have untethered us from seeing the natural world as our life support system and that what the climate crisis really requires is a huge change management process.

## 'I'm not smart enough for this'

Faced with a need for such an enormous revisioning of the way we live on the planet, it is unsurprising that the last decades have been characterized by debate about whether climate change is real and really something we need to address. As Peter Stott explains in *Hot Air*,[1] scientists and environmentalists have spent years trying to present data to a world that wants to obfuscate, either in the form of vested interests or self-motivated governments. It is understandable then that for those leading the science, rigorous findings and accurate data should be front and centre. Unfortunately, those who wish to stall the pace of change have become adept at creating opposing information. Caught in the middle of this tussle are the public, and indeed anyone outside of the debate, for whom the complexity of the problem and disagreement over the solutions, has created a feeling of confusion at best and incompetence at worst.

While data is crucial – you will not see us arguing against rigorous scientific findings – it *alone* cannot change behaviour. To do that, we need to build relationships. In the face of public apathy, those who seek change often reach for more data on the basis that if people understood better, they would act.[2] While this seems logical, the data is often landing in the laps of people who have convinced themselves that they are not equipped to comprehend it. We want problems that we can identify with and solutions that we can hold in our hands. The scale and complexity of the climate crisis has interfered with the effectiveness of campaigners and researchers, leaving those they seek to influence feeling 'too stupid' to truly hear them. In our training sessions, people frequently tell us they do not feel knowledgeable enough to act on climate change. This includes those with a background in environmental science.

To be clear, those who took a stand for evidence-based research and who for years doggedly applied pressure to those with the power to make change, were doing the right thing. What the rest of us need to understand now is that each of us can hold a piece of this jigsaw without needing to hold the entirety, and that solutions will constantly evolve. Every aspect of how this crisis unfolds comes down to the decisions that we, our organizations and our governments make on a daily basis; everyone counts. How we frame an issue has implications for everything that comes next. The British charity Climate Outreach describes how people do not:

> form their attitudes or change behaviours as a result primarily of weighing up expert information and making rational cost-benefit calculations. Instead, they are influenced by stories that 'feel right' – narratives that resonate with their values and identity, presented by people they trust, and made acceptable by ... social norms.[3]

## In practice

### Why data does not change behaviour

*Irina Feygina, facilitator and behavioural science consultant for climate and clean energy*

*I felt called to work on climate change from early on, stemming from my personal history as a refugee from Russia to the USA, which inspired a desire to contribute to tackling complex societal issues. I was puzzled and troubled by society's paralysis in the face of climate change. Despite an ever-growing body of knowledge and evidence, denial of climate change was spreading, especially in the US, and society was becoming increasingly polarized around the issue. As a social psychologist, I have focused on exploring the underpinnings of this counterintuitive and destructive response. In my work, I have been involved in bringing psychological insights into state and federal government climate projects, as well as climate communication and engagement efforts, and have witnessed the centrality of relationships to their success.*

*At the root of our resistance to acknowledging and addressing climate change lies a deeply seated need to protect and uphold our identities and the socioeconomic systems that we inhabit and are dependent on for survival and well-being. We strive to maintain a sense of confidence and trust in these systems and to see them as legitimate, stable and benevolent. This powerful motive is driven by our deepest needs to feel safe and secure, and experience a sense of belonging and connection. Climate change profoundly threatens established systems. It challenges modern faith in technology and progress, and our sense of control over nature; it brings into question the sustainability of industrialized society; it underlines that a globalized economy cannot thrive, or even exist, without a healthy ecological foundation; and it makes clear that prioritizing short-term comfort will undermine our ability to thrive in the long term. In short, that indefinite growth – on which a capitalist economy is based – is unsustainable.*

*When our systems are threatened, people often experience reactance. We may try to defend the status quo and resist the possibility of change or deny the problem altogether by dismissing it and blaming the messenger. This is exacerbated by extensive disinformation campaigns perpetrated by special-interest groups aimed to protect their profits. People latch on to the lies as a means to maintain faith and certainty in the status quo, which, ironically, undermines the ability to actually protect what we care about. To influence people we have to understand what drives their beliefs and behaviours. The need to belong is our most powerful motive because our ability to survive and thrive depends on relationships. Climate change responses are driven not by analysis of data or facts, or intellectual or economic capacity, but by the need to feel safe and secure, and see ourselves, and the groups and systems we identify with, in a positive way. Only by working with these underlying needs can we build relationships across the divide.*

*How can we harness these motives toward responding to climate change? One example is by helping people recognize that climate solutions help to protect what we love, rather than threaten it. For example, clean energy can increase energy independence and improve the stability and resilience of energy supply for businesses and homes; and mitigating climate change can protect our health and preserve cultural and family traditions. Former President Obama and Pope Francis often used this approach to engage their constituents with climate issues. The goal is to help people see the interdependence, rather than conflict, between ecology and economy, between environmental protection and personal well-being.*

*Importantly, we need to focus on the solutions, not the problem. While threat and fear are debilitating, solutions can foster a sense of efficacy, empowerment and hope, especially when in connection with other people. Solutions are more engaging if they align with the norms and values of groups, and are consistently communicated and endorsed by those whom we trust and look up to. For example, the greatest predictor of whether people install solar panels on their houses is how many neighbours have done so, and saving energy is best motivated by competition with similar households. These approaches respond to our need to belong and be in a relationship with each other and the world.*

*In addition, we need to shift from one-directional communication of information via media outlets towards bi-directional dialogue, in which we can practise listening and holding space for people's lived experiences. In one example, a public administrator had organized a community event with a group of distrustful constituents aimed to engage them with a sustainability-oriented activity. To his distress, on arriving at the location, his team discovered that the technological setup for their presentations had malfunctioned. Afraid that attendees would leave prior to the tech being repaired, the team headed into the crowd and started talking to people, connecting over sports and kids and pets, and eventually opening up about their lives and challenges. Though the tech was eventually repaired, they barely used it, instead spending the time to listen and deepen this precious connection, which later served as the foundation for building trust and collaboration with this and other local communities. The connection-first approach is now a best practice in this agency.*

*Focusing on building relationships not only helps us reconnect to the humanity in ourselves and each other, it fosters a sense of community and empowerment. We need to feel each other. We need to tell personal, visual, compelling stories and create a narrative of hope and engagement in community. As we engage together, we shift the social norms around us and create new identities, which is the most effective way to bring about change – not only in individual behaviour, but that of politicians, who are beholden to the communities and interests they represent. Their capacity to function is also dependent on relationships – with constituents and each other – and we can harness this need toward climate action.*

---

### A shared catastrophe is not a shared experience

In this book we will regularly use the phrase 'existential threat' because climate change poses such a threat to humanity. However, it is very important to recognize that for many communities on our planet – for whom current daily life contains an all-too-present element of threat, such as those living in oppressive regimes, minorities in countries that persecute them and many others – the climate crisis does not represent the first threat to their survival that they have experienced.[4] Some peoples and communities have experienced existential threat multiple times before and may have collective or individual trauma or developed coping mechanisms for managing those feelings. This can include those who experience ongoing and widespread racism; citizens of any country that has experienced civil conflict or exploitation such as indigenous communities, descendants of slavery and descendants of genocide; and those who are already living with the felt effects of climate change. So when we write about the loss of control, or threat to lives and livelihoods, we recognize that we may well be standing in our own white, Northern Hemisphere privilege, and that while some are fortunate that the climate crisis is their *first* reckoning with that illusion of control, for many it is very much not.

---

## Seeing ourselves in the story

While seeing ourselves in the story is crucial, it is noticeable that the climate crisis is often imaged as an environmental not a human problem. Although human activity is central to the cause of every environmental crisis that we currently face, high-emitting humans are conspicuously absent from much of the representation of its impacts. Until recently, when wildfires and floods have devastated the US, Australia and parts of Europe, images of the climate crisis traditionally fell into three categories: the natural world as isolated (a landscape sometimes with animals, never with people); humans as victims (featuring indigenous or low-carbon-emitting communities from the Global South, living lifestyles we do not identify with); and the system at scale (imposing man-made structures shown at a scale that few of us can individually control, such as coal-fired power plants or aerial views of teams of bulldozers ploughing through a forest).[5]

This was because, until relatively recently, the first effects of the crisis were felt not on us those of us in the Global North but on the Global South and the natural world. As the scientific research, from which the headlines are derived, reflected what was happening in those places, the images accompanied that research. However, when we do not see ourselves in the climate narrative, we do not connect with how we are part of the problem *or* the solution. For this reason, Climate Outreach created a specialist image library, 'Climate Visuals'[6] to encourage a new visual language that includes people-centred narratives and positive solutions. In our increasingly breathless 24-hour news cycle, where many of us are inhaling information rather than truly digesting it, images and headlines also have a huge influence on our sense of agency. As desperately sad as it is to see a polar bear adrift on a floating patch of ice, it is not necessarily galvanizing. Our empathy is triggered but immediately accompanied by deep powerlessness. We want to help and yet cannot, which can lead to avoidance.[7] While the early downstream problems of global heating are felt most acutely on the natural world and Global South, their origin is upstream with us, the highest emitting humans, who have failed to live symbiotically within the limits of the planet's resources.

The story of 'living within our means' is not one that has guided us for much of the industrializing era. Instead we have lived a story of domination and excess, in what many social commentators call the Anthropocene.[8] Here we can simultaneously be islands of individualism or nation states, while also being globalized when it suits, creating an unhelpful illusion of individualism, competition, growth and control over others. Its opposite, to be in a relationship, is to be collaborative, and accepting that we are out of control. Until challenged by the coronavirus pandemic of 2019, there was a pervasive false sense of security (certainly, though not solely, in industrialized countries) first that humans were in charge and second, that we were insulated against the crises of others. Those of us in countries with rank and power, largely due to economics and the windfall of our domineering histories, hubristically felt that we not only held the levers of ultimate power but that we were also were inoculated against the plight of others (animal, plant and even human), even as we lived in an intricately globalized, interlinked world. Covid-19 pulled that rug swiftly from underneath us.

This unravelling of our sense of dominion and lack of control is made even more terrifying when confronted with the existential nature of climate change. With each unfolding crisis we are reminded that rather than controlling the tides, we are but driftwood washed in and out of shore. Many of us struggle with this idea because we have been socialized to *expect* control, and that we must be in control in order to act. If we believe that we need to know each detailed step of the journey before we set foot on the road – then it follows that our actions become stalled when we are faced with uncertainty. Both climate change and being in the natural world require us to engage with something uncontrollable and uncertain, with quite literally a life of its own. Reframing our control to being 'in a relationship' rather than 'in dominion over' means a shift to a collaborative mentality, and the, perhaps radical, pursuit of accepting that we are out of control and do not alone determine our destiny.

## In practice

### The power of a compelling story

*Angela Natividad, Advisor to Creatives for Climate, writer at MusebyCl.io, author of Generation Creation and editor at Atelier Insights*

*In older times, shamans were often tasked with gauging whether a person's ailment was physical or spiritual, individual or collective. The prescription made could be a mix of herbs, or perhaps a story. Like medicine, stories should be carefully chosen and dosed. They have the power to heal or harm.*

*Joseph Campbell's 1949 film* The Hero with a Thousand Faces *introduced Hollywood to the Hero's Journey. From* Star Wars *to* Mad Men, *many modern stories follow this format—a hero goes on an adventure, learns a lesson, triumphs over adversity and returns changed. Most of us also hail from a linear storytelling culture, a means of grouping stories that emerged in Palaeolithic times. In such mythologies, a god creates the universe, and at the end of the arc, the world is destroyed. (Usually there's a fight with a dragon in the middle.) This contrasts with circular mythologies (going back 65,000 years at least), where the universe existed before us and goes on after us, death forever yields life, and humans are not the only protagonists – there are coyotes, foxes, even bears running off with women. In these stories the world is vividly alive with non-human, yet equally legitimate, agendas.*

*Stories matter when it comes to understanding our place in the universe, and on this planet. Linear storytelling cultures are always waiting for the end of the world – demonstrated by how often we refer to ourselves as destroying the Earth. Equally so, many apocalyptic films centre around the single strong leader, who rescues humanity from the world's end. It is almost impossible to think outside the template of the Hero's Journey, where all the action happens in service to a single human protagonist. Yet other stories exist that might help us act more regeneratively. These stories remind us that we are neither stewards nor enemies of nature but are inextricably connected, in a collaborative community with living, vibrant systems.*

*Coaches play a critical role in interrogating stories – their own, the culture's, their clients'. The coaching currently popular among millennials often advances some promise of redemption or success – classic Hero's Journey stuff.[9] This narrative suggests that, like technology, we require constant optimization, and our lives are something we can succeed or fail at, pulling us further from our bodies and natures, and fuelling anxiety and shame. It also encourages us to believe everything exists mostly in service to our own trajectories. You can see how such logic, passed down over generations, impacts how we design our economies and results in an entitled relationship to the planet.*

*It is a very Hero's Journey mentality to ask, 'What can I do to make real change in the world?' We make change all the time, simply by being alive. We exhale, trees inhale. When we die, mushrooms reincarnate us into pulsating life. The circular worldview reflects the more-than-human kinships of interdependence necessary for life in shared space. The world does not need a messiah.*

*Instead, it is more useful to ask, 'How can I change the nature of my interactions?' Removing the framework of the hero, and our compulsion for definitive ends and streamlined answers, provides room to manoeuvre. It connects us to where we can be most effective.*

*When the stories we hold are investigated, coaches can be integral to undoing corrosive narrative threads within clients. You can return people's bodies to them, helping them reconnect to their own inner compasses. People will not only begin to settle into their skin; they can act with conviction and focus without constantly punishing themselves when they misstep. This is what storytelling can do: yield enough such shifts, and paradigm change occurs at scale. But this does not happen when people are too panicked or anxious to truly listen to themselves.*

*As Clarissa Pinkola Estes writes, 'It is not by accident that the pristine wilderness of our planet disappears as the understanding of our own inner wild natures fades.' Paradigm shifts are not guaranteed to happen in our lifetimes, certainly. Humans are notoriously terrible at predicting our impact on larger system dynamics. Hence the necessity of learning to trust the messy myth logic of one's life, instead of seeing our failures as personal ones. The storyteller Martin Shaw once said, 'Stories are spirits'; to transmit a story is to transmit something living. What if, in seeing ourselves as interconnected beings among others, be they trees or animals, we could bring more curative stories to life – for ourselves and for the world?*

## When nature feels unnatural

In the industrialized world, our tendency to delineate the world in time and space, to create order and to organize has led to great technological advances and developments but has also meant that our culture sees itself as separate from nature and not belonging to it. This enables, and even encourages us to dominate, exploit and plunder the planet's resources without acknowledging our fundamental connectedness. It has also enabled us (in the Global North at least) to become increasingly individualistic and socially disconnected. Populations in industrial and industrializing societies are increasingly moving into the built environment of cities and (as a result of leaps in technology, ready access to high-speed internet and higher-skilled populations) behind screens. This is only set to continue, according to the UN, with 68 per cent of the world's population projected to live in urban areas by 2050.[10] We have eroded our ability to be in really deep relationships with each other as well as impeding our ability to be in a healthy relationship with nature.

For many, untamed nature is either 'other' or a place to fear. Where once children played outside without adults, now they are supervised at play parks or on outings to managed forests that turn nature into an attraction not a part of the landscape. The idea of walking in the countryside feels unsafe for many people, especially women and people of colour, who either fear or have actually experienced harassment.[11] Women especially are frequently reminded by the media that open countryside is unsafe for them, a place in which help cannot easily be found and in which predators assault lone joggers and walkers.[12] Our food supply chains have grown so long, and our food so processed, that some children have trouble distinguishing a tomato from a potato,[13] and very few of us work outdoors or even have access to outside spaces. For many, our daily lives and work do not rely on the stability of the natural world in the short or even medium term. While our forebears closely watched the changing of the seasons for reassurance of their survival, the majority of us now only check the weather report when we are on holiday.

Perhaps it is no surprise that at the same time that we as humans are becoming more remote from nature, so nature is receding and disappearing as a result of our farming and industrial practices. We may speculate that it is odd that we no longer have to wipe insects off the windscreens of our cars on summer evenings,[14] or that the hedgehogs are no longer there that as children we saw tottering around our gardens.[15] Our access to nature is diminishing, just as nature itself is in terminal decline. This is quite simply heartbreaking for everyone, from those of us who remember nature as something in which to tumble and explore, to those who have never had the chance, and especially for those who have long-predicted that this would happen, and who have been watching a slow car crash take place for decades. Heartbreak is not an emotion that we enjoy being with, and we will later look at the impact of all of this pain on our response to the crisis. At the same time that we have lost our felt connection to the natural world, environmentalists and scientists have been attempting the Sisyphean task of making us care about it. For anyone working on this issue for a long time, this is a frustrating and a deeply saddening experience.

## A human change challenge

What is really required is for humans, our economies and the structures we have created to run them, to change. If we reframe this problem to be not an environmental catastrophe, but rather a failure of humans to change our individual, organizational and systemic behaviour, then we find a different set of tools with which to get results. Instead of needing an army of people who understand polar ice floes, we can instead start to see this as a change management project on a global scale. As all of us have undergone some kind of change, we can relate our own experience of change in our lives at a micro level to the systemic shift that we want to create at a macro level. Climate change coaching is useful here because it not only helps people connect these dots, but it also focuses upon the nature of our relationships – be they people or the wider world, or even the climate crisis itself. Our tendency to see ourselves as separate has meant that we are often not in a meaningful relationship with the people, structures and entities that might help us to address big problems, but we can be.

This requires a new mindset, and making the shift to it also means acknowledging that instead of feeling galvanized to act, we have often found ourselves unable to identify with the issue, disconnected from nature and feeling intellectually ill-equipped to create solutions at a useful scale. Naming our old mindset as 'powerlessness' helps us to recognize what we do when we feel that way. As powerlessness is not a feeling that humans enjoy, the behaviours that flow from it tend to be suppression and distraction. We are convinced, however, that our inaction is not because people do not care but rather that we are operating from a mindset that says we do not know what to do for the best in the face of an entrenched system and set of norms that are causing damage. As we hope to show you, a combination of head, heart and hands, or data, emotion and action, can help people to change their perspective around climate, their connection to nature and their role in creating positive change. First we have to understand the nature of the psychological problem with which we are dealing.

# 2 How change works

## And what that means for the climate crisis individually, relationally and systemically

Changemakers frequently tell us it is not the technical aspects of their work where they encounter their *greatest* problems but convincing other people to try new things that is the greatest challenge.[16] Indeed, if creating and maintaining behaviour change were easy we would not be writing about it and neither would the numerous researchers who have explored it.[17] In this chapter we will explore how humans approach change more broadly and the specific challenges that the climate crisis presents when it comes to behaviour change. We will outline an individual change journey while showing its relational dimension and explaining how all of this links to change in our organizations and communities.

From the field of addiction treatment through to studies of organizational change, there is an enormous body of work exploring how humans move from not engaging with the need to change through to being fully committed and active, and remaining on track. One useful model from this work is by Elizabeth Kubler-Ross,[18] who described what have become famous stages of change. First we learn new information that shocks us; not everyday information we can easily assimilate but something that requires a big shift. As we start to come to terms with this, because of its magnitude, we may deny it, get angry, bargain with ourselves and others, and become depressed or confused. Finally, when we reach some kind of acceptance, we can regain motivation, set a direction and start actively problem-solving. As neat as that sounds, Kubler-Ross herself warned that this was not a linear process, rather a set of behaviours that we may employ at any stage, and cycle through many times. Human change is a messy, dynamic, iterative journey. Confidence, morale and effectiveness are affected as we go through change – internally it can be an uncomfortable process that we may wish to turn away from. We often experience feelings of loss and sadness for what was before we begin to accept and feel familiar with the new reality. This can make us turn away from change, and we will look at how we resist change. To begin, let us look at the first component of how we introduce and react to new information.

### New information: shocking knowledge about climate change

With its ability to give us a powerful 'shock', climate change should easily get us engaged with the need to change. Indeed many of us try to shock when we talk about climate change with others because we believe that people need to 'wake up' to the problem (and often because we are operating from our own threat response). The stress created by a threat can be useful when we are under real attack, to produce a rapid immediate response in us. If you wake to find your house on fire, your threat response will get you out of bed and onto the pavement far quicker than your logical mind can. Similarly, this is when you would be right to order 'Get out NOW' to your family. However, using this behaviour to engage others around climate change can be very unhelpful.

The existential, systemic nature of the climate crisis magnifies the existing ways in which change unsettles us and requires us to embrace the unknown. This gives us many good reasons to back away. The facts of the climate crisis are also a direct challenge to the world view of many people,[19] calling into question the predictability of the world and with it their sense of control. All change is a foray into the unknown, and being out of control makes change hard. Feeling that we have a locus of control over our lives makes us feel safe, and when we feel safe we behave in helpful, skilful ways, and we are nice to be around. When we feel out of control, we can become defensive, aggressive or brooding and unresponsive. Change that is externally imposed necessarily takes control away from us, and reminds us that our control was only ever an illusion. The larger the change, and/ or the less we have initiated it, the more unsettling it feels. It is no wonder then that climate change is a huge challenge to us psychologically, and that we have manifested a range of avoidant behaviours towards it.

While with our logical minds we accept the need to embrace change for the sake of our survival, our emotional selves can derail us. We may feel that the change confronts our sense of identity and self ('I will always be someone who flies every month') or that it threatens a long-held belief ('A proper meal includes meat'); or it may threaten our values ('I value spontaneity too much to swap my own car for a car share'). There are also big differences in how each of us responds to change. Some people may naturally be early adopters, while others may hang back. Equally, the same person may find one kind of change easy and another hard. For example, someone who considers themselves to be an early adopter of new technology may be slower to make changes to their diet.

It is easy to understand why we shout 'fire' when the house is burning around us – and let us be real, it is. The problem with the climate crisis is that it does not look like a burning building to everyone … yet. Instead some people may think you are being hysterical – 'Calm down! You're being over the top!' – or try to minimize it – 'It's just a little fire, grab a bucket, no big deal' – or even fail to see the fire altogether – 'Hasn't the house just always smelled of smoke?' The more we argue about this, the more attached to our individual positions we become and the more gripped we are by our individualized sense of threat. This makes us more likely to criticize and blame, which only damages the very relationships that we seek to influence.

When we feel threatened, our bodies release adrenaline, our breathing becomes shallow, our blood pressure increases and our heart rate speeds up because our stress response is an unconscious physical process by which our bodies ready us to respond to danger.[20] There is an increase in the activity in the primal part of our brains (the amygdala) designed to react to threat with the 'flight–fright–freeze– appease' response. Imagine an alarm going off loudly in our internal house – all we can think about is getting everyone to safety; we are less able to reason, think clearly and collaborate until the alarm stops. This is exactly why people who work in life and death situations are trained to overcome their natural inclination to panic.[21] From surgeons to firefighters, each knows that staying calm under pressure yields better results than being emotionally hijacked. They also know their safety and success relies on great relationships.

To create the amount of change that we need, and at the speed that we need it, it is important to bring greater numbers of people into change and help them to remain there, rather than frightening them into 'backing off'. While people may well need new information, they need to reach the later stages of change where the real benefits are felt, and that requires our support and empathy. There are key points at which we risk 'quitting' the change process. These tend to be right after learning new information, when we are deciding whether or not to do anything about it, and also later when our motivation levels wane and a sense of doubt creeps in. At these moments, and indeed throughout change, support is a key success factor. If your best friend goes on training runs with you, chances are you will do that marathon, but if they keep inviting you for coffee when you should be training, that likelihood reduces. Support, validation, a sense of being 'good enough' and of belonging are fundamental human needs. Loneliness can be a powerful driver of people disengaging with change, and we have coached many people who feel isolated and for whom a coach was their main support.

This relational aspect is a hugely important part of the overall human response to climate change. How the millions of micro interactions that occur in relation to climate change play out will have an effect upon our overall human response to the challenge, and whether or not we succeed in responding on the scale we need. The effectiveness of these interactions can in turn have an impact on our own ability to remain in a relationship and in agency with climate change, as well as possibly our relationship with the other person. To more successfully do this we need to support people to de-escalate their feelings of threat (and to do the same to our own).

There is another good reason to lessen the grip of threat when we are starting to engage in change – besides feeling happier and more comfortable – and that is that we are much less productive. When we feel threatened, we are not as creative or imaginative as we are when we feel optimistic and capable. As we have said, when threatened there is an increase in activity in the primal parts of our brain, but there is also a corresponding *reduction* in activity in our prefrontal cortex, which changes the way we think; we become more 'black and white', less able to respond to nuance and more likely to perceive threat and danger in our surroundings. We may even go blank. Additionally, if the threat comes in the form of being blamed by someone else, we instinctively look for the safest ideas, not the most inventive or innovative, because we want to reduce the chances of additional blame coming our way.

OVER TO YOU: Consider a big change that you have initiated in your own life. How successful were you? What helped and what hindered the change?

## In practice

### Be a communication hero not a narrator of doom

*Asher Minns, Executive Director of the Tyndall Centre for Climate Change Research at the University of East Anglia*

*Over the years I have seen climate change research evolve from a mathematical topic of atmospheric physics to what it is now, the umbrella of everything.*

*Climate change is now far more famous than The Beatles, not that I remember The Beatles. In the beginning, it was a pub band.*

*The history of climate change in the public and political world has been awareness raising. Most of the world is now very aware of climate change, of what it is, and that more of it is bad, and the worst of global warming has to be avoided. I have known for most of my career that people want to know about climate change, but they mostly do not know how or with whom to have conversations.*

*Research shows that fear headlines and disaster visuals do not engage or create agency. Early research by Saffron O'Neil and Sophie Day showed that fear is generally an ineffective tool for motivating genuine personal engagement[22] and that non-threatening imagery that links to everyday emotions and concerns are the most engaging. In 2006 Tom Lowe assessed the pro-environmental publicity around the Hollywood global warming disaster movie* The Day After Tomorrow, *a film that was promoted as a popular tipping point to engage the general public with the urgency and consequences of global warming, and that was lauded by large parts of the environmental movement as the film that would change everything. Serious panel discussions were held at its national premiers. If you do not know it,* The Day After Tomorrow *is the story of a father–son rescue because of a very abrupt ice age – over a matter of days – due to the sudden shutting down of the Gulf Stream. It is the highest grossing Hollywood film ever made in Canada. Tom showed that while the movie increased anxiety about climate risk in viewers, it decreased belief in the likelihood of extreme events. Exactly the opposite of what the publicity and the welcoming environmental movement assumed would happen. Like most movies, many viewers had forgotten they had seen it when asked about it a few weeks later.*

*These research papers have contributed to a paradigm shift in the communication of climate change by researchers and then NGOs. They were influential in my own organization because they were done within a body of 180 climate change researchers. The organization Climate Outreach was subsequently born based on the need for evidence-based climate change communication by applying these types of understanding from the social sciences. The science of climate change communication draws on a broad church of published literature from studies within the fields of climate change communication, science communication, social sciences, marketing, PR and public affairs, cognitive psychology, as well as museum studies, arts and data visualization practice as appropriate. You could more neatly summarize it as 'how you are heard is not what you think you said'. A game-changing framework is the Universal Values Map of Shalom Schwartz, which visualizes audience values. Engagement is not through science facts, or economic arguments, or other sender–receiver one-way models of communication. It is what we believe as much as what we are told that matters. This is not humans as a rational actor. Politicians and PRs understand this, scientists less so.*

*The biggest climate change communication trope that misinforms the public is 'false balance', the fairly lazy journalistic practice of simply quoting opposing views when reporting on climate science (and many other topics). It is only relatively recently that BBC editors began to change this practice for climate stories, beginning in 2018, just ahead of the release of the IPCC 1.5C Special Report. Nevertheless, disaster headlines and imagery persist, and one impact of these and of social media narratives appears to be climate anxiety, in particular among young people.*

*Doomism climate change narratives do not fit with any robust and carefully peer-reviewed published science. There is no* Day After Tomorrow *disaster scenario, from climate science or anything similarly apocalyptic. I worry that young people are being gaslighted by some environmentalist adults, who then follow-up by asking kids how bad their climate anxiety is, creating a self-fulfilling circle. My communication is all about theory to practice, wherever possible, and when I train researchers in the science of climate change communication, one of my key take home messages to them is 'Do not be a preacher of doom be a climate change hero.'*

*Those of us in the climate change communication field can maximize the environmental interest of the public, of those in policy and business, and of the media, not in preaching doom but in connecting. In the UK we could see a social tipping point because people are genuinely hungry for information about climate change. The end of the climate silence is hopefully nigh, the end of the world is not.*

## The relational nature of change

The threat that the climate crisis poses can surface some emotions that leave us feeling overwhelmed and deeply disempowered, making it very easy for us to lose faith in our ideas and our actions, because none of them will solve *all* of the problems, *all* at once. The crisis raises some particularly difficult truths, namely:

- It is impossible to solve quickly, and as humans we have a predilection for the short term over the long term.
- It has many moving parts, which often conflict with each other. There can be no simple, perfect answer, and we prefer a silver bullet.
- It requires a shared vision and then collaboration between people, organizations, communities and nations, and this requires a herculean effort of diplomacy.
- Fundamentally, it requires us to be in trusting relationships with others, which feels a bit icky. We find it easier to imagine a single hero (the kind who can use that silver bullet).

The urgency, combined with the complex and existential nature of the crisis, further amplifies our disempowerment. Any change involves doubts, but unlike a personal change process, climate change leaves us powerless on two dimensions: individual and systemic. We doubt not just our own ability to create the necessary change but also whether our systems have the ability to change. We ask ourselves both 'can I do it?' and 'can we/humanity do it?' This is what makes climate change a relational issue as much as an individual or environmental one. We can unpack this concept further by looking more closely at the different levels in which we are 'in a relationship' with ourselves, others, the system and with the issue itself.

In Fig. 2.1 we have mapped some of the middle stages of the Kubler Ross change process, but also added in the difference between not engaging with climate change (stage 1), relating to it but feeling powerless (stage 2) and finally relating to the change and feeling a sense of agency (stages 3 to 4).

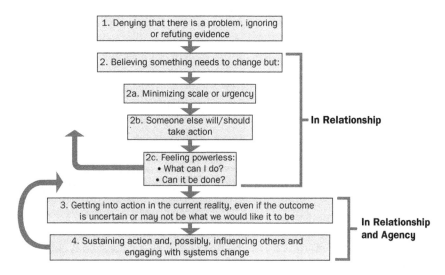

**Figure 2.1** The flow chart of climate action

Stage 1: Our starting point is not being engaged (or not 'in a relationship') with the climate crisis. Here we may still be attached to an old personal or social identity, or still arguing about whether we should believe in the climate crisis or not. 'I am not a tree hugger'; 'None of my friends and family believe this stuff'; 'The planet has been warming for years.' Some people stay here, and some gradually move into a relationship with the problem, at stage 2.

Stage 2: Here we are engaged with the climate crisis and recognize the need for change. However, we do not feel a sense of personal agency. This is a crucial point when we may get stuck and disengage if we do not find a way to believe that we can make any difference.

2a. Minimizing: Here we may seek to downplay the scale or urgency of the issues.
2b. Shifting accountability: When we lose faith in our own or others' ability to change we often shift responsibility, feeling our own actions will not make a difference – we say 'I won't, they should'.
2c. Feeling insignificant: We may be unsure of what to do or where to begin, and therefore not take action – 'Who am I to do this?', 'I'm not an expert', 'I don't know where to start'. We may also cite perfectionism or a fear of getting it wrong, or get stuck in a mindset of scarcity ('there's not enough time!') or one of overwhelm ('there's too much to do!'), from which it is very hard to move forward. If we deny our feelings we are more likely to disconnect from the process. We may 'come out of relationship' with the need to change, and return to stage 1.

Stage 3: The next key transition is when we begin to have a sense of personal agency and enough empowerment to get into imperfect action and overcome hurdles and challenges as they arise. This is the place in which we also have acknowledged the complex emotions that arise when we face up to the climate crisis and create change in spite of those feelings, even when we believe that we may be too late and not able to make the change we really want to.

Stage 4: By this stage, change is sustained and we may be influencing others. However, we may also find our motivation and belief waning as we get tired, experience setbacks or if we feel alone. At this point, we can backtrack to stage 2 if we feel overwhelmed or exhausted. That may happen because we receive new information (or a new take on old information) that leaves us feeling hopeless again and diminishes our sense of agency. Here it is vitally important to look after ourselves, giving permission to recalibrate goals, and connect with others for solidarity and support. In Part D we explore ways to do that so that this loss of agency is a temporary pause, rather than a permanent disengagement.

It is important to say that the process is rarely this linear. As with all change, we often cycle through the later stages several times as a new reality appears that we need to adjust and respond to. As the situation is constantly evolving, the final three stages in particular are often cycled through repeatedly as the size and gravity of the crisis becomes apparent with new information and awareness. We frequently return to stage 2 as the context evolves, or as we develop a different perspective on something familiar. Both ingredients – being in a relationship with the issues and having a sense of agency in ourselves and each other – are required for us to take ownership, overcome obstacles and enrol others to join us in change.

## Two types of agency – individual and systemic

The difference between everyday individual change and change for the climate is that to succeed in catalysing change for the planet we have to scaffold our personal sense of capability *and* our belief in the system. It is not enough to help people feel individually capable. We have to help them believe in those around them as well. Let us consider an example of how these two types of empowerment work in practice with one of our fictitious characters. Monica is a young activist and works as an administrator to pay the bills. She wants to apply for a new job, but although she's downloaded the application form, she cannot bring herself to fill it in. 'I want to apply for that job, but I know I won't get it. I'm too inexperienced.' This is a lack of personal belief. While Monica doubts her own ability, she would never say: 'I don't believe that *other people* can get new jobs.' She knows that other people can, she just doubts that she can. However, when it comes to environmental action, Monica is likely to express a systemic type of disempowerment as well as an individual one. She might say 'I want to do something about plastic waste on our local beach, but no one else cares. If I clean it up people will still go on littering, so what's the point? I'll just end up losing momentum and stop too.' Here, Monica lacks systemic and individual belief. She may feel confident that she can grab a sack and start picking up plastic, but before she even does so, her lack of belief that her community will collaborate defeats her individual belief, and so she does nothing. She is likely to say to herself: 'There's nothing anyone can do. Litter on the beach is a fact of life.'

Note that even when there is some factual accuracy to the things Monica feels, it is not what we say but *why* we say it that is important. In both cases Monica is about to embark on change when her mind intercepts with a note of doubt about her own or her community's capability. This is an assumption, remember. No real research has taken place; Monica has not spoken to the HR department advertising the job or asked her community if they will help keep the beach clean with her. Instead she is allowing her voice of doubt and assumptions to hold her back. This is simply Monica's mind trying to keep her safe from risky change. To stay safe, she undermines her sense of self-efficacy and delays getting into personal action. The sad thing is that continually undermining ourselves like this, over time, dismantles our sense of self-esteem and our belief in the goodness of our societies.

Just as individuals need to cultivate an empowered relationship with climate change, they must also attend to their relationship and engagement with systems change. Stage 2 in the flow chart (when we believe someone else should act) is often the stage at which we give away personal power to the system – believing that systems such as businesses, governments or other countries are the problem and it is for them to solve. Equally, if we assume that the structures of power in our lives (for example, our governments) are dependable and hold the answers, reckoning with the failure of those systems can impair our capacity to act. Feeling that the system of which we have been a part has been destructive or does not have the answers can create a feeling of grief and dismay that can encourage people to disengage. Our first task in galvanizing change has to be dialling down fear and dialling up our own and others' sense of faith and belief that the change is (a) worth doing and (b) likely to be successful. We also have to accept that change is required of us, that we are grappling with a complex problem and that we cannot control everything.

OVER TO YOU: Listen for how people talk about climate change, in the media as well as in your daily life. Start listening for signs of both individual and systemic doubt and belief, and consider where that person might be on the flow chart.

# 3 Empowerment and belief

## A model for resonant action

We have so far explained that the climate crisis is a human problem that requires human change. We have shown you that the process of human change can bring up lots of feelings of doubt, and that when we are facing globalized challenges, our doubt can be both in our own abilities and in those of the system. We also described how a sense of agency is key to taking action. So here we want to unpack that idea of agency further and offer you a model of change that brings together the dimensions of doubt and agency (or disempowerment and empowerment), with their impact on our motivation and ultimately our action.

## The two siblings of empowerment and disempowerment

Empowerment and disempowerment can be called many different things: agency and doubt; resonance and dissonance; self-belief and insecurity. Most commonly we are talking about whether we feel capable or incapable. We can think of empowerment and disempowerment as closely related. Often our doubts signpost us to our most dearly held beliefs. For example, we may feel 'stupid for not knowing enough' about climate change, which might indicate that competence matters to us. If we can find a way to feel competent with our current level of knowledge, or define what we additionally need to feel competent, we can return ourselves to a state of empowerment.

Unlike the crushing presence of powerlessness, empowerment is the more fun-loving sibling. Like good and bad angels on our shoulders, empowerment is the cheerleader saying 'You can do it!' and is the sense of confidence that accompanies us when we feel secure in ourselves or in our chance of success. We may or may not also feel happy and content, ambitious and driven, courageous and adventurous; these are possible but not given. What we will not feel is wracked with doubt and indecision or distrustful of ourselves or others. When we are empowered, we are often also more at peace with ourselves and our contribution. Perhaps we do not notice feeling empowered as much as we notice feeling disempowered, because the voice of doubt is much louder than the softer voice of courage. While there are certainly times when we feel 'on top of the world', and very aware of our empowerment, there are far more times in which we are using muscle memory for tasks that we have done before, at which we naturally feel capable. We would not describe ourselves as empowered when we turn up to an environmental group at which we know everyone, but we may well have felt the nervous jitters of disempowerment the first time we went knowing no one.

Wouldn't it be great if we could just tell ourselves that we are powerful and make it so? Unfortunately, for most of us, simply telling ourselves (or others) to feel empowered does not work; we have to *feel* it. We have coached many people who have told us 'People tell me I'm great at my job, but I don't *feel* good at it.' We humans are very good at talking ourselves out of our self-belief. When we receive praise, a disempowering voice may verbally deflect it with 'Oh it was nothing', or internally reject it ('If only they knew how stupid I am they wouldn't say that'). In other words, there is a disconnect between someone's true capability and their *feeling* of it. To build someone's sense of capability we have to employ a range of tools, and in Part B we will offer you a whole toolbox that moves people from a place of doubt (disempowerment) to a place of motivation (empowerment). First, we want to show you the impact of each on the action someone takes.

## The link between feeling empowered and taking responsibility

Empowerment is a word that is much loved in the self-help section of the book shop, so you may have some baggage about it. You may think it is a nice-to-have and 'Why cannot we just tell people what to do?' If your experience of 'empowerment' is that it is about making people feel nice then you may come to this with an eye roll of 'don't tell me I'm a little sun beam!' However, empowerment is not just a feel-good state, it is a key component of responsibility-taking. You may yourself feel a tremendous sense of responsibility and feel angry that others can appear so careless or apathetic. The chances are that in addition to feeling responsible, you also feel a degree of empowerment that your efforts make a difference, and this helps you act on your responsibility. While there may be people who feel capable of acting on the climate crisis and choose not to (or who have not thought about it that much), there are vastly more for whom their powerlessness holds them back from taking responsibility, and therefore, action.

Empowerment and responsibility can be seen as two sides of the same coin. If we want people to take responsibility, then first we have to understand what is stopping them. As we said in Chapter 2, in many cases, this comes down to a deep lack of belief: 'I can't', 'I don't believe I can work out how' or 'it won't make a difference'. When someone feels responsible but does not believe acting will change anything, it can feel very lonely and disempowering. Before people get into action they have to believe that the action is worth taking. When that belief is self-belief, it is called empowerment. Rather than judging people for not acting, we can find out why they do not believe anything they do will count, by asking 'What is telling you that nothing will count?' or 'What if it did count, what would you do?' To explain this, let us turn to one of our fictitious characters, Rotimi, an IT manager and the leader of her company's green team. The team is hard-working, but Rotimi is sick of having to make decisions. She complains:

> I wish for once, someone else would make a choice! It's always me deciding everything. They're really hardworking, they come to meetings, but they lack oomph. This week I discovered that Cathy had stopped work on a project because she was waiting for someone to make a decision she was perfectly

capable of making herself! I can't follow them around like a kindergarten teacher! Honestly I've tried everything, I don't know what else to do. I think it's just that there are people in this world who don't care enough, and unfortunately they're in my team!

It is easy to feel like Rotimi: out of ideas and exhausted. Yet, ironically, both Rotimi and her team have the same problem: they all lack confidence in their ability to solve the situation, and so are giving up their responsibility. Rotimi is expressing a lack of belief in her ability to lead. She may think she has 'tried everything' but she almost certainly has not, instead her usual ways of managing people (the ones she feels empowered doing) are not working, and she is doubting that she has what it takes to lead. That feeling of incompetence makes her feel insecure, from which vantage point her options are to blame herself or her team. She chooses to blame the team, which is Rotimi passing the responsibility for the problems in their relationships to them. This pushes the problem away, but it does not make the disempowerment go away. Rotimi will still be making the decisions, and now she will feel resentful as well as incompetent. That will make it harder for her to lead. At the same time, Rotimi's team may also lack self-belief. Cathy may have worried that she would not 'get it right' or 'get it right the way Rotimi likes it' and so stalled making that decision even though Rotimi thinks she is capable of making it. She may have had a bad experience of taking the initiative in the past and it backfiring. On a deeper level Cathy may have been brought up not to share her opinions and does not see it as her place to do so. We have a clue to all of this because we know that they are committed and hardworking; it is just initiative and decisions with which they struggle.

Remember that disempowerment is another word for fear, insecurity and doubt. It can be sneaky, and may not sound like a straightforward inner critic such as 'nothing I do will make any difference'. Instead, you might hear:

- 'That's more than my job's' worth' = 'I don't feel confident to stray outside of my role' (perhaps for fear of management reprisal).
- 'We've got to be sensible about this' = 'This makes me feel insecure because it could damage my reputation/team/ability to win support.'
- 'No one needs to hear my opinion' = 'I don't believe in myself.'
- 'What's the point when X country is such a polluter?' = 'I doubt my significance.'

When people do not make decisions, it is very often because they lack confidence in their decision-making; put another way, they do not feel empowered to make them. They may otherwise be dependable and conscientious but find self-initiated change hard. Figure 3.1 breaks this down into stages that create a virtuous circle. Whether we start with action or belief, we need both to sustain change. Now notice how often, when it comes to the climate crisis, otherwise conscientious people turn away from action because they feel incapable of making the right decisions or of taking action that seems worthwhile.

Think about what it would be like if everyone felt confident and even excited to make decisions that favoured the planet. Wouldn't it be inspiring? Wouldn't it be contagious? Wouldn't you be mad not to get involved with something that exhilarating? Empowerment enables us to make decisions for ourselves and others, feeling not just the possibility of success but confidence that if we fail, we will

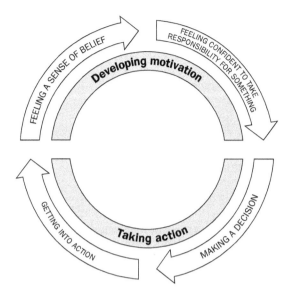

**Figure 3.1** The interplay between belief and action

handle it. Decisions lead to action, so if we want people to be more active on this issue we would do worse than to focus on their sense of belief.

OVER TO YOU: Have you ever met someone who was like Cathy in Rotimi's green team? What might you have done to help them feel more confident?

## You gotta have faith – the power of hope and belief

It is not fashionable to believe in things. It can sometimes feel that we are supposed to be cynical or to expect to be let down. As climate justice writer Mary Annaïse Heglar said in her 2019 article 'Home is Always Worth It':

> … the thing about warming – whether we are talking about the globe or an oven – is that it happens in degrees. That means that every slice of a degree matters. And right now, that means everything we do matters. We, quite literally, have no time for nihilism.[23]

Yet those who believe are often deemed naive or gullible, despite the fact that belief is the secret sauce of change. People arrive into our coaching practices all the time asking for help with changes that they are easily capable of making on their own or for help finding solutions that they could research solo. They are smart enough to know, however, that when left to their own devices, their minds intervene and start protesting: 'You'll never manage this. This change is not for you. What was wrong with the old way?' To make change we have to dial down doubt and dial up belief. The power of the mind to subvert change is the reason that data alone does not change behaviour. Someone who is thinking of becoming vegan can find numerous websites sharing easy, delicious recipes. However, if when they reach those websites they hear their voice of doubt chirp louder than their belief that they can become vegan, they will shut the browser.

Belief matters when it comes to personal goals, where we can also call belief 'confidence'. It matters even more when we want to take collective action, where another word for belief is 'hope'. Hope is a verb not a passive act, and we would be mistaken if we think that we can just cross our fingers and the rest will happen by magic. As Rebecca Solnit said 'Hope is not a lottery ticket you can sit on the sofa and clutch, feeling lucky. Hope is an axe you break down doors with in an emergency.'[24] Hope is the fuel for motivation and, when accompanied by impactful actions, becomes a renewable resource. Hope plays strongly into the success stories of every team that wins against the odds. Hope enrols others to join and be inspired by us and to create powerful relationships.

We know that individuals making small changes alone will not do the job here and that we need large-scale systems change. However, if we dismiss individual change entirely, we ignore the fact that it is not just about net impact but the way people feel when they act and the ripple effect that occurs when powerlessness is transformed into a belief in our collective ability. Connected people can change systems by sheer force of numbers, especially when, in their belief, they behave like the future is already here, and we cover this more in Chapter 18. This can influence the creation of new policies and products because it indicates a new direction of travel for the system, which is then reinforced with legislation or innovation. Another term for this belief might be 'trend data', which is gathered by organizations like Ipsos Mori[25] in the UK and used by politicians and companies alike in order to design policies and products that appeal to voters and customers. This is how people feeling a sense of individual and societal belief can forge relationships with each other and with climate change itself.

## Building motivation before action – Being and Doing

In coaching we talk about two aspects of human behaviour – motivation or Being, and action or Doing. Both are important for successful, sustainable human behaviour change. Key to all of this is a feeling of empowerment or resonance, which is the internal motivation that comes from something holding meaning for us. When we feel empowered, we feel capable and find wells of energy. We may also feel more congruent and able to express how we feel. When we feel disempowered we may not allow ourselves to acknowledge our dissonance, or lack of meaning, and act out of fear, pushing the negative feelings onto others or things. To determine our state of empowerment or disempowerment in relation to climate change, it is helpful to unpick the difference between what we do and how we feel.

### The 'Doing' of human behaviour

'Doing' is the most measurable, visible aspect of human activity. We may measure not just what we have done – 'Our organization has planted 30,000 trees' – but what we have *not* done – 'I haven't eaten meat for 12 years'. In both cases we can put concrete measurements around the Doing. As a result, we can see a link between our Doing and our output. In conversations about climate change, tangibility and measurability is often where the focus lies, for example 'What should I/we do to reduce our carbon footprint?' The trouble is that the Doing is also where scrutiny, judgement and criticism abound, and where guilt and self-criticism can arise. It is often where disagreements and impasses occur when people hold opposing viewpoints and where uncertainty about what is the 'right' choice can result in inaction. While the Doing is clearly crucial (this is, after all, a book about climate action), it is not the complete story. If we wish to successfully create *sustainable* human behaviour change at individual, relational and systemic levels, we need to take into account the Being as well.

### The 'Being' of human behaviour

A simple way to think of Being is as the thing that differentiates a person from a robot. Being is about how we feel and whether or not an activity holds meaning and resonance for us. Being matters because it relates to the amount of internal energy that we bring to an activity, and how likely we are to continue and succeed. Because the Being aspect is unseen, it is much less tangible, harder to quantify and measure, and more difficult to explain than Doing. Yet, arguably, it is just as powerful, if not more so, because meaning has the ability to enrol others. If we only focus on the visible, measurable Doing, we are also less likely to recognize when someone's motivation is waning. Our being is also crucial to our well-being and sense of fulfilment since spending time in a prolonged state of dissonance can have severe consequences for our mental health.

To illustrate the Being and Doing, let us invent a fictitious couple, Petra and Rahim, who both want to reduce the carbon footprint of their food. Petra thinks an easy solution is to become vegan, so she asks Rahim, who does all the cooking, if he can cook plant-based dinners. If Rahim is as motivated to go vegan as Petra then he might relish the challenge to cook this way. A meal cooked with love, however it tastes, will always be appreciated. But let us imagine that Rahim resents cutting out the food he loves. His bad mood festers as he's cooking, so Petra keeps clear of the kitchen. The resulting meal is eaten in brooding silence. Rahim has little internal motivation, he is just 'keeping Petra happy'. If, however, Rahim has his own intrinsic motivations (high energy or Being) for going vegan, his entire attitude towards cooking will be different, the mood in the kitchen will be pleasant so Petra will hang around. In both examples Rahim's Doing is the same – he's cooking a vegan meal – but his Being as he does so will affect their experience of the meal, their relationship and their future choices.

## A useful model of sustainable change

Figure 3.2 shows the Flynn model of sustainable change[26] that Sarah developed to illustrate how our Being and Doing interacts and affects our response regarding a particular climate related activity. You can use this model to clarify your own motivation levels as well as to assess the motivation of those you seek to influence. This can help you decide what your most important intervention should be to help someone become highly motivated and sustain action. The model plots motivation (Being) on the Y axis, on a scale of dissonance (disempowerment) to resonance (empowerment) and action (Doing) on the X axis on a scale of inactive

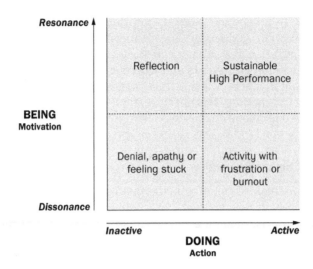

**Figure 3.2** The Flynn model of sustainable change

to active. To bring the model alive, think of a specific climate related activity that you are considering undertaking and hold this in your mind as you explore the model. Take a moment to familiarize yourself with the diagram, and, even before we fully explain it, think about where you might plot yourself right now.

Let us look at the four quadrants in turn and see what happens to us in each state and what we can do to support change there.

### Bottom left – little motivation, little action

When we score low on both the motivation and the action dimensions, we are usually feeling little meaning and not doing much. That can look like us being a bit flat and not having much energy for activities. It can be a passing thing, when we are feeling a bit tired, or it can be a longer-lasting state, for example when we have experienced a transition or a crisis but have not yet found a new rhythm for how we do things. This can, however, also be a place of renewal if we are tired or if something major has happened in our life (such as a bereavement). We can also move here if we have stayed too long in the bottom *right*-hand quadrant (low motivation and high action) and are exhausted. In industrialized countries we often force ourselves out of this quadrant by blaming ourselves for being 'bad' or 'lazy'. That can mean that we miss the healing that comes from being here. Just as trees cannot bear fruit all year round, so we also need a period of wintering. People here deserve our kindness and may benefit from being reassured that inaction does not mean that they do not care or that their active days are over or will never start. Other people might be in this quadrant because other things are crowding the climate crisis out. For example, they may be concerned about the climate but highly active in caring for elderly relatives with no capacity for anything else. They may alternatively not yet have found an anchor that resonates around the climate. Equally, if someone is an active climate change denier but feels unsafe to voice their views publicly they may also be here, because they are silencing themselves and that feels dissonant.

### Bottom right – little motivation, lots of action

When we have high amounts of action, but low motivation, our actions can feel meaningless and mechanical, and are likely to be unsustainable in the long term. This can be a quadrant in which all effort feels Sisyphean – as if we too are condemned to push a boulder up a mountain, only for it to roll back down every day. Life always contains some tasks we do not find motivating, but when the balance tips too much this way, our well-being and productivity suffer. If we spend too much time in this quadrant our activity will tail off as we move back to the bottom *left* to recover. In the bottom right we are likely to have little intrinsic motivation, being motivated more by external factors, operating out of guilt or fear of others' disapproval. Negative feelings may kick start action, but unless we discover meaning in that action, we will not sustain it for long enough to attain lasting results, and we may struggle to lead ourselves and others. People can find themselves in this quadrant because they previously held strong intrinsic motivation, but became burned out, depleted or overwhelmed and have lost connection with their belief. People who have been doing a great deal for some time may still be doing a lot, but their sense of positivity will be waning due to exhaustion, lack of impact, or a sense of scarcity. They may be hiding this from others and feeling isolated, or alternatively taking it out on others. Some may be in this quadrant because environmental action is expected by those around them, such as green schemes launched at work, which they do not find meaningful.

### Top left – high motivation, low action

Being highly motivated but low on action is typified by reflecting, thinking or dreaming. This is the state of wanting to take climate action, and looking around for where to start. This can also be a stage of feeling galvanized and excited to do something and making plans, just not yet putting them into action. This is often an undervalued state in cultures that emphasize productivity, being busy and 'doing', but it is an important aspect of being human. This is a place for finding inspiration, creativity and imagination, and for reflecting or taking stock of our mental well-being. Action without this time for reflection can become automatic and send us into the bottom right (low motivation, high action). Often people use coaching sessions to sit in this quadrant and think more clearly about what really matters to them. Sadly, many only realize the importance of reflection after they burn out. The health warning on this quadrant is that if we stay here for too long without acting, we can feel dissatisfied, bored or frustrated, and slide down to the bottom left as we lose enthusiasm. Overwhelm of information or a sense of perfectionism can send us back down into the bottom left and leave us feeling disappointed that taking action on a particular thing is not part of our identity after all.

### Top right – high motivation, high action

Think back to a time when you were part of a great team, working on a project that really mattered to you, and you felt like you could conquer the world.

Remember that clarity of purpose that helped you to overcome hurdles? That is what it is like when we are high on both motivation and action. We have more energy, we can see past obstacles to achieve our goals and we have a deep sense of purpose. Here we are in our most creative and high performing state, in which we thrive and flourish. Coaches often work to help clients to reach and remain in this quadrant in order to achieve a goal that matters to them, and many conclude coaching here because they have discovered their motivation, got through the pain of change, or overcome whatever was holding them back. This is likely a place of solidarity and collaboration with other climate actors, where our individual motivation becomes collectively contagious. We have connected to a deeply meaningful goal and found a way to take action that speaks to us personally. We have aligned our Being and Doing by putting structures in place to keep us going and we have some measurement of our impact, if not individually then collectively. Here we also have support in place for when our motivation dips. It is also important to know that we cannot be in this quadrant forever. We also need rest and recuperation. Just as for high performing athletes, so for all of us: rest and recuperation time is as much part of our success as the stretch training and competing.

OVER TO YOU: Which quadrants on the model are most familiar to you? Use it like a map to work out where you are, the reasons you might be there and which quadrant you might move towards (if any).

There is no perfect place to be on this model; it is a guide to help you understand what you and others need to support your climate action. There is nothing 'wrong' with being low on the motivation scale and feeling dissonant. The model simply helps us to notice where we are. On a typical day (or even during a conversation) we can oscillate from low to high on the motivation scale. Motivation is highly individual, with different activities holding different levels of meaning for different people. We are not suggesting that we should, or could, always be in a place of high motivation or that organizations can or should be filled with people who are in a permanently energized state. What we are suggesting is that we can only get so far in one dimension (either motivation or action) without the other, and that we need both to create sustainable behaviour change.

## The search for meaning – resonance and dissonance unpacked

When action meets motivation, resonance is created, so we can also think of this model as one of resonant action. This is when an external need or reason for action connects with our internal world – what matters to us, what we enjoy, what we are good at, our identity and self-concept. It is a key part of creating continuous, resilient, internalized change. In Chapter 2 we talked about being in a relationship and agency as important precursors to action. Resonance is the glue that helps bring all these elements together and keeps them connected when challenges arise. 'Resonance' is our psychological and emotional energy, while its counterpart, 'dissonance', makes our energy fall flat. Someone who is experiencing dissonance may tell themselves that they 'have to', 'should' or 'ought to' keep going, regardless of how bad it feels. Yet sustained action comes from feeling great, not bad, so we need to help ourselves and others steer back into the more motivating waters of resonance, even when that means leaving behind an action that we feel we 'should do'.

To illustrate this, let us return to Rahim and Petra. Imagine first that Rahim goes along with veganism because he loves Petra. But every lunchtime at work he buys a cheese steak sandwich. He feels guilty while he's eating it (dissonance) but he also dreads dinners at home (more dissonance) because he doesn't enjoy vegan food. Now imagine that Petra notices this and has the courage to ask him what is wrong. Rahim and Petra have a pretty good relationship, so he feels able to tell her:

> I've tried to go vegan, but I don't think it is for me. Everything I enjoy has dairy in it, and I really like meat! I don't want to kill the planet, but going vegan is making me want to eat more meat than before, and it is making me miserable.

Notice that it is easy to step away from something when it feels dissonant – by buying a cheese steak sandwich – and disrupting the process of behaviour change. There is also a ripple effect to Rahim's dissonance in several directions. Until this point Rahim has not been honest with Petra, which is affecting their relationship (Petra asked because she thought something wasn't right), and because Rahim is feeling dissonant about veganism, he is actually *more* motivated to eat meat than before. Rahim's dissonant action is not only unsustainable, it is actually causing setbacks.

So what should the couple do? Well for a start, Rahim could either find a way to make veganism resonant for him or stop trying to do it. It is far better to make a resonant choice to stop doing something than to continue with a dissonant choice until it damages a relationship. Instead they can revisit their original *resonant* reasons for trying veganism; their aim of reducing the carbon footprint of their food. Petra, who suggested veganism and is enjoying it, can stick with it (she has the motivation to keep going), whereas Rahim might find it fun to challenge himself to reduce their food waste by 50 percent or their food miles to a 40-mile radius. Once he feels good about food again, it is completely possible that Rahim will start to cut down on his meat consumption (perhaps when he finds out through his research about the carbon miles of animal feed). One day, Rahim might turn to Petra and say 'I'd like to give veganism a go again', or become vegetarian. Then it would then be a resonant choice, and he would stand a much better chance of sustaining it. The trick is not to stay wedded to a particular action that feels bad but to find ways to make that choice motivating or return to the first principles of what you wanted, dig back into that resonance and find other ways to bring it to life.

## *In practice*

### *Climate change coaching to support someone through dissonance*

*Elizabeth Bechard, coach and author of* Parenting in a Changing Climate

*The first words out of my client Allison's mouth when we met over Zoom for an initial coaching session were an apology for the state of her hair: 'I'm sorry I*

*look like such a mess – I haven't had a chance to take a shower yet today!' I reassured her that she looked lovely, noting that she seemed a bit frazzled. A mother of two young children, Allison had recently given birth to the second one, and I knew it had not been easy for her to make time for a coaching session. I asked her if she wanted to take a few centring breaths together, and she gratefully agreed. After a short mindfulness practice, she seemed calmer and was ready to begin.*

*'What brings you to coaching with me?' Allison let out a heavy sigh. 'I've been anxious about climate change all the time since the baby was born,' she said.*

> *It seems like there's a new climate disaster in the news every day, and I know I should be paying more attention, but sometimes I just want to tune it all out and pretend it is not happening. I'm trying to make eco-friendly choices in parenting but the amount of plastic in our lives is overwhelming. I know I need to do more about the environment. I know I should be a climate activist. But I'm exhausted all the time and I just don't know where to begin.*

*There was emotion in Allison's voice – it was clear that she cared deeply about making environmentally conscious choices in her family's life, but the idea of doing more than she was currently doing seemed overwhelming. I knew that we did not have time to tackle everything she was concerned about in a single session, and after reflecting back her concerns, I invited her to narrow in on a smaller, more manageable focus: 'If you could choose just one small piece of this puzzle for us to focus on in our time together today, what might that be?'*

*Allison sighed again, this time with a sense of palpable relief. She admitted with a chuckle that sometimes she forgets that she does not have to tackle everything.*

> *One thing I know we want to be doing is composting. I used to compost before getting married and having kids, but we just have not made it a part of our family life yet. And I know it makes a difference.*

*I wanted to understand how composting fitted into Allison's values: 'What makes composting important to you?'*

*In college, Allison had volunteered in a community garden and knew quite a bit about the impact of food waste on the environment and climate change – it turned out that sustainable, regenerative food systems were an issue especially close to her heart. Her posture shifted as she spoke, her voice became more energetic than it had been before. I could see that something in her lit up when she spoke about it. As the conversation continued, Allison admitted that while she was enthusiastic about the idea of composting, she also wondered if it was a worthy contribution to the cause. Real climate activists, she said, were the people showing up to climate marches, organizing boycotts and protesting on the front lines of injustice. We explored the 'should' that had come up at the beginning of our session: 'I should be a climate activist.'*

*'What does it mean to you to be a climate activist?' I asked. Through an exploration of values and limiting beliefs around what it meant to be an activist, Allison realized that she could give herself permission to broaden her definition of 'activism' beyond the narrow stereotypes that she had used to hold about it,*

*expanding her view of climate activism to include the way she and her family lived on a day-to-day basis. The idea that composting in her home could 'count' as climate action felt empowering and motivating, and as we came up with a clear goal for starting the process of getting a compost bin set up, Allison seemed energized.*

*Towards the end of our session, I invited Allison to name the insights she was taking away:*

> *I know I'll eventually want to do more around climate action, but the idea that composting is a good enough place to start makes me feel like I can actually do this. I hadn't realized that the way I was thinking about activism was setting me up to feel like a failure all the time – it feels really good to know that I can find ways to be an activist in the life I have now.*

*When I checked on her several weeks later, she and her family had successfully set up a composting system at their home. Allison also reported that she had started making calls to her representatives about climate change on a regular basis. While climate change still seemed overwhelming at times, Allison had built her confidence in taking small, doable action steps that aligned with her current stage of life. It was a win for Allison, and a win for the planet.*

## Supporting change with structures, processes and systems

Is it possible to get into action, without there initially being much intrinsic motivation at all? Yes and no. We can certainly support our behaviour change in a range of ways, both individually and organizationally, but our sense of resonance is how we translate an external reason to act into an internal drive to do something. In other words, resonance sustains change, so while we can start new things first and discover the resonance second, the whole change endeavour will crumble if we do not find the resonance at some point. If we do try to form habits for which we are poorly motivated, there are a few external ways to help make them stick. We might incorporate a new activity into an existing routine to make us less reliant upon personal willpower – like putting a bike helmet instead of the car keys by the front door at night to make it easier to choose the bicycle over the car in the morning. We can also consciously attach meaning to tasks when our motivation is waning, for example by reminding ourselves what doing the activity will give us – 'Even though it is raining, if I cycle to work, I'll feel clear-headed when I get there' – or by adding something enjoyable as we carry out the task, such as being able to stop for coffee on a cycle route on which we would not be allowed to stop in a car.

We know from the fields of behavioural science and behavioural psychology that receiving immediate positive reinforcement has a powerful effect upon our behaviour, even when motivation is low. If we cycle to work on a sunny day and arrive feeling recharged, we will likely do it again. If, on the other hand, we receive an immediate negative consequence, such as being splashed by a car as we cycle in the pouring rain, we are less likely to repeat that action. However, as long as the burst of motivation lasts, we are likely to carry on, and this is true even when we 'force' ourselves to do something a few times, provided that we start to see some results. For example, we may notice our midriff getting slimmer because we are cycling to work, and that might keep us doing it on mornings when we are tempted to drive.

Our behaviour can also be modified externally without there being strong personal motivation. Everything from our supermarket shopping and online browsing choices are examples of how our actions are shaped without our conscious choice or motivation. However, when processes police us but fail to truly influence us, we do not continue with the behaviours after the processes are removed. This is passive engagement rather than ownership. A better way in which we can be externally supported to take action (even without intrinsic motivation), is by having friends who encourage us. You may not be sold on going vegan, but yet give Veganuary a go because some of your friends suggest you do it together. Without the anchor of intrinsic motivation however, even a small bump in the road may reset you back; we may go back to eating meat in February when the social support has disappeared. Resonance helps us connect to our internal motivation to change rather than being 'pushed' from the outside, and in turn helps us form a habit, where our self-concept makes a shift from 'someone who does not like cycling' to 'someone who cycles'.

Aside from making their new behaviours stick, you might argue that it doesn't matter whether someone cares about climate change, as long as their impact upon the environment is reduced. In one sense that is probably right; however, this sense of meaning or resonance is important when we consider the ripple effect from person to person, because we are less likely to positively influence others to change if we are not connected to a meaningful reason for it ourselves. Imagine your friend asking 'Clare and I are doing Veganuary, do you want to join us? I'm doing it because Clare won't stop going on about it, and you know, it's only a month.' Now consider this instead 'Clare and I are doing Veganuary, do you want to join us? It's a great way to get healthy at the start of the year, and it really makes you feel like you're part of the solution.' Which of the two would interest you more?

If powerlessness makes us feel bad, and feeling bad makes us avoidant, hard to be around and resistant to change, then empowerment makes us feel capable, chirpy and able to take on a challenge. This matters when it comes to systems change, because whatever mood we choose to put into the system is contagious. Think of a tipping point in human history, where something shifted from seeming completely impossible to being an inevitability – the success of the women's suffrage movements perhaps. In each case the law changed when it was felt that there was more of a groundswell of support in favour of the change than against it. More prosaically, not many years ago the major car manufacturers said that electric cars would not catch on, but now all major brands have at least one model. In each of these examples, innovation and policy change clearly played a big role but so too did a generalized sense of confidence that the paradigm was shifting.

We began this chapter by talking about belief. When as a system we connect to an idea and we believe in the possibility of change succeeding, it is a self-fulfilling prophecy. The same is also unfortunately true when we stop believing. In economics this is called 'market confidence' when investors suddenly lose belief and markets collapse. So the same is true for other, non-financial aspects of life, from the way we will now more easily talk about our mental health to the way in which it is now accepted almost globally that girls have the right to be educated as well as boys. The good news is that when we all get together and believe, change happens. This is where a coaching approach comes in.

# 4  Coaching for climate action

## What is climate change coaching?

Climate change coaching brings together the practical skills of coaching with the knotty problem of our disempowerment about climate change. We have so far explained that powerlessness can hold us back from taking action and that empowerment is the key to being more imaginative about change, sticking with it when it is hard and inspiring others to join us. In this chapter we will look at how a coaching approach can help you to make that shift from disempowerment to empowerment and the specific differences between climate change coaching and traditional coaching.

## What has coaching got to do with it?

You can think of coaching skills as a practical toolkit through which we apply the theoretical ideas of psychology to our everyday lives. At their most simple level, coaching skills are a way of developing self-awareness. There is never a bad time to increase your self-awareness; however, coaching is most useful during times of change and uncertainty, when we can acquire the emotional muscles to not just survive but to thrive. In our fast-paced world, in which we barely have time for a sandwich let alone any proper thinking, the simplicity of coaching questions combined with the reflective pace that they necessitate can offer us a doorway into a different part of our brains and dramatically alter the way we think. Married to that, coaching purposefully includes a large element of championing which deescalates the sense of threat and overwhelm that this subject can generate. Dialling this down makes us less wary of change and more open to influence.

Coaching skills can help us to move people from a sense of doubt to one of agency, but this is not the only work to be done. Beyond achieving a proactive mindset, we also want to catalyse action. This is again where coaching can be helpful because it naturally has a dual focus on thinking things through *and* getting things done. Coaching marries two crucial aspects of the problem we face in galvanizing climate action, as we show in Figure 4.1, in that it creates both the time to reflect and work through our feelings and a focus on action as a means of making progress and building mental resilience.

The extraordinary thing is that when we put emphasis on the left-hand side (reflection), the right-hand side (action) often takes care of itself. Whereas when we focus immediately on action, we either miss the reasons that someone has stalled or we arrive at bland ideas.

Unspoken or unconscious disempowerment can take many forms, from straight up fear of change, to believing the change crosses our values, to a sense of not being able to speak up for ourselves. We can also agree to things when we

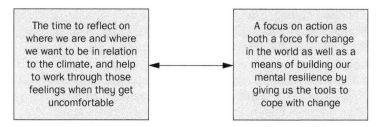

**Figure 4.1** How coaching balances reflection and action

feel disempowered because we fear disapproval or because we have not thought them through properly and our inner critic tells us to 'get on with it'. When we do that, we often come unstuck. Just as Rahim could not articulate a good reason not to go vegan, we are not always consciously aware of the underlying beliefs and feelings that drive our actions. If you manage or are collaborating with someone, it is better to have their concerns surface and be examined at the outset, than to have those concerns derail the process (and the other people depending on it) later on or for that person to 'go quiet' and avoidant. Coaching skills can help you to start to surface these concerns, by opening the discussion up to what the person thinks or feels about the idea/project/new direction. When we help someone to get clear on what would make them feel comfortable to move forward (even when it is an activity that we would like them to do), they feel safer and more capable. It is no surprise really; when we feel secure and confident, we are much more open to influence. That makes us more likely to take resonant action, which in turn builds our confidence further, making us open to future influence; creating a naturally virtuous cycle. Where disempowerment can disconnect us from our own capability, damage our relationships and cut us adrift from the issue of climate change itself, so empowerment can create the conditions for connection, innovation and collaboration.

## What is climate change coaching?

Climate change coaching has several components that may feel different to more traditional coaching. First, it encourages us to be in relationships with all levels of the system (from ourselves, to others, to systems and nature and to the crisis itself), rather than on individual actualization alone. Second, climate change coaching recognizes the powerful role of psychology in tackling climate change, and rather than putting the climate into a technical box that is not for coaching, it works with someone's emotional needs as a result of the crisis, rather than assuming that knowledge alone is required. Finally, and most importantly, climate change coaching addresses not only someone's individual sense of powerlessness but their lack of belief that the system can change too.

The biggest difference between climate change coaching and more traditional coaching, however, is that we ourselves are part of and vulnerable to the systems that we are coaching within, at the biggest level of the system: the planet. This means we have to do much more work on our own relationship and sense of agency, in order to help others with theirs. A manager using a traditional coaching approach with a colleague may get triggered by them sharing an organizational issue, because they are in the same company, but will probably be able to set that aside long enough to help. However, when that manager hears 'I'm so worried about the future for my children', they may find it harder to stand apart if they share those existential fears too. This makes it all the more important that we work on our own feelings and triggers. For this reason, we offer 'over to you' opportunities to help you understand your own relationship to each of the ideas that we cover. It can also help to try to adopt the powerful mindsets of acceptance and compassion and we discuss those in the next chapter.

## Solving the right problem – working with the who, not the what

Whether you have been coached, are a coach or are a professional considering using a coaching approach in environmental work, you probably already know that coaches do not give advice. You might be baffled as to why they don't when they are being paid good money to 'help people'. It all comes down to recognizing what we are helping with. While it is true that some people naturally jump into action, it is more often the case that we are not sure how to start, or that we are taking action that we feel we *should* but don't enjoy. When someone is in this dissonant place, it is very easy to become an unhelpful thinking partner, by focusing on action without understanding the feelings lurking behind. When someone feels capable (empowered), focusing on action becomes a fun brainstorming exercise. But when they feel stuck (disempowered), starting with action pretty quickly becomes a frustrating process of you suggesting perfectly achievable things while they tell you all of the reasons they 'cannot and will not'. Many of us have been socialized to offer advice: 'Have you tried this? What about that?' We may also want to be helpful, or perhaps we ourselves find the very idea of being stuck a bit terrifying. The trouble is, suggesting solutions usually does not work and leaves us feeling frustrated or dismissed. This can damage relationships because it can take a lot of vulnerability for someone to share their concerns about the climate crisis.

When we give unsolicited advice, we are often focusing on the wrong issue. The real problem is lack of belief, not lack of creativity. None of us truly lack creativity, we just sometimes disbelieve that we can access it. This is an easy trap to fall into, in which we focus on someone's problem rather than on their *relationship to the problem*. When someone tells us 'I feel really stuck about my frequent flying for work', instead of reaching for a guide to sustainable travel, we can ask 'What do you usually do when you feel stuck?' This is moving away from what they could do and towards who they are and their existing resourcefulness. You will be surprised by the answers you get when you make this switch. This focus on the 'who' is especially important when it comes to climate change coaching because it can be very easy to be triggered by the problem (or the 'what'). Maybe you are also wondering about flying, and that could cloud your ability to help. But while you share the same concern, you do not share the same body, mind or life experiences.

When you shift focus to their *relationship* to the problem, a whole suite of ideas opens up before you both. This is great news for all of us because it frees us from needing to be subject experts about climate change (even those of us who are). Many people shy away from simply talking about climate because they feel like they do not have enough technical knowledge or do not want to lecture people, when instead if we share our lived experience of the crisis and the blocks it creates in us, then it is possible to create real transformation in our thinking, as we realize that we do have the resourcefulness we need to take action.

The other reason to focus on transforming someone's perspective around a subject, rather than telling them what to do, is that when someone feels positive and capable they have far more access to the creative parts of their brain. They are therefore much more likely to devise fantastic solutions. As we have said, when we feel threatened we pull up our mental drawbridge and stick to what we know. When we feel safe we try new things. It is not that the person is the 'kind of person who cannot come up with ideas' or 'doesn't know where to start' (even when they tell you that is the case), it is simply that they are not in the right mind-set to do so. We would never dream of telling a child that they are not imaginative, but yet as adults we think it of ourselves and others all the time. A climate change coaching approach can help you create the conditions in which others can rediscover their imagination in the face of complexity.

Climate change coaching is best suited to solving the problem of people feeling powerless either individually or systemically. While someone may also need to go and find a technical solution to their problem, almost always when someone says they are stuck or unsure, there is some coaching work to be done to help them believe they can do it and hold them to account. Information and action are a big part of the picture but not the only part. For example, if Monica wants to learn about ocean acidification as part of her beach-cleaning action, coaching cannot teach her that, she needs to take a course. However, Monica might start a master's in marine biology and need a friend to help her believe that she is 'the kind of person who does a master's'. Even when the problem genuinely is a lack of knowledge, a coach or someone using this approach can play an important role in encouraging belief, helping us to stick with it and holding us to account, especially when our voice of doubt is working overtime to convince us otherwise.

Coaching is the one of the few modalities in which someone is supported to find their own answers and put them into action. Climate change coaching is not, however, a single silver bullet and a coaching approach does not solve everything. If you are reading this book in order to have more impact then you are likely to be considering a range of different ways to influence people. Changing systems requires a multi-pronged approach. At the same time, it may well be that these ideas of self and collective belief are among the most undervalued components of successful change. Here's what a coaching approach can help with, and what it cannot.

### What a climate change coaching approach can directly help with

- Create the space for someone to properly consider climate change in more depth and what it means for them personally.

- Help someone feel understood when they are isolated or lonely and think no one else understands.
- Support someone to unpack their self-doubt and connect to their motivation and purpose.
- Help someone decide how to act – what they want to do or what part they want to play.
- Create accountability for an action (or for researching an action).
- Maintain a mindset for sustained action.
- Make it safe for someone to surface their concerns or resistance to change.
- Manage our numerous complex emotions and create a space to process big emotions when we are upset, angry or experiencing climate grief.
- Support ourselves and others to stay resilient and encourage us to care for ourselves.

**What a climate change coaching approach cannot directly help with**

- Providing technical information.
- Lobbying to make your views heard.
- Telling people what to do regardless of their needs and wishes.
- Trying to force someone to care about climate change when they do not or take specific actions that they do not agree with. We are not sure, however, that there is any methodology that can do those things!

You will notice that climate change coaching supports people who are concerned or engaged in the crisis; however, do not dismiss this as being a tool only for those who are already deeply committed. Many people are anxious about climate change and not acting, yet feeling guilty on the sidelines, unsure of what to do. As Elizabeth Bechard's story in Chapter 3 demonstrated, coaching skills can be used to open safe, non-judgemental conversations to help people engage safely. It is often the case that someone does care and we just have to find out what exactly motivates them. While you would not be coaching if you were doing the things on the second list, you may, however, use coaching skills alongside the first two of those things. For example, a sustainability consultant may present a technical report to a team and then listen well and ask powerful, open questions to gauge how it has landed. The best lobbyists are also those who are highly skilled at building trusting relationships, such as Jill Bruce does in Chapter 6. A coaching approach can be integrated into many aspects of life and work, because at its heart coaching is about listening without judgement and asking great questions that galvanize action.

**The difference between holding coaching sessions and using coaching skills**

There is an ethical difference between formally coaching someone in a coaching session and using coaching skills in everyday conversations. If you are a trained coach holding a coaching session you are ethically bound to respect the coachee's agenda – meaning if they do not want to talk about the climate crisis, you do not pursue it. In Chapter 7 we look at ways you can discuss this with clients in a coaching session. If you are a professional coach wanting to use your skills, it is important to be clear what role you are playing at any one time. Imagine that a coach wants to lobby their local councillor, who is also their coaching client, to save a local woodland from being cut down for housing. It would be unethical to raise this with the councillor in a coaching session because the coach has an agenda of their own. As a resident however, they could visit their councillor in the council office and use coaching skills such as recognizing how hard it is for the councillor to please everyone (the skill of acknowledgement in Chapter 5), as well as being clear on what they want the councillor to do for them as a resident (the skill of contracting in Chapter 7). It would also be important that they made it clear they were there as a resident, not as the councillor's coach. This could get sticky if the councillor themselves wanted to bring the woodland development as a topic to a coaching session. The coach would then want to be open about their own agenda and if the coach did not feel that they would be able to manage their own views while coaching this topic, they might legitimately tell the councillor that they cannot coach them on the topic. While the best path is to be able to manage and 'put down' our own thoughts and feelings during coaching, it is not always that easy. It is better to be open and say no to a topic in service of the bigger coaching relationship than to unintentionally lead the coachee towards our own agenda because we are very attached to it. If we have not agreed with a client to bring climate into our coaching then as coaches we must be mindful of not being 'an activist in coach's clothing', and do our activism outside of our coaching. As agents of climate action, a coaching *approach* can be used to great effect in our normal life and in our activism, where we can inspire others to join us, take action or change their organization.

## In practice

### How I use climate change coaching in my work

*Sarah Taylor, senior specialist climate change adaptation, Natural England*

*I first experienced coaching a few years ago, when I was lucky enough to be offered some free sessions by a brilliant trainee coach. I used the sessions to think about the different parts of my life and how I could maximize the things that feel the most important, worthwhile and fulfilling to me and explore some of the things I was finding difficult. I was thrilled with the insight these sessions gave me on a personal level, and it also sparked an idea about how I might be able to apply this in my job. About a year later a member of our internal coaching network got in touch to talk about using coaching skills in our approach to climate change. It was an exciting idea, which led to some of Natural England's coaches being trained in climate change coaching, and a series of group coaching sessions for colleagues who wanted to create more climate action in their roles.[27] This gave us new ideas for how we approach climate change at Natural*

*England but they also showed us how we can address the huge emotional and professional challenges of climate action, while looking after ourselves to ensure we can stay active and effective.*

*For me personally, learning climate change coaching added to my positive experience of being coached and helped me build on some of the realizations I had gained. I know that I'm inclined to be directive in the way I call for action, both of myself and others. My role as a climate change specialist requires me to gather evidence and share it in a way that my colleagues and our partners can use, and I know how much we need to embed climate change across our organization. Presenting information in a way that galvanizes action is core to my role. But I was also starting to see that my tendency to direct (or demand) action, coupled with my own feelings of scarcity, overwhelm and urgency, were in danger of creating unhelpful outcomes. Coaching was prompting me to ask what the best ways were to create change and what else I could try. I used to think that taking care of the thinking for my audience, made it quicker for them to act. In my head I said 'I have the information you need, no need to figure it out for yourself, I can tell you what to do, that'll save us all time.' If people didn't then leap into action, I assumed I had communicated the facts poorly, and would just say it again. My experiences of coaching are helping me realize that my 'shortcut' might not be the best way to galvanize action and I could develop more supportive and effective communication skills.*

*Coaching empowers people to come up with and own their actions. For climate action, we really need this kind of positive energy. Action needs to be evidence based, but in that we can also understand and utilize our emotions and our connection to the natural world and to each other. Our purpose and enthusiasm can be part of the story too. It is why we do what we do. Coaching can also help when our emotions become a barrier, and we get stuck. Nature is so often an inspiration, but our relationship with nature can bring difficult emotions too, when we think about what we have done to the natural environment and the scale of the challenge we now face. I often feel a huge range of emotions, which can negatively impact my motivation. I have learnt that coaching can help us work with these feelings and devise actions that help us find our place and make positive change in the world. Discovering a coaching approach has inspired and invigorated me. I have been so enthused by the revelations I have had so far that I am now training to become a coach myself. I know the things I am learning will help me in my approach to life in general, help me be a better ally, actor and advocate and will also make my work for climate change adaptation and nature more effective.*

*This learning journey will continue for me as I now take my coaching training and start designing the next steps for climate change coaching with my colleagues within Natural England. I'll be thinking about how my role as a climate change specialist can be informed by what I have learnt, and how I can share the coaching approach more widely. We will look for the best way to support and grow the climate change coaching network that has developed out of the training we received, so we can expand our effectiveness as a group. I cannot claim to have completely stopped using my 'shortcut' or demanding people take action, but I do notice when I am doing it and try to change my approach. I'm still on the first steps of this journey, but I see there are useful changes I can make and I'm excited to try them out.*

## An emotional focus

Until now we have talked about transforming someone's feelings around an issue as if we are impartially helping them to work through an issue unconnected to us. If you are a professional coach, this fits with the principle that the coach does not have an agenda of their own. However, this is also a book about how all of us can use a climate change coaching approach to influence others to change, and by that definition we do have something that we want them to do. Coaching skills can help us just as much there too because they put choice in the other person's hands. When someone feels cajoled or pushed into doing something, they either verbally push back or worse, they agree but feel resentful or anxious. The task may then be done poorly or not at all, which can damage their self-confidence and your relationship, and diminish their likelihood of being open to future influence. Whether you want to help someone engage with climate change itself or to influence someone to do a specific activity, in many cases what is standing between you and the change you seek to make is not only the person, but their emotional response to change.

Regardless of who has initiated the action – whether you are wanting to influence someone to do something for you, are talking to a friend about something that matters only to them, or you are a coach with a client – the process is in fact the same:

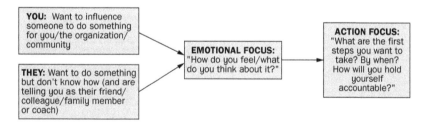

**Figure 4.2** The role of the emotional focus in generating action

This makes it look smooth when, of course, it is often a process of back and forth or 'circling the airfield' with someone getting close to action (landing) and stepping away (back up into an emotional holding pattern). Different people also need more or less reassurance before they will risk taking action. What may seem easy for you may seem edgy for someone else, because of differences in your skills, experiences and approaches to risk. For some people any kind of change feels risky, while for others, standing still and *not* changing feels much more confronting. A coaching approach gives you the tools to keep building someone's confidence or resonant reasons to act, and we will give you lots of specific ways to do that in Part B. You'll know that someone is feeling resonance if, when you ask them what they want to do, it feels like you only had to lightly tap them for an answer to appear. If it feels instead that you have to push and needle to get an answer, you are not there yet and you need to do more to dial down doubt (dissonance) or dial up motivation (resonance). This can feel slow to begin with, but the good news is that once someone feels confident, the action part of the conversation can happen very quickly. If we return to the model of sustainable change (Fig. 4.3), we can see that coaching skills focus on the Y axis.

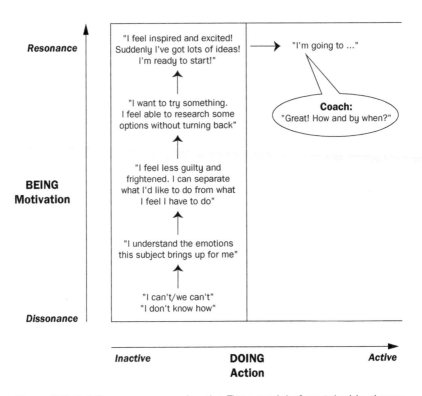

**Figure 4.3** Building resonance using the Flynn model of sustainable change

Using coaching skills we can noticeably move people from dissonance (the 'low motivation/being' bottom of the Y axis) to resonance (the 'high motivation/being' top of it). This often makes coaching skills the missing component in innovation processes. If we look at the model, we can see where the greatest innovation is likely to emerge – in the resonant top half when people feel energized, positive and ready to act. There is no point trying to brainstorm ideas when we are feeling dissonant (in the bottom half of the graph) because our dissonance will blinker us, and our doubt will shut down every suggestion. This is also the place in which we can feel threatened or unsafe. When we feel that way we will generate low-risk, low-return ideas that do not stray too far from the norm. By contrast, when we are feeling resonant and empowered, we are likely to see opportunities, not problems.

When someone reaches a place of resonance, they naturally start to generate ideas and will naturally want to move into action. In fact, just try and stop them! At that point our job is simply to offer some accountability – for example 'What will you do first? By when?' As an experience in conversation it can feel that we are just lightly nudging someone as opposed to pushing them into a change. Professional coaches will offer formal accountability, by asking clients how they want to check in, but anyone can use a looser version of this, for example: 'Can you tell me how it goes?' or 'Do you have a sense of when you might get started or what's the first step for you?' Rather than having to do lots of hard thinking about the best solution someone could employ, with a coaching approach we instead improve the quality of the other person's thinking and belief so that they can come up with the hard answers themselves. This is as true when we have a fixed outcome in mind and are trying to influence someone else to decide how they will do it, as it is for a professional coach who has no attachment to the result.

Put like that, shifting someone into an empowered state sounds not only crucial but quite a trick. Yet coaching skills are powerful because they are simple and transform someone's perspective with such a light touch that it can almost feel like magic. The good news is that you already have the ability to do this. Coaching skills are innately human. We all have the capacity for empathy, for cheering each other on and for comforting each other when we are struggling. Our children remind us of this every day. You do not need a certificate to be a human being. So while we will teach you some pretty powerful magic, know that you are already a magician; you just possibly forgot how to use your wand.

## Reconnecting our inner magician to our inner professor, balancing our thinking and feeling selves

Despite the fact that coaching skills are so human, it is sadly true that as we 'grow up' we are subtly taught to value our thinking selves over our feeling selves and to prioritize the verifiable over the instinctive. While we are not for one moment arguing for the abandonment of logic or intellectual knowledge, we do suggest that our thinking and feeling selves be allowed to come more into balance. Kimberly Nicholas framed this as: 'I've realized that giving space to my feelings gives me more empathy with what others are going through as part of the shared human experience and helps me connect with them more deeply.'[28] This balance is not only useful because we know from behavioural science that data alone rarely changes behaviour, but because we also know that the coming decades will expose us to more, not less, uncertainty and unpredictability, and that it is within our feeling selves that we will find the self-trust to more successfully cope with that.

Let us go back to Rahim and Petra to look at this in practice. Rahim knew in his heart of hearts that he didn't want to go vegan, but his thinking self may have said 'Look, Petra wants you to do this, and you can't think of a good reason why not. You should agree.' Later, when he was munching on his guilty cheese steak sandwich and finding ways to blame Petra for it, he may have had another thought: 'I knew veganism was a bad idea! Why did I go along with it?' Whether we call this Rahim's gut instinct, his intuition or his heart speaking, this is a great example of us not listening to our feeling selves until it is too late. On the flipside, Petra may have had this thought process: 'I just *know* veganism is going to be good for me. I haven't got any proof yet to convince Rahim, but I know that if we try it, I'll love it.' After a month as a vegan she might then have thought 'I *knew* I'd love this and I do! And only now do I realize how bored I was of what I was eating before.' This is an example of Petra trusting her feeling self to guide her when she did not know enough to logically explain why. Notice in both cases their feeling selves 'knew' all along. This chimes with Daniel Goleman, Richard Boyatzis and Annie McKee's work on emotional intelligence and the way that over our lifetimes they observed that people's 'brains had quietly picked up … accumulated lessons. Though their logical intellect was still stumped … their intuition told them what to do based on … lessons learned.'[29]

The funny thing is that when we act from a felt sense of rightness, we rarely regret it, even when the decision turns out to be a bad one. If Petra had hated veganism, unlike Rahim she would have been likely to say 'Well at least I gave it a go and tried. I'll just have to find another way to reduce my carbon footprint.' When we listen to our feelings during decision-making we are more likely to voice our concerns, and also if we later turn out to be wrong, we are less likely to engage in blame (of ourselves or others). Blame is another example of shifting responsibility away from us because we feel powerless. This means that just giving some airtime to our feeling selves can help to diminish our disempowerment, because we create space to articulate what is going on for us, and not brush those feelings aside (as we showed you in Fig. 4.3). We will cover blame in more depth in Chapter 13 when we look at anger.

Just as it is important to allow space for emotions, so it is also important to rebalance towards logic if we operate too much from feelings. When we lash out at others because they represent a systemic problem (for example, when we see someone glorifying fast fashion), we are not in balance either. When we act with too much feeling and not enough thinking we can also find ourselves exposed to negative consequences. At a systemic level, polarization, after all, breeds on triggered selves not engaging in logic. An example may be when we open our social media and get triggered by a viewpoint that challenges something we care about. For example, a friend says that they are 'So excited to be flying to Bali this spring!' when we feel strongly about the damage done by air travel. Our triggered self may reach for the keyboard to give them a piece of our mind about the fact that they 'are stealing our children's futures from them'. If we act on that triggered feeling and leap in with harsh, disproportionate judgement, we will damage the relationship and most likely later regret it. If we checked in with our thinking selves at that moment we could step back and examine our reaction, and maybe also stand in their shoes. That does not mean that we will not say anything, but perhaps rather than dashing out a very public dressing down, we would send them a private message to say how concerned we are about flying. Maybe then we'd find out that this is their first flight in 3 years and maybe by discussing it with them calmly, we may plant a seed for how they think about flying in the future.

Notice that in both cases it is our internal voice of disempowerment and doubt driving us, not our voice of reason (thinking self) or values (feeling self). This is why a coaching approach does not abandon logic but rather allows logic and feeling to come into equal relationship with each other, so that they can both assist us with decision-making. When we can balance our thinking and feeling selves, we are better able to both manage change and to deal with uncertainty. This is part of what we mean about being in a positive relationship with ourselves. As many of us exist in societies that have downgraded the value of the feeling self, in this book we will speak more to how you can expand that side of your range. A coaching approach, which is based on working with the whole person, not just what they do and not just what they know, can help us to strike this equilibrium.

## In practice

### The head, the heart and the hands

*Dr Sarah Higginson, research knowledge exchange manager, Centre for Research into Energy Demand Solutions (CREDS)*

*In my work with energy demand researchers, I have learnt that we need to bring our whole selves, or our heads, hearts and hands, to our endeavours. Our heads champion facts, rationality, strategic planning, research, evaluation and monitoring. Our hearts drive our passion and compassion, allowing us to bring our values into play. Our hands invent practical solutions, build solid new structures, and bring an applied approach and real-world solutions. Working singly, the head, heart or hands are less effective than if they work together, when they can offer checks and balances for each other.*

*The head, heart and hands can be blended within ourselves but also by working across sectors or in multidisciplinary teams. For example, and clearly generalizing greatly, academics and many businesses tend to operate from the head; the 'public' and media from the heart; and NGOs from the hands. This is why collaboration between different stakeholders is so important to achieve impact. Rather than competing for scarce resources, we can tap into our shared resourcefulness and work together towards transformation.*

*Of course, the head, heart and hands can also sabotage our efforts. Unchecked, heads can become cold, calculating, dismissive and distant, caught up in intellectual arguments. Our hearts can spiral us into despair, grief and overwhelm, eventually getting angry and resentful. Our hands can lack direction, fail to achieve impact and become impatient. It is clear that this will quickly lead to exhaustion, which is why inner work is so crucial to our cause. There are many in our field who are operating out of balance, who may value the cause more than the people around them, or 'results' more than their relationships, or even themselves. This can lead to unfortunate side effects like workplace bullying and burnout. Doing the work on ourselves is essential because, ultimately, our inner realities are reflected in the world we create around us.*

*I have worked with a coach over the years, who has helped me to understand my overall purpose and how to bring my head, heart and hands into better balance. With her help, I have learned to harness the logic of my head. She has helped me to deepen my values, find my authenticity and courage, and step into my heart. With her I have learned to rest in my breath and reside in my body, helping my 'hands' to develop and facilitate powerful processes that engage those with whom I work. When I am able to bring all three to my work or life, I am effective and things seem almost to manifest around me. My own transformation has allowed me (on good days) to 'show up' completely, to bring my whole self and to be authentically present in the work I do. This is not to say I do not have my share of hurdles. One of the risks of bringing our whole selves, displaying our full power or being deeply authentic is that it is threatening to the system and to many people, which can result in a backlash. For the most part, though, coaching has enabled me to hold this lightly and stay engaged.*

*For me, apples offer a powerful image of what I mean by bringing our whole selves to achieve impact. I have several apple trees and every year, apparently effortlessly and without harm to anything, they produce more apples than I (or the creatures who share my garden) can possibly use. Rather than fear and lack, apples speak of an unbelievable abundance born of deep connectedness with their purpose and with their role as part of a whole. Like apple trees, we can all flourish abundantly if we remember our connectedness to ourselves, each other and our whole purpose.*

## The importance of being in a relationship

It may be money, as the song goes, that makes the world go round, but it is friendship, as Woodrow Wilson reminded us, that is 'the only cement that will ever hold the world together'.[30] We have learned through successive crises in this century alone that our individual and collective resilience – our ability to weather the storms that we have experienced – is built on our sense of connection or undermined by our feeling of isolation. We cannot hope to tackle the problems that will be created by climate change without standing and working together. As easy as it is to write that, it is much harder to do.

It might sound odd to suggest that we need to make a conscious shift to being in a relationship with others. 'I've got plenty of relationships' you might say, and we hope you would be right. Yet how often do we step out of a relationship with things, organizations or people because they upset us or we do not agree with them? Polarization has become an unfortunate characteristic of our societies, in which battle lines are drawn around everything from politics to fashion. There may be times that you find it very difficult indeed to stay in a relationship with another person, especially when they refuse to see your point of view or do as you want them to. You may also find it terrifying to be in a relationship with climate change itself, and feel a strong desire to bury your head or to avoid the feelings that haunt you.

Coaching relationships are fundamentally based on trust, to create a relationship that can withstand challenging conversations, times of hardship and the tricky road to change. For the relationship to work there must be mutual belief that the coach is genuine and that the person they are coaching is inherently capable. Climate change coaching is no different, where we must also start from a position of seeing everything (including people, systems and the crisis) as worthy of being in a relationship with and being capable of evolving. You may already be frustrated with the results that come from a transactional, command

and control approach or from perceiving the person you are trying to influence as being uncaring and hopeless. When you adopt the mindset of a coach it is easy to build trusting relationships with others, systems, our planet and fundamentally with ourselves, because your starting point is that everything is resourceful and worthy of respect. Crucially, these kinds of relationships are the foundation for fostering more open, transformational conversations that lead to real change.

## Developing healthy relationships with everything

While some relationships repel us, others may ensnare us, and we can find ourselves in unhealthy relationships; sacrificing ourselves for causes or work to the detriment of our health or families, or tolerating toxic behaviour from others because we doubt our ability to be on our own. We may ourselves play the obsessive partner, in an unhealthy relationship with 'all the problems', addicted to every new development and glued to bad news. Ideally we want to be in a healthy relationship, with ourselves, with others, with systems and with the planet. Healthy relationships are safe – because we do not feel threatened – and so provide fertile ground for more skilful communication. To be in a healthy relationship does not mean agreeing with views which we find abhorrent or tolerating abuse, but it does mean being compassionate rather than contemptuous in return. Sometimes the healthiest way for us to be in a relationship with ourselves, is by stepping out of a direct relationship with someone else, quitting a job that is ruining our health or taking a break from climate activism.

When we do decide to step back from a relationship it is important to reflect on *how* we walk away too. Blame itself can be a damaging emotion that can pull us off course from our broader intentions. This may mean shifting perspective to understand how the other side is 'good and right' not 'bad and wrong' and the same goes for seeing ourselves that way too. If we can do this, it helps us to then embrace new relationships wholeheartedly, rather than just hovering at the periphery. That might mean leaving an activist group because you do not agree with their new strategy, but at the same time recognizing that this does not make them bad or you a quitter, but that rather you are just not right for each other any more and are better apart. Equally it might mean agreeing to disagree about sustainability with a relative for the sake of having harmonious family gatherings.

The benefit of this approach is that it generates a sense of belonging to ourselves and each other, that enables greater change to happen. When we stand in a relationship rather than in opposition, numerous possibilities open up:

– We become more collectively resilient. We recognize the need to 'fit our own oxygen mask first' so we can go the distance rather than burning out.
– We build bridges and find ways to understand other viewpoints.
– We see ourselves as part of the problem, on the pitch with everyone else, rather than externalizing it and blaming others.
– We take a share of the responsibility for the issue or the group, rather than being separate to it or leaving it for someone else to sort out.
– We access more creativity in our teams when we work more effectively together helping new ideas to emerge.

Maintaining healthy relationships can be a tough shift to make in our Anthropogenic world of individualism and control, where separation is so often encouraged. To create change in the world we have to be in a relationship with it – to engage emotionally not hurl mud at it from the sidelines. It sometimes helps to think of all of our relationships as being like a marriage: we are in this for the long haul and by choice, not trapped or not noncommittal. We enter into the relationship hopefully, expecting to go the distance together, even during the times when we want to turn away. We will work on the relationship, and in doing so we will learn multitudes of things about the world, and about ourselves.

This means checking in with how well we relate to each level of the world around us, and it means moving from being *out* of relationship to *in* it. Here are some examples of how we can shift to being in a more helpful relationship:

– An individual who burns out because they find it hard to listen to their body …
  to being someone who makes time for rest and reflection.
– A friendship in which politics is a no-go area … to one in which politics is discussed without blame or judgement so that new ideas can bubble up.
– An organization in which the boss makes all the decisions … to one in which decision-making is co-created and collaboration can occur.
– A population that has become disconnected from the planet on which it lives … to one that recognizes and acts for the symbiosis of all elements of the planet, so that regeneration is possible.

When we step into a relationship, we step out of full individual control. This letting go is important, because if we truly want to influence others, then we need to recognize that to be in a true relationship with ourselves, with others, with systems and with the planet is to *share* control. This is when we truly collaborate. All of us can build better relationships, whether that is with each other or with the climate crisis itself. And all of us can create conversations that are catalytic rather than guilt-inducing. From the sustainability professional who is landing their technical knowledge into a company, to the community campaigner lobbying local politicians, to coaches supporting leaders as they consider their legacies, to parents at the school gates sharing ideas on reducing waste; to the flowers that bloomed in the ashes of the Australian bushfires: each of our talents are needed in concert with everyone and everything else's.

If you stop here and you do not use any of the tools in the next chapters, know that being a friend to others rather than going into combat with them is a big step towards influencing change. If you also spend time reconnecting to the natural world, then nature can support you just as you work for it. If you also make a conscious decision to cultivate self-compassion and be in a kinder relationship with yourself then you will be helping to protect your mental health as you do this work and ensuring that you are still doing it long into the future. If you are keen to read on, prepared to be an even more active part of our global change journey, and ready to be a magician, then turn over to Part B, and we will teach you some tricks.

# Part B

# Transform the way you communicate

# 5 | Climate change coaching basics

What do you want to change about the way that you talk and work with others when it comes to climate change? Perhaps it is that you want to understand better what they need in order to become engaged or to hear why they feel powerless and incapable. Or maybe you want to be able to challenge people when they appear to be making excuses, but you don't feel that you have the tools to do that. Then again, it could be that you suspect that your approach and mindset could do with an upgrade when it comes to helping others embrace the climate challenge. Whether you come at this wanting to soften your approach or better hold people to account, a good starting point is to consider how you would like to be treated in the same situation. We humans generally appreciate someone taking an interest in our thoughts and ideas, and want to feel included, validated and secure. We prefer to be treated as adults that are given the benefit of the doubt. Many of us also like to be stretched and thought of as competent and capable of taking on a challenge.

In this chapter we offer you what is taught to all coaches at the beginning of their training – the mindset that coaches are required to embody and the practical skills that can transform your conversations – with tools and examples of how these operate in a climate change context. You will know that your coaching approach is working, because ideas and solutions will appear when there was no sign of them before, and sometimes a whole world view can shift on its axis in the space of just one conversation. We have divided this chapter into the key principles of climate change coaching:

- Developing a climate change coaching mindset of trust, positive regard, compassion, acceptance and non-attachment, presence and pace.
- Listening closely to everything, including the other person's needs and concerns, and also their verbal and nonverbal signals.
- Normalizing and acknowledging feelings to reduce isolation, scaffold someone's self-belief, and recognize them for who they are as much as what they do.
- Asking great questions that stretch and challenge someone to find new answers to old problems and create a fruitful thinking environment.

We will unpack all of these in this chapter, to show you how to connect more sincerely with others and create a safe environment where people are open to the possibility of change, and in which we can offer stretching questions that show how capable we think them to be.

## Adopting a climate change coaching mindset

The shift to a climate change coaching mindset requires four key components. Many of us are already cultivating these things in our daily lives: for example, self-compassion as a form of self-care. However, in the context of the climate crisis, the mindsets that we think we have mastered, or don't think we need, come into play more usefully because of the large-scale complexity of the issue. In climate change coaching, we do not only extend these mindsets to other people but to all levels of our relationships; those with ourselves, with others, with systems (teams, organizations, communities, governments), with the natural world and with the climate crisis. Let us explore them more deeply.

- **Mindset 1 – Trust and positive regard** – that everyone and thing is capable of change
- **Mindset 2 – Compassion** – continuously cultivating love and forgiveness
- **Mindset 3 – Acceptance and non-attachment** – that change may not happen as we want it to and that we can be unattached to specific outcomes
- **Mindset 4 – Pace and being present** – slowing to create deeper reflection and better progress

### Mindset 1 – Trust and positive regard

If relating well to each other is the foundation for transformational conversations, then trust and positive regard are building blocks. As the educationalist, Laurence J. Peter so humorously said 'You can always tell a real friend: when you've made a fool of yourself he does not feel you've done a permanent job.'[31] It is very easy to define people by their circumstances or actions, and for that definition to stick. While we would not want to be written off so easily, we are often guilty of doing it to others. We can categorize people into 'capable' or 'useless', 'easy-going' or 'up-tight', 'dependable' or 'flaky' based on something they have done or a situation in which they find themselves. Yet we are failing to recognize the role of their systems in shaping them. Someone who is scatterbrained in one organization may be a valued creator in another, where the environment better nurtures their talent. People also allow circumstances to unhelpfully define them, for example: 'The last two people that I tried to influence about climate change did nothing! I'm just *not the kind of person* that can influence others.' These are all signs of fixed mindsets, which makes it imperative that we employ a growth mindset.[32]

We can instead believe that people can transcend their current situations, to grow, develop and change. This may take constant work (both to remind ourselves of this potential, as much as to encourage a growth mindset in others), but if we steadfastly believe in someone's ability to be brilliant, they stand a much greater chance of becoming so. This is not always easy to do. People enrol us in their disbelief, citing lots of good reasons that change is impossible. Sometimes, although we think someone is fantastic, they doggedly try to convince us of their lack of ability and beg us for help because they have lost their self-confidence. This is seductive because they will often compliment us in the process, and it is very easy to slide into giving advice. 'You know best, what should I do?' When someone is resisting a change that *we* want them to make, it is even easier for us to see them as stubborn or difficult. In each of these situations, when doubt arises (in them or us), our best approach is to remind everyone in the conversation that they are capable. Phrases that begin with 'You are the person that ... [insert life experience]' or 'I know you to be someone who ... [insert values or behaviour]' can transport people out of this problem and into a memory of successfully overcoming another one. If we return to Rahim and Petra, Rahim might tell a friend: 'I just don't know how to tell Petra I don't want to be a vegan.' Rather than give advice, his friend could remind him of his experience and values – 'You know Petra better than anyone, and you really care about her. What would you *like* to tell her?' – or he might point Rahim to their shared resourcefulness: 'You guys have been through lots of things together; you can handle a change of diet.'

Much of this is traditional coaching, but to do this in a climate change context we need to believe not just that each of us but also that the wider system is capable of change. This means demonstrating trust that change is possible at all levels, including believing in the inherent resilience and creativity of nature. If we think of nature as a fragile bird with a death sentence then we respond differently to believing in regeneration. Similarly, we can develop fixed mindsets about systems never changing, when we also have plenty of examples of when systems have dramatically changed, and can also point to smaller changes occurring daily that could one day create those dramatic shifts.[33] When we see ourselves, other people, organizations and the planet as fixed in our current downward spiral, we are more likely to give up on change. Even when someone expresses doubts that their friends, organization, government or world population can change, we can hold a steadfast belief to the contrary. Climate change coaching requires that we consciously and continuously look for the good in people and systems, and put aside blame, contempt and doubt, which are manifestations of our own insecurity. The complexity of the problem can naturally create a sense of overwhelm in us, so to help others we need to be able to put that down and stand slightly apart. This doesn't mean being cold, quite the opposite, it means empathizing about how muddy they are without getting stuck in the mud with them. We can believe for them, until they can too, that each component of the world has the potential for greatness, the capacity to solve problems (or to find allies who can help to do so) and that the circumstances in which we, organizations, communities and even the planet currently find themselves are just that: circumstances. And circumstances can – and do – change. Indeed change is *the* only fact of life.

### Mindset 2 – Compassion

'Compassion' seems to be a real trigger word for a lot of us. We often experience pushback when teaching compassion because it seems to be a word that suggests lying on the mat and letting others wipe their feet on you, or letting ourselves off the hook of working hard. We are sometimes asked 'How am I supposed to forgive this awful behaviour? And how will wrapping them in cotton wool help? Surely getting tough is the only way to get people to shape up?' Not only does compassion not mean false praise or pulling punches, but it is also worth noticing what an Anthropogenic mindset this is, in which domination and exploitation (of others and of ourselves) is preferable to love and coexistence. Instead of seeing compassion as weakness, we can ask ourselves why we think a punitive approach helps people to change. When we feel valued (are held in positive regard) and are given the benefit of the doubt, we relax and accept challenges, not when we experience some form of bullying. We are rarely open to change when we feel judged or blamed.

Compassion is also not about letting people 'get away' with bad behaviour; it can be compassionate to call someone out. However, we can only succeed with the challenges we issue to others when we start by showing that we accept their weaknesses as well as their strengths. The key here is to separate the person from their acts, and to hold someone with compassion, even while we are unhappy with their behaviour. So this can mean telling someone that what they are doing is damaging, because it is damaging for them as much as everyone else, not because we are trying to protect ourselves or the world *from* them. This is about calling them back to what Buddhists would call their Buddha-nature, or in Christianity might be framed as 'hate the sin but love the sinner'. Self-compassion similarly need not be about allowing ourselves to give up on our ambitions but rather recognizing the damage that those ambitions of the mind might have on other parts of our system, like our bodies, or our families and friends. The tough side of self-compassion might also be about setting and enforcing boundaries, and knowing when it is time to end a relationship. People are more open to influence when they believe us to be compassionate towards them, and when we are being self-compassionate we are less likely to engage in self-blame with ourselves, which can in turn lead to defensiveness with others. We look again at compassion and blame in Chapter 13 when we look at coaching anger.

Compassion is not about having an easy life, it is about having a loving life. Holding a compassionate mindset means finding ways to understand and empathize with the reasons for someone's actions, and looking for the sense behind them. This changes the way that we approach conversations, because we start from a principle of 'this will make sense somehow'. Whatever the issue, we can almost always find something to feel compassion for, either for the circumstances they may have experienced that led them to those conclusions or actions, or for the values that underpin them, even when we disagree with the way they are expressed or enacted. This does not mean colluding with views with which we disagree or censoring ourselves into silence around them. Rather it means neutrally observing and naming what we see. From those tiny seeds of understanding, more can grow. We may never want to be best friends with a climate change denier, but if we were in a conversation together we would stand a better chance of altering their perspective if we approached them with love, rather than disgust. No one is beyond the capacity to change, and no one *will* change by being shamed or made to feel unwelcome. Whatever your starting point, try to cultivate compassion, forgiveness, for yourself, others, the organizations and communities that make up our systems, and even for the planet on which we are hurtling into the unknown. When we make a decision to be in a compassionate relationship with the world around us, much more possibility opens up as we see the world and those around us through more understanding, loving eyes.

### Mindset 3 – Acceptance and non-attachment

Acceptance (and with it non-attachment) is the ability to hold things lightly, not tightly. It is a state of not clinging to specific outcomes, actions or timeframes. It can bring with it a tremendous sense of peace and a much greater capacity to be in a relationship with all aspects of our world and its peoples. Zen Buddhist teacher and author David Loy goes to the heart of the matter in his piece in Chapter 19 here when he says:

> we [can] accept that the way things are is the way things are and do our best to respond appropriately … Acceptance is accepting that there is a problem that has to be addressed, rather than hiding from it, or fighting it.

Acceptance is about coming to terms with where we are and how we got here, which for some of us means less about avoiding the subject and more about putting down our anger and blame. We can waste a lot of valuable energy in holding on tightly to our guilt and frustration. When we can come to peace with 'what is', we can channel that energy into action and collaboration. This means not being triggered by thoughts, but rather accepting those thoughts and letting them go.

We can usefully look to the modality of Acceptance and Commitment Therapy here, which encourages us to recognize how we become fused with our thoughts and are then more reactive than responsive. As renowned psychologist and author in this field, Russ Harris says:

> Acceptance does not mean putting up with or resigning yourself to anything. Acceptance is about embracing life, not merely tolerating it. Acceptance literally means 'taking what is offered.' It does not mean giving up or admitting defeat; it does not mean just gritting your teeth and bearing it. It means fully opening yourself to your present reality – acknowledging how it is, right here and now, and letting go of the struggle with life as it is in this moment.[34]

This means noticing our thoughts and the attachments we have to things being a certain way, and questioning their usefulness. For example, does becoming attached to the idea of our sustainability strategy being a roaring success help us to engage with staff around it or does it incline us to push rather than listen or insist on the strategy being implemented a certain way? Scrutinizing the thought that 'this *must* be a success' can help us to loosen our grip.

The route to acceptance is not easy or quick, and we may be advised not to seek mastery but, rather continually *pursue* acceptance and non-attachment. Otherwise, trying to master acceptance could lead to us becoming attached to the idea of 'I *must* attain acceptance and non-attachment'! Instead, expect these to be within your grasp one moment and snatched away the next by your very human instinct to doubt or control. When we *can* hold onto them, acceptance and non-attachment can free us from stress, anxiety and friction in our relationships. Engaging in the process of acceptance means that you will have to reckon with your own disbelief, denial, anger and grief. Baked into this is the uncomfortable truth that we might not change or change fast enough to survive. That can be very hard to accept when we care so deeply about our beautiful world. Acceptance might feel like giving up on the world, your loved ones and your ambition, but you are not; you are recognizing that we are part of a vast system that we cannot individually control. Coming to a place of acceptance about this may be the single best thing that you do for your mental health when working on the climate crisis. Equally, we may also want to avoid attachment to individual control and certainty, because it is after all the belief that we can steer the planet that has got us to this point. Acceptance still allows us to work hard and set ambitious goals, but to do so knowing that we alone are not responsible for their success or failure, and cannot control the outcome, because we are part of an ecosystem.

OVER TO YOU: Take a pen in your hand and grip it as tight as you can. Try writing with this ferociously tight grip. Now what happens when you relax your grip and write with the pen held loosely? Think of the issues that you hold like the pen tightly gripped. What makes the attachment so tight? What do you need to accept to loosen your grip?

### Mindset 4 – Pace and being present

If you want to galvanize people to take sustained climate action – the kind that goes beyond turning down the thermostat – the first step is to slow down and pay proper attention. Just like when we drive fast and our field of vision narrows, when we speed through an interaction we miss important information about someone's thoughts and feelings. That can be hard when many of us feel like racing vehicles, speeding along because 'The clock is ticking and there's so much to do!' Slowing ourselves down is, however, in service of getting things done well and surprisingly it can actually lead to faster as well as more sustainable outcomes. That is because it is not about how many minutes we spend talking, but about how present we are while we do it. Many people assume that because we talk about slowing down *an interaction*, coaching must be time-consuming. In reality, you can have a very effective coaching interaction in a matter of minutes, when you are present to what is really going on and can respond to that. We have all had a rushed conversation with someone who said they were 'fine' while their eyes showed anguish, and only later on our own do we wish we had asked more. If we slow our internal engines down, we can have those transformative and supportive moments together. The trick is not how long you spend in the conversation but your sense of pace when you are there.

Coaching is often also quicker than problem-solving because we don't need to waste time listening to the ins and outs of the problem but can instead go straight to the emotions behind it and so get to the heart of the matter much faster. For example, imagine that a colleague says they have a problem, and explains a piece of work that is bothering them. Rather than asking about a project that we have no need to understand, we can short circuit the backstory and gently ask 'what's at the root of all this for you?' You might be surprised by how often the answer has nothing to do with the work and all to do with a feeling, perhaps in this case some insecurity towards a colleague with whom they have to collaborate. When we can help that person identify the real, emotional problem, we can give them an understanding of not just how to solve *this* problem but future problems that involve the same emotions. For example, perhaps this person feels their colleague dismisses their expertise but has not realized that they are triggering deeper anger at being dismissed. A conversation about how it feels to be dismissed can unearth a new approach to avoiding being triggered that they can use when they feel dismissed in *any* circumstance, not just with this colleague. This is how coaching skills increase resilience because they equip people with emotional self-knowledge so that next time they feel a specific emotion in relation to a *different* situation, they can recognize it and have tools to respond.

This is in stark contrast to helping someone solve a specific issue, which they may never experience again, and is how coaching helps people both feel more confident and learn about themselves so that they can solve their own problems long into the future. Even where this *does* take a little more time now it will pay off handsomely later.

Slowing down and being more present could also be a feature of the new paradigm for which many of us are yearning. Pace is a factor of our current paradigm and is therefore implicated in the climate crisis. The fast pace of most industrialized societies means that we make quick decisions about consumption and waste that are short term and poorly considered. Life can feel breathless, and our solutions are often rushed. To satisfy our economies' emphasis on unending growth, we have created a time scarcity that encourages us to consume, and in turn pushes our workers to endure working long hours for poor pay (often then encouraging yet more hours worked in overtime), so that we in turn can buy more things to keep the whole merry-go-round moving. Perhaps if we drew more on nature's wisdom, we would hurry less, as Ancient Chinese philosopher, Lao Tzu, wrote 'Nature does not hurry, yet everything is accomplished.' Many are dissatisfied with a life lived this way, and the Covid-19 pandemic brought this into sharp relief, as we recognized the value of community, friendship and of slowing down and having time to be present. If you are still finding it hard to accept the need to slow down in order to create change, perhaps this perspective can help you. If the world that you would like to create includes behaviours such as caring for each other, or making time for relationships, or of *not* engaging in the fast modern economies such as fast food and fast fashion, then you can be a part of creating that shift right now. We do not need to wait for the structures of the new paradigm to appear, to model its behaviours. Being present, as we shall reiterate throughout this part of the book, is also a way of de-escalating a lot of the fears that people experience around the climate crisis, from scarcity, to overwhelm to anger.

## Listening closely to everything

Listening is the simplest, most underrated skill of all. We all have the ability to do it, and we all know how great it feels to be truly heard and given enough time to express ourselves. However, whether we interrupt someone before they finish their sentence, listen superficially while rehearsing our reply in our heads, or spend the entire conversation silently judging the other side; it seems easier to listen badly than to listen well. We can categorize listening into three categories: transactional listening, person-focused listening and attuned listening. Unfortunately, in everyday life, the majority of us engage in transactional listening, which is not really listening, more waiting for a pause to interject. The comedian Jimeoin described it brilliantly in a 2010 Edinburgh stand-up routine:

> Normally what happens is, when people are telling you their story, half way through their story it suddenly reminds you of a story. So the whole way through their story you are not listening to a word they are saying. You are just thinking 'hurry up and finish that stupid story' … mine's next![35]

People give us clues to their interests, passions and concerns all the time, it is just that most of us are not listening. A big reason for this is that we believe we do not have time to listen deeply. As we said above, this stops us being present. Yet think about the most effective salesperson you have ever experienced, how well they listened and how much they let you tell them. Listening builds bonds between us, and when we trust, we buy, whether it is a product or the need to act on climate change. Yet often when we discuss change we rush people for a decision. Whether we are deep thinkers or fast actors, we all like to feel that we have got a handle on something before we agree to it. In many cases, deep listening is the tool we can use to hasten that process of decision-making. Moreover it is relatively easy to cultivate. It simply takes discipline and practice to hold your tongue and 'get over there' with the other person and their thoughts, leaving your own thoughts for later. If you are prone to listening to 'radio me, myself and I' while talking to others, start to consciously tune it out so that you can really tune in to what the other person is saying.

### Transactional listening

This is listening superficially, often while listening to our own thoughts and opinions. This often involves relating someone's experience to our own: 'You are thinking of giving up flying? Good for you. I took the bus the other day instead of driving … So liberating.' This is often an unskilful attempt to show empathy, except that it has the effect of giving the other person the impression that you are not interested in them at all because it shifts the spotlight away from them and over to you. There are lots of times when we might use this type of listening, for example when we are chatting with a close friend about something of not much consequence and are sharing back and forth. Transactional conversations are also much more likely when we are rushing. That is fine when we are in a relationship that can withstand transactionalism, such as with a family member to whom you might cut across to ask 'Can you grab some bread while you are at the shops?' It doesn't merit a big discussion, and they know you well enough not to be offended by how short you are. However, it is a pretty disastrous approach when we don't have a close relationship or are speaking to someone about a charged subject like the climate crisis, or a change related to it. Then, you would want to have a proper conversation, in which you really hear the other person's views.

### Person-focused listening

In contrast to transactional listening, this is listening closely to the other person and the particular words they say, while tightly managing your own internal monologue and external reaction. This could mean 'You're thinking of giving up flying? I notice you also said that you're feeling conflicted. What's conflicting about it?' The discipline here is not so much listening to them as not listening to yourself, requiring you to 'get over there' with them and tune out the stories of your own, of which theirs remind you. This is often the type of listening that the people we train find the easiest to master, because it involves picking up on specific words, replaying them back and asking open questions about them. A feature of transactional listening is often responding to others with a comment, usually from our own storybook, so a good starting point for using person-focused listening is to consciously make yourself ask a question whenever someone tells you something. With time you can either ask a question, or paraphrase or summarize what they have said. Paraphrasing is creating a concise version of their words, while summarizing covers everything that they say. An example of paraphrasing might be: 'You want to stop flying but are conflicted about your annual summer trip.' Meanwhile, summarizing might be: 'You are thinking of stopping flying, and the family are keen to try it, but you are conflicted about what it will be like to have a summer break closer to home.' Because it is easy to slip into our own stories from these two skills, we suggest that you start with purposefully asking a question first, until you find it easy to really be with the other person and not look for a place to interject your own narrative. The good news is that with relatively little practice this type of listening can be mastered. It is straightforward to do and can dramatically change the results that you have.

This type of listening is also helpful for spotting verbal clues, in which the words we say, give us away. There are some specific words that express dissonance. When someone has handed power to their inner critic they will use words like 'should', 'ought to', 'have to' and will talk about not having a choice, being overwhelmed ('there's too much to do') or gripped by scarcity ('there's not enough time'). They may also give away imposter syndrome – 'What will other people think of me if I mess it up?' which is a reformulation of 'Who do I think I am to do this?' and 'They'll find out I'm a fake.' A great question when someone has told you all of the ways they have to/ought to/should is: 'What would you *like* to do?' People often don't recognize for themselves that these words are doubts, and so by drawing attention to them we can help them to distinguish between something that feels truly motivating and something they think they 'should' do, but is actually dissonant.

### Attuned listening

The most skilful listening is that which is highly attuned, not just to what the person says but also to their non-verbal cues. This includes the animation in their eyes, faces and bodies as well as their posture when not moving, which are all signals to whether someone is truly motivated or deeply fearful. The documentary maker Valerie Kaur explained her process for eliciting her subjects' stories as a process of dropping into this type of listening:

> When I really want to hear another person's story, I try to leave my preconceptions at the door … I consciously quiet my thoughts and begin to listen with my senses … I try to understand what matters to them, not what I think matters.[36]

We can yield the most astonishing results because this type of attunement enables us to notice the congruence (or incongruence) between someone's words and their body language and to receive messages that are more intuitive and physical (what Sarah Higginson called the head, heart and hands in Chapter 4). Congruence is the lining up of all parts of ourselves, including our minds with our emotions and our bodies. When something really resonates for someone we see their words match their body language and tone of voice. When it does not, the dissonance shows up as a mismatch in what they say and how they behave. Importantly, here we are not listening for signs of *happiness* but of resonance. It is OK to sound sad if we are talking about feeling that way. For example, if someone is talking about a relative who has died they may be deeply sad and yet also experience the resonance of remembering that person with love. Dissonance is not the feeling of difficult emotions, it is the state we find ourselves in when we *avoid* our true feelings.

Coaches look for signs of 'not being with' an emotion because it is this dissonance that causes incongruence to show up. While the mind can lie to itself (and to us) with words, the body will tell the truth. Often people don't know that their words are incongruent with their bodies, and helping them to see this can shift their self-awareness. Someone who is self-soothing by stroking their arms as they tell they are 'honestly happy to stop flying' may not realize that they are harbouring bigger doubts, but their body is signalling that it feels unsettled and needs to be soothed. Being able to say 'I notice that you are stroking your arms as you say that. Is there anything else we need to think about?' could not only open up the conversation so that their anxieties can be addressed. This type of listening allows us to bring in the wider environment in which someone exists, even in small ways. 'As you said that, I noticed you brushed a leaf from your jumper. I wonder what the tree that the leaf came from might say about this?' or the even simpler 'I just heard a siren wailing; what might that siren be signalling for you?' This, of course, does not give you an answer from the tree or the siren but allows the other person to go somewhere else in their minds for an answer and to consider a different perspective. Listening in this way can create a much more trusting relationship between you, because the other person feels deeply seen.

OVER TO YOU: In your next conversation test out either the second or third types of listening and see what happens to the quality of the answers you get and the conversation you have. If you can, look out for signs of incongruence and dissonance, where the body and tone of voice don't match the person's words.

## In practice

### The power of listening to everything

*Rakel Baldursdottir, climate change coach*

*An environmentalist came to me as a coach because he wanted a clearer view of how he could do more to contribute to climate action. He had worked in the sector for some time and didn't feel he had done enough but did not know quite how he could contribute more than he was already doing. I felt a scarcity in the tone of his voice and by the look in his eyes there was some hesitation so I suggested that we look at some ideas.*

*We began by looking at some of the possibilities he saw in the situation and considered the pros and cons. But I felt there was something more to it and his body language indicated that too, so I asked: 'What is it that you really want to do?' My client thought carefully and then said: 'I want to put targeted education into schools about climate change and solutions they can use.' That felt like a breakthrough and a way of narrowing down the topic. 'So is that what you want to look at in this session?' I asked.*

> *Yeah. I would like to explore that to see if this is what I really want to do. I have been giving this a lot of thought but (and here he paused in silence for a moment) … somehow I'm just stuck.*

*Before that silent pause, he had sounded so determined and actually grew in the chair, so I asked: 'How important is this for you?' He replied, 'This is very important to me, but I don't know if it is possible. I'm not sure I'll get the support to do this within my job. So this is probably not going to work.' As he said this, my client had a little hopelessness in his voice.*

*At that moment we had a barrier that we needed to look further into so I asked, 'How about we look at what this would look like if you had full support at work and no barriers to putting this targeted education into schools?' My client leaned back in his chair, closed his eyes and allowed his thoughts to flow. 'Yes, that would be exciting and enjoyable. I would be using my knowledge and passion to educate others. I would have come up with a teaching programme that works well and I would see a butterfly effect taking place.' I could see the client was really visualizing his future and visibly enjoying that vision, and I gave him space to stay there for a while, by being silent myself. Then to build on his sense of empowerment I asked: 'Tell me more about that butterfly effect.' 'Well, I would be going from one school to another and leaving something behind in every place which would keep on growing and having more impact.' 'How do you feel when you are there?' I asked. 'I feel really good! And I feel like I'm in the right place' My client said with relief and a smile on his face.*

*To ground this great feeling I asked: 'What did you do to get into the right place?' Suddenly my client said, 'By following my intuition!' As he did, I could see that something new was happening; a new discovery. That felt like another breakthrough. I said, 'Really nice work there. Can you summarize this?' My client looked up with a distant gaze in his eyes, took a deep breath and summarized his exploration, from his scarcity, to his values and purpose, and passion for his solutions; his strengths and how he could embrace them. I listened and when he had finished I asked, 'What do you have now?' 'I have more confidence,' he said. 'You have more confidence, that's so great! What was missing?'*

*He paused and then said, 'Support, I suppose. I feel like I can now present my idea to my director. Maybe I could even get a grant from the community to do the project. That is something I really hadn't thought of.' To remind him of his earlier breakthrough I added, 'So if you follow your intuition, you will gain more confidence to promote the idea and have the strength to make a teaching program that empowers others?' 'YES!' he beamed. 'This is what I really want to do, and now I know how to do it and what it means for me.'*

*In the creation of self-awareness so much can happen. With powerful questions in the right places, my client found the way back to his heart. Once there, he had a very easy time setting up an action plan. With this client there was a sense of trust from the beginning. I felt how much he wanted to get out of his scarcity and his ambition to find his own purpose, and I stayed empathetic and responsive by listening and staying calm and remaining silent while he thought.*

*At the end of the discussion, when I asked for feedback, he said that what worked so well for him here was that it was such a good space to explore his own thoughts. He said:*

> *I've often thought about this, but now it's so clear. When I said it all out loud, I had this conversation with myself as much as you. I feel so much relief and freedom and I feel an increased sense of purpose and excitement.*

## Acknowledging and normalizing feelings

If we could encourage you to do nothing else it would be to use the skill of acknowledgement. It is a true super skill, that soothes doubt and loves our fear into feeling safe. As climate change coaches, whenever we sense that someone feels insecure, uncertain or outright fearful, this is the first skill that we deploy. That is because acknowledgement is another word for recognition, and it is a way in which to tell someone 'I see who you really are, even while you are telling me that you cannot, or aren't good enough, or are frightened. I know that you can, you are good enough, and that you can find your courage.' Acknowledging someone for who they truly are is how we remind them that their current fears, behaviour or circumstances are situational, not a part of their identity. When we use this skill we see people bloom, growing taller in their seats and more certain in their views. This validation makes us feel secure, which in turn makes us more likely to try new things and grow.

Recognition is a fundamental human need, and yet a shocking number of people report[37] receiving next to none in their everyday work and relationships. Instead they are repeatedly told what they have done wrong, while not being told about the positive attributes that they have. Again this could be a feature of our current paradigm showing up in our relationships. In our desire to progress and improve, we forget to celebrate. Recognition is something that keeps us engaged when everything is plodding along but in change it takes on a far more important role. When someone is feeling powerless, acknowledging their capability is the most effective way to revive their sense of agency. It naturally follows then that when someone is feeling powerless on behalf of the system or the planet, in climate change coaching we need to acknowledge those parts as well as revive their belief in them too. While with an individual an acknowledgement might sound like 'You have shown real commitment to this relationship/project/idea', systems acknowledgement could be 'This is an organization that follows through on its pledges', or 'the forest has a really extraordinary ability to recover from wildfires'. There is a health warning with acknowledgement, however: to work, it must be sincere and honest, and it must be valued by the person you are giving it to. Telling someone who cares about their ideas but not their time-keeping that the thing you value most about them coming to a meeting is that they are always on time will fall flat, and offering platitudes about an organization that are palpably untrue will just get you an eye roll. You will know if the acknowledgement has landed well with someone because their mood and body language will change, and they may pause to consider the new perspective that is opening up for them.

While 'saying what we see' is a quick and powerful form of acknowledgement for most people, for those with a chronic lack of self-esteem, a more long-term approach may be needed. People who have been taught to believe that their ideas are unworthy, or have been overlooked and their self-esteem diminished, may readily tell you how insignificant they are. This creates a vicious circle, in which they convince others to believe and overlook them more. Acknowledgement can help them to see themselves as an equal, but it will need to be sustained and specific to soothe an entrenched inner critic. You may acknowledge their choices – 'I really respect you for bringing leftovers as lunch to reduce waste' – or how their ideas have influenced your thinking – 'I thought changing our energy supplier would be hard, but when you mentioned doing it easily in your last company, I decided to put it on the list for the COO' – or simply and regularly ask for their ideas and show how much you value them. While these techniques may sound basic, when you have convinced yourself that you lack value, it is not as simple as someone opening the door for you to walk through, you also need help to believe that you have a right to walk through it.

Another form of acknowledgment, which you need to use slightly more cautiously, is normalizing the way that someone feels about the climate crisis, which can be very useful given that the situation is as grave as it. We can easily tell ourselves that we are pathetic, weak or an outlier for feeling anxious and frightened. Yet in this context, to be alarmed is not only widespread but proportionate. Nevertheless, many people feel isolated and ashamed of admitting to these feelings, and too often when they open up to someone, the other person responds confidently with facts, which makes them feel even more unusual for feeling overwhelmed and out of their depth. When we normalize something we are communicating that 'you are being human by feeling this way'. We can use a range of phrases to normalize including 'that sounds like a really human response to this situation', 'I imagine that very many people feel as overwhelmed as you', 'I feel that way too', 'that seems like a sensible response, given how frightening it is'. The important thing is the tone with which you use these phrases, and you may wish to speak more softly and leave silence afterwards. A tone of care and compassion says 'I hear you, and you are not alone' and sounds very different to a brash, dismissive tone that implies 'don't think you are special!' Done gently, normalizing begins to dial down the volume on someone's inner monologue of self-criticism and lays the ground for dialling up their motivation. We also help people feel less isolated and that they can turn to us in the future.

## Asking great questions that stretch and challenge

In the context of climate change we encounter the kind of disempowerment that can easily get us stuck in unhelpful thoughts and where bold questions can bring us back to our senses. Great questions can shake us out of our current, stuck perspective and encourage us to see things differently. They can also help people get clear about their resistance so that you can move forward together. Children have a great way of asking big, powerful questions that stop us in our tracks, such as 'What do you want?' So one way of formulating great questions is to imagine what a child might ask. But do not assume that this is an easy or weak approach; these are typically open questions that will get you more than just a 'yes' or 'no' answer, and in conversations where we are working with big fears they can often be quite challenging. The trick is to not imply the answer in the question, which could influence their response, getting you confirmation of your assertion rather than a true answer. As we have said, people tend towards being agreeable and respond as much with what they think the other side wants to hear as with what they truly believe. You may notice this when you listen to radio journalists interviewing the general public who often agree with the interviewer's assertion.

In a world in which we are so used to broadcasting our thoughts on social media, one of the hardest questions for many of us is 'What do you think?' For those of us who have been raised to be self-sufficient and impervious, this is a question that is really edgy because it implies that we need help. Yet this is powerful when we want someone to collaborate with us. Some people (typically extroverts but not always) will automatically volunteer their opinions without being asked, but many others will not. If you are one of those that happily share their thoughts uninvited, you may assume that someone who does not share with you does not have anything to say, but beware this presumption. It is very possible that they are waiting to be asked. In climate change coaching, our questions can also move beyond the individual (and their life or work), to include the system in which they operate and their relationship with the planet itself. Here are some useful examples:

- When someone has shared a sense of being out of control: 'What do you want / want to happen?'
- When someone has told you that things just have to be this way: 'What if it were different?'
- When someone has explained all of the things they are no good at or the world is bad at: 'What are you great at?' 'What does the world/company/community do well?'
- When someone believes that the natural world is helpless: 'What does nature know about change?' 'What would nature say about this?'
- When you suspect that the problem is that someone's principles or values have been crossed, or that they are holding onto a deeply held belief: 'What's the most important thing about this/what matters about this?' They may well name values in answering these questions – 'It's about respect.'
- When someone has told you that there's no point because their community/company/country will not join them in action: 'When has your community/company/country acted to do the right thing?' The example may be unrelated to climate, it is more that we want to demonstrate the system's ability to act. Similarly you could ask 'When have they surprised you by doing the right thing?' to the person who is world-weary and cynical.
- When someone tells you they'd love to help but … : 'What help would you feel comfortable giving?/What would make it easy to help?'
- When someone sounds unsure and worried: 'What's your biggest concern?' and possibly also 'What would help you feel at ease with this?'

In addition to broad, open questions, an equally powerful form of powerful question is a customized question that uses someone else's words within it. For example, if someone were to say 'I feel so conflicted about whether to stop flying,' we could ask 'If you weren't conflicted, what would you be instead?' And when they answer 'Well I'd feel sure-footed,' we could follow up with 'What could you do to become sure-footed?' The reason to do this is because when someone is considering such big topics, they are engaged in an internal thought process that we are not able to see. We do not therefore want to distract them from that with a question that satisfies our nosiness. If it seems like this is not working hard enough, don't worry; the hard work is managing our own desire to shift the spotlight to ourselves. If it were easy to ask such big, tailored questions, everyone would be doing it, and we know that is not the case.

OVER TO YOU: Tune into a morning news show on your local radio station and listen to how the presenter does or does not use open questions with their interviewees, and the effect it has on the quality of the information they get back. Write down the open questions you find it easy to ask in conversation. Now write down the questions you find it hard to make in conversation. Try to use two easy questions and one hard question this week.

We have introduced you to the basics of climate change coaching. If you read no further then this alone is a powerful toolkit. Working on the mindset you adopt about others and the climate challenge, will alter the way you interact and increase your ability to connect. Listening deeply, in a way that is attuned to the other person's hopes, fears and congruence, can help you to give them the information they need to make more joined up, confident decisions. Normalizing and acknowledging how difficult this situation is and how capable we are in stepping up to it, can dial down our understandable doubts and remind us that we have all faced challenges before. Finally, we can challenge the perspectives we hold with big, open questions that send us to different parts of our minds for newer answers. We have given you the foundations of great climate change coaching, next we will apply this to influencing, outside of a coaching setting, to take these skills out to the world.

# 6 How to make and break influence

While inside a formal coaching session the client sets the agenda, in the world beyond, our coaching skills can be put to excellent use to influence the climate agenda and galvanize people to take action for the environment. As more of us begin to recognize that all lives are now lived within the context of the climate crisis, everyday interactions become an opportunity to influence others in service of the planet. This might be asking other parents at the school gates to join an environmental group, enrolling colleagues in your projects or sharing individual solutions with family members. Whatever your sphere of influence, coaching skills can be the bridging link between information about the crisis and ideas for what we can do about it.

Holding a coaching mindset, combined with the skills we shared in the last chapter, can be transformational in helping others overcome their fears and embrace change for the planet. When we have conversations in which we demonstrate care for others, and in which we help people to feel safe, voice their needs and begin to feel capable, it can dramatically increase our capacity to influence. Showing that we care also sets off a mutually reinforcing cycle of positive regard that allows a really solid relationship to develop, in which we are more likely to take suggestions from each other in the future and become trusted advisors.[38] In other words, when we influence well, we open the door to future influence and collaboration. When we do it badly, that door slams in our face and it is much harder to rebuild that trust.

## Trust as a lever for change

Just as it is human to resist change, it is also natural to want to sway, inspire and influence each other. In his 2013 book *To Sell Is Human*, Daniel Pink wrote:

> The ability to move others to exchange what they have for what we have is crucial for our survival and our happiness … The capacity to sell is not some unnatural adaptation to the merciless world of commerce. It is part of who we are … selling is fundamentally human.[39]

In the context of climate change we are often asking people to give us their time, not their money, because we want them to take conscious action, either to lobby for change, alter something in their team or organization or simply switch energy providers. We can feel uncomfortable with the idea of 'selling', believing that it requires us to hoodwink others for our own benefit, yet as Pink says, almost all of our day jobs now involve some form of collaboration or influencing, and in the context of climate change, it is to all of our benefit to influence people to change. The good news for those of us who are squeamish about this subject is that the foundation stones of great influence are not the ability to cheat, cajole or bamboozle. Rather they are mutual respect, trust and care.

The first place to look for these qualities is in our friendships. It is no surprise that if you ask any marketeer about the most effective way to sell something, they will tell you it is through friendships. This is because friends have the power to influence our 'buying' choices, whether it is a new kind of washing powder or backing a campaign to change the law, and is the reason for the explosion in the use of peer-to-peer marketing using social media influencers and platforms. This sense of friendship as a lever for change is even more powerful during times of crisis, when our instinct is to retract from the world and be wary of those trying to influence us. That is because when we experience scarcity (in this case, of money), we are much more risk averse. We need a deep sense of trust to make change at moments of fearfulness. Trusting relationships are therefore a key source of influence, particularly for those who are gripped or challenged by the climate crisis. This may be because of existential fear about the crisis itself or because their life or role is threatened by proposed changes in favour of the planet. Influencing others necessarily means that we are inviting them to change, either in what they do or in how they behave. Just as if we were planting a flower bed, we would first prepare the soil, so with influence before we launch into what we want to grow, we need to create the right conditions for the conversation. This is a combination of a relationship mindset and the climate coaching skills that we described in the last chapter, plus a few important ingredients that we will share here. As is often the case, the starting point is with ourselves.

## Opening ourselves to influence

As Tolstoy so famously said 'Everyone thinks of changing the world, but no one thinks of changing himself.'[40] We would all rather avoid change, while believing that other people definitely should. However, in well-functioning, productive relationships both sides are open to influence. The American relationship psychologist John Gottman, drawing on his decades of research into successful marriages, suggests that when spouses accept each other's influence they strengthen their friendship 'not just because the absence of frequent power struggles makes the marriage more pleasurable, but because … [they are] open to learning'.[41] Being open to each other demonstrates respect and trust, which creates a virtuous circle of mutual appreciation. Gottman's research shows that spouses who feel secure are more likely to cede influence to their partners. He also explains the reasons for and consequences of not doing this: 'He [a husband] will not accept his wife's influence because he fears any … loss of power … [but] because he will not accept influence, he will not have very much influence.'[42] While Gottman is writing about marriage, this easily translates into any relationship in which people choose to cooperate – a manager and a member of their team, for example. When we start from a position of wanting to maintain control, we turn cooperation into a power struggle. In an organization, if we are closed to influence we become known as someone who seeks to maintain their power or who will not yield, damaging a culture of collaboration and limiting the ability for the wider system to change. Clearly influence is not a one-way street in which we influence others and are done with it. At its best, it is a reciprocal process where we ourselves also need to

be open to influence so that multiple change processes are happening simultaneously. This is how many small ripples of change can shift systems.

To be open to influence we must first be aware of our own behaviour. While in
the last chapter we used the word attachment to signify holding tightly to our
own agenda, we could easily replace it with the word insecurity, because the less
secure we feel, the more tightly we cling to ideas, identities, job titles, even relationships as life rafts for our sense of self. Those who are more secure are
comfortable with what they don't know and haven't achieved. Whether they are
aware of it or not, they are demonstrating the qualities of non-attachment. While
we want to galvanize others to act, often (because we hold tightly to a goal or
agenda), we can inadvertently draw on unskilful behaviours that undermine this.
In many cases this shows up as asserting rank over others to shore up our sense
of self, but it makes it hard for others to interact as peers.

One example of this is that we can claim greater (or the greatest) experience
or credibility over someone. Without being asked, we might say how many years
we have been working on this issue/topic, or how many credentials we have – 'I've
been campaigning on this since the 70s' or 'I have a PhD in ...', which can be intim-
idating. Another is that we can assert an ability to know the future better than
others by painting a dark vision as an inevitability. For example, 'We aren't going
to survive this. There'll be wars, mass migration ... And we've definitely only got
50 good harvests left.' There may be elements of well-researched fact amid these
statements, but they are not used to inform. Instead we are trying to create certainty for ourselves as much as to assert rank. We know that claiming that it will
all be OK is unrealistic, so we swing to the other extreme, as if in voicing the
worst we can prepare for it or because (as we said in Chapter 2) we believe that
shock tactics are enrolling. Again, we are wielding knowledge as a form of rank
– in this case 'I've learned more/done more thinking than you have' making it
harder for people to connect with us. It is very important to recognize that no one
has a crystal ball, and none of us can know the future, however many reports we
read. As Rebecca Solnit wrote after the disappointment of COP26:

> People who proclaim with authority what is or is not going to happen just bol-
> ster their own sense of self and sabotage your belief in what is possible. There
> was, according to conventional wisdom, never going to be marriage equality in
> Ireland or Spain ... The future is not yet written.[43]

Instead of asserting rank, we can reference our sense of purpose, values or
beliefs, which are not only more effective at making us feel secure but offer a
more accessible way to connect with us. 'I care deeply about the justice of this
issue, which is why I've dedicated four decades to it.' We can also model the ability to be with uncertainty and say how it makes us feel: 'I don't know what the
future holds but what I read really worries me. So I want to do whatever I can,
while I can.' Or even 'I'm sorry, I'm just so worried about this that I find myself
wanting to shock people into doing something about it. I know that doesn't work,
and I didn't mean to upset you.' If you find that you use these behaviours, don't
beat yourself up; they are a very common response to threat. Instead ask yourself, would this enrol me to join someone? If the answer is no, change tack. Try
the alternatives here and adapt them to feel more natural to you. If all else fails,

focus on building a great relationship instead, because the safety we feel when we connect can dial down our own insecurities.

OVER TO YOU:

1. In what circumstances do you allow yourself to be influenced?
2. Choose someone you already know and who you would like to influence. Rate how open you have been to *their* influence on a scale of 0–10 (10 being 'all the time' and 0 being 'never'). What can you do to improve or maintain your score?

## Consciously choosing how we behave in high-stakes conversations

We have talked about the importance of relationships to achieve collaboration and action. However, as we have said, we can be guilty of putting the relationship second to the issue (and hope to persuade people with facts not friendship) or alternatively we can silence ourselves about the climate crisis for fear of upsetting others. All too often we do these things unconsciously, entering into high-stakes conversations without considering the dynamics of the relationship and how we can best respond. This can lead to us becoming disconnected and becoming an island, either because we isolate ourselves by being too challenging or from censoring ourselves and feeling less than authentic, which can in turn damage our well-being. To understand these behaviours and improve collaboration, researchers Kenneth Thomas and Ralph Kilmann developed the Thomas-Kilmann Conflict Model as a framework for handling high-stakes conversations (see Fig. 6.1). The model considers the relationship between how much focus we put on asserting our needs (or how much we value the issue) and our level of cooperativeness (or how much emphasis we place on the other person's needs). They assert that in any conflict situation, we may respond with one of the five behaviours below: competing, avoiding, accommodating, compromising or collaborating. In the context of climate change, we can see all of these behaviours in play.

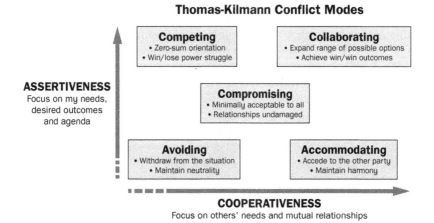

**Figure 6.1** The Thomas-Kilmann model

If we choose to 'compete' we engage in a zero sum game of win or lose. This can be at the expense of others, and of cooperation, and is undoubtedly a systemic feature of our climate-destructive paradigm. This may, however, be useful when we need to make a quick (or emergency) decision but will only antagonize others during open discussion. While 'avoiding' may sound irresponsible, it can be employed when the issue is trivial or when we have little chance of succeeding. In climate terms, this may be the conversation you have at a bus stop with a climate-denying stranger who you will never meet again. We may decide that for our own sense of self-esteem we will let this one go; however, we don't want to use this style when we need to get something done with a key stakeholder. Similarly, 'accommodating' someone else's needs too much may well be at our own expense, and work against what we need to achieve. It is perhaps the case that many of us have been doing this by not talking openly about the climate crisis with friends for fear of sounding judgemental or even CEOs who do not push the climate agenda for fear of shareholder revolt. 'Compromising' is something that many of us are encouraged to do; however, Thomas and Kilmann see this as a lose-lose in which neither side gets what they want or need because it is mainly useful when we need a temporary solution, not a long-term approach. Compromising may be a default for a lot of us because it requires a degree of both assertiveness and cooperativeness, which makes it feel safe for those who fear conflict as much as for those who find challenge comfortable. 'Collaborating' is seen as the ideal position, in which we allow the other person to influence and input ideas as much as we do, and in which we define parameters together, expanding what we ourselves may have considered. Collaborating, of course, requires the most trust of all of the positions, and improves our ability to build relationships.

Conversations about climate change can easily go in one of two ways: (1) we see that the other side is uncomfortable and we change the subject (accommodating at best, avoiding at worst) or (2) we are so tightly attached to the issue and our need to change minds on it, that we become overly assertive and fail to listen closely to the other person's perspective (by competing for airtime and arguing about facts). The Thomas-Kilmann model shows how, if we can demonstrate that we value the relationship while also asserting the importance of the issue to us, we can achieve collaboration, giving us the greatest chance of sustainable action. In the face of this systemic, complex issue, collaboration is more essential than ever. This means that the starting point is to connect well and build trust. By encouraging us to listen well, respect and be compassionate to each other, a coaching approach helps us to create the kind of honest and successful collaboration in which people feel included, consulted and at choice. As we shall show you, these are skills that can also help bring your staff with you during organizational change, because they create a space to process the emotions that come with big change. While many leaders talk about the importance of a burning platform to galvanize action, it is not just the fire but also the strength of relationships (particularly a sense of solidarity) that supports transformation. Investing in our relationships also, of course, helps us to expand our networks, which can then lead to further collaboration.

OVER TO YOU: Pick someone you want to influence and plot your current behaviour towards them on the Thomas-Kilmann model. What do you need to do to become more collaborative?

## In practice

### Creating a trusted relationship to lobby politicians to vote with the environment

Jill Bruce, lead climate ambassador, Federation of Essex Women's Institutes

*I have been lobbying my local MP, Sir Bernard Jenkin, for over 6 years. I started in summer 2015 when I went to see him at his surgery. At that time he argued that climate change was not a serious concern, and he always voted with the government on environmental issues, often to the detriment of the climate crisis.*

*I had never spoken to an MP about climate change before, and I was not an expert on the subject. But I am frightened we are creating a world we cannot survive in. My grandfather served in the First World War as a stretcher bearer and my father served in Burma in the Second World War (while my mother, aged 21, was sheltering her first baby alone under the kitchen table during the Blitz). Those generations selflessly gave so much so that future generations could live in peace and prosperity. I am a baby boomer, born in 1950. My generation have enjoyed previously undreamt-of opportunities. Yet we have filled the sea with plastic and the air with carbon, wrecking the world for our old age, and for all future generations. While, at first, we did this naively, unaware of the consequences of our actions, now we know exactly what we are doing and we have little time left to save ourselves from the consequences of our own actions. I became a Women's Institute volunteer Climate Ambassador and decided that the most useful thing I could do would be to lobby parliamentarians.*

*When I first met Bernard, I introduced myself as both his constituent, and a WI member because the WI is seen in the UK as respectable and conservative, and I thought this would make it feel less confrontational. I took in a YouGov poll of 1,000 WI members' views on climate change, who expressed concern about local wildlife and flooding, which was very relevant to our constituency, which is rural with a coastline already under threat. I did not expect to completely change Bernard's views in one visit, so my objective was to introduce the topic, pick three or four points from the Royal Society literature and say why I was concerned. I really wanted to hear his views and had an idea to get him to chair an (as yet unplanned) public meeting. I was friendly, but I was also persistent! At the end of the meeting he agreed to the meeting, shook my hand and said 'very good lobbying'.*

*That was the start. In November 2015 we held that meeting. I had heard that MPs would otherwise drop in briefly and leave meetings, so I asked him to chair because that meant he had to arrive early, listen to all the speakers and stay till the end. We have held a public climate change meeting together every autumn since then. I decide the agenda and invite the speakers, and I always give Bernard an opportunity to speak. I have found an astonishing array of world-class speakers from well-known organizations, such as Greenpeace, WWF and the Committee on Climate Change who are prepared to come for no fee, in order to convince Bernard and our audience of the need for climate action. Bernard attends because his constituents are there, his constituents attend because speakers from the organizations they support are there and the speakers attend because Bernard is there. The end result is that over the course of seven annual meetings he has listened to many hours of knowledgeable people talk about climate change, and he has taken tough questions from the public. Bernard and I both now know far more about the need for urgent climate action than we did 6 years ago.*

*Before I retired, I worked in sales and marketing, which proved to be very useful training for lobbying. Right from the start I have always treated Bernard with respect and listened as much as I have talked. I never attack. I simply hold him to account. As a result, we have developed a professional but friendly relationship. I email him whenever I feel he should be taking action, copying in his team, who have also become very helpful. I congratulate and thank him whenever he does something that I believe will help, and express concern when I think he has missed an opportunity.*

*In the summer of 2021, Bernard had an epiphany. He made a public video about his conviction that climate change is already adversely affecting the UK, and chose to launch it at our public meeting. In it he says that 'he will dedicate his next years in politics to climate action'. This is his own conclusion, and I believe he will be more committed to seeing it through for that reason. I do not see this as 'job done', however. Fossil fuel lobbyists will be ratcheting up their efforts to persuade him to water down, and slow down. I will still keep asking him to back pro-environmental policies, to speak out and to write to ministers on government policies that I believe to be wrong. We cannot use up any more of our world's resources, and must try to restore what we have already destroyed, if we hope to have a habitable planet for the generations after mine.*

## Allowing people to come to their own solutions

There's an old joke: 'How many psychologists does it take to change a lightbulb?' Answer: 'One, but the lightbulb has to want to change.' People change when they want to, regardless of what we want. Jill Bruce's story shows that rather than preaching or solutionizing, we can create safe conditions in which someone can change. This is particularly important in the context of climate change where judgement and blame are rife. Yet when someone voices their powerlessness, and especially when it sounds like not taking responsibility, we can easily be triggered into offering unwanted advice. Most of us have experienced someone trying to get us to act one way or another about the climate crisis, and much of the time it serves to entrench us more firmly into our original position. In the context of climate change, it is very likely that the friends, colleagues and family members of the person you are speaking to have already tried to 'fix' them with solutions, and you may inflame existing frustration. Yet although we have said that coaches don't offer advice within formal sessions, we would be glib to suggest that advice is never helpful. There are specific conditions in which we accept advice, however, that are worth bearing in mind when we are trying to influence someone, which we outline in the table below.

Notice that if we want our suggestions to be taken on board, we need to either be held in esteem by the other person, appeal to something that already interests them or demonstrate empathy for them. In other words, we need to create a relationship of mutual respect, trust and care. Holding back from suggesting solutions can be challenging when we think we have the answer, or when we feel

**Table 6.1** Circumstances in which someone may accept your ideas

| Circumstances in which someone may accept your ideas | Circumstances in which someone may NOT accept your ideas |
|---|---|
| **They have asked for advice** from you because they judge you will give correct information / advice. For example:<br><br>They ask: 'I know that you know a lot about this, what do you think?'<br>They may still not take this advice, but as they are information-gathering, they are open to influence. | **When they are feeling fearful about the change** to which the ideas relate (in this case you will hear a lot of push back). For example:<br>You ask: 'Why not just call them and ask?'<br>They say: 'I would but I've got a lot on and I don't know the number' (hiding their fears) OR<br>'I cannot do that! They'll think I'm an idiot!' (open about their fears). |
| **They have asked for advice** because they think you will help them think through the problem in a thoughtful and intelligent way:<br>They ask: 'You've been through a similar thing, what do you think?' OR<br>'You've helped me think about things like this before, could you help me decide about this?' | **When they do not respect or trust you**, or think that you have not thought about their problem properly. For example, before they finish their sentence you jump in with an off the cuff solution:<br>'Stop right there, I totally get it. What you need to do is …' |
| **When the solution confirms their existing ideas** or chimes with an approach they were exploring. For example:<br>You ask: 'Have you thought of researching how other people have done it?'<br>They say: 'Yes! I didn't think that seemed proactive enough, do you think it does?' | **When the ideas step on their values** or feel out of character. For example, when you tell someone who avoids conflict and values harmony that:<br>'You should go right in there and give them a piece of your mind!' |
| **When you empathize** with the way they feel either about their current stuckness or the subject to which the advice relates, they are more likely to subsequently listen to your advice, though they still may not take it. For example:<br>You say: 'I can understand how anxious it must make you feel not to know what to do for the best.'<br>OR<br>'I can imagine how maddening it feels to see the beach covered in plastic.'<br>OR both together<br>'It must be upsetting to see the beach covered in plastic and not know what to do.' | **When they feel misunderstood**, dismissed or belittled. For example:<br>You say: 'Oh that's easy, I cannot believe you are worrying about that!' OR<br>If you do not have a close relationship, or the other person does not feel confident to speak openly, they may also not tell you that they feel dismissed and instead change the subject. For example:<br>You ask: 'So this is just a case of choosing the right scheme and going for it?'<br>They say: 'Yeah, kind of. Don't worry about it, I think I know what to do.' |
| When you share ideas as a form of **brainstorming**. For example:<br>You say: 'Here's a thought that you've probably already had, have you tried … ? Oh great you have, so what else do you think you could try?' | When they feel that **the advice is being forced upon them**, that you are trying to make them take your advice or that you have an agenda. For example: |

(Continued)

**Table 6.1** (Continued)

| Circumstances in which someone may accept your ideas | Circumstances in which someone may NOT accept your ideas |
|---|---|
| Note that this works best when the other person is feeling motivated, not incapable. If they are truly stuck, brainstorming can reinforce their feeling of being incapable of solving their own problems. | 'Honestly it will work! Try it! No don't tell me you cannot, of course you can! I've done it and it worked for me.' This begins to feel like haranguing and triggers a suspicion that they are being 'sold to'. |

the urgency of the crisis, but rushing into action can slow us down in the long term. There are myriad ideas for how we can adapt and mitigate the worst effects of the climate crisis, but before we can use them, we have to override our human fear of change. In climate change coaching we meet with some pretty powerful fears, and often the worst thing we can do is offer our solutions, because their fear will always outsmart us. Notice how often when you offer advice you unwittingly engage in a tussle with the other person's fear. What is more, when someone is unsure or doubtful, offering advice only serves to undermine their confidence. If we could hear someone's internal monologue at that moment it might say: 'I am confused, and you are reminding me that I cannot come up with any ideas on my own.' If we attempt to brainstorm when someone is in this place, it reinforces their sense that they are incapable. In many cases, we need to make sure that our excitement, passion or urgency don't spill over into us prescribing how someone else lives their life or does their job. Whether you are a coach or someone wanting to use coaching skills, it is important to remember your scope of influence. We aren't in their shoes, living their life or doing their role. Even if they are a colleague, you most likely will not be with them when they start taking action, so it is critical to help them develop agency and confidence about it.

OVER TO YOU: Think back to the last time that you offered someone solutions when they expressed difficulty. Did they put those solutions into practice, and if not, what do you think stopped them?

## In practice

### Talking more easily about climate change

*Dr Charlie JR Williams BA DPhil FRGS, climate scientist and research fellow, University of Bristol*

*I am a climate scientist, and have been a university academic for almost 20 years. During that time, I have conducted research into various aspects of climate change, published extensively, have lectured around the world and taught countless undergraduate and postgraduate students. Although I suffer massively from impostor syndrome, it is probably true to say that I know a fair bit about climate and how it has changed in the past, is changing today and might change in the future. While I do not consider myself an expert, my friends and family often do, which sometimes makes for some interesting and challenging conversations. Here are three approaches that I use.*

### Avoiding lecture mode

*When climate comes up in the news, I am asked questions by my friends and family, and I am always mindful not to go into lecture mode. There are two good reasons for this: (1) I really despise arrogance (undoubtedly because I was fairly arrogant myself in my early 20s, and can now see this vice for what it is) and would never want to come across as a know-it-all; and (2) nobody likes the dinner-table bore! If I were to deliver a 20-minute monologue on atmospheric physics, I probably wouldn't be asked again. I am also very mindful to explain things in an engaging way. There is a very fine line, sometimes difficult to tread, between oversimplifying and therefore patronizing, and over-complicating and watching as their eyes glaze over. I tend to tackle any questions by turning them into a conversation, either by answering their question with another question or by answering the question directly (if indeed I know the answer, which is not always the case) and then asking the person what they think. I want to engage others with the subject on an equal footing, and avoid them feeling they are back in school.*

### Finding common ground

*This is just as relevant to those who are very much on-board when it comes to tackling anthropogenic climate change as it is for those at the other end of the spectrum, who are ardent climate sceptics. It is particularly true for those of my friends and family who are somewhere in-between, i.e. questioning whether their actions make a difference. Rather than diving straight into the science of climate change, I usually try to find some common ground. Again, I try to turn this into a conversation rather than a lecture (when dealing with climate sceptics in particular, throwing numbers or hard science around tends to cement their opinions, rather than helping to change them). An example of this might be talking about how much someone enjoys taking their grandchildren to the local park and how it would be sad if that park were to deteriorate; or talking about hobbies, such as sailing and how it is a shame that because of factors such as pollution and human waste, our marine ecosystems and biodiversity are rapidly declining. I have also discussed the issue of meat consumption by taking the natural world out of the discussion and considering how a diet high in red meat is not great for our physical health. Although none of these things are directly related to anthropogenic climate change because they are all indirectly related, more often than not, we end up talking about climate change.*

### Reining in the temptation to be overzealous

*While enthusiasm and zeal are definitely needed during the climate crisis, sometimes it is very easy to go overboard and my approach is to be slightly more cautious. My attitude is that my role is not to preach and tell people what they should be doing, but rather to provide the facts and help people to think about what they could be doing. Of course, when it comes to our own families, we can all be occasionally 'over-persuasive', and I am certainly guilty of trying (so far unsuccessfully) to persuade my parents to follow a plant-based diet two or three times a week. But even here, I try to use gentle persuasion rather than showing downright disapproval. If I were to be constantly judgemental about the choices and lifestyles of my friends and family, I probably would not have many friends, and having friends to let off steam with and rely on is something that helps me to do my job. Whether this makes me a lukewarm climate activist is for others to judge – if they must!*

## Influence the destination not the roadmap

Most commonly when we want to influence someone, we have a specific destination in mind; however, problems can arise when we get too attached to how they reach that place. A more useful approach is to work with them to co-design the route, which is often the key to unlocking someone's willingness to go on the journey. This requires a certain amount of detachment on our part because when we use coaching skills we co-design the roadmap as we go. This is powerful because it signals to our conversation partner that they are equally in control, while modelling the ability to be comfortable with uncertainty. We can show those around us that we don't need to anxiously grab the reins or flex our expertise or knowledge in order to control the world. As we do this we also help others to develop their own psychological flexibility to operate in a context of uncertainty.

To demonstrate this, let us look at the fictitious story of Jonathan, a professional coach by day with a background in marine biology and a love of the ocean. As a coach Jonathan is fantastic at asking open questions and listening really deeply to his clients, who love working with him for the high positive regard that he shows them in their sessions. Outside of coaching, he can sometimes forget his skills and allow his passion to take over. Jonathan is involved in a local environmental group and wants to encourage Melissa, the head of the group, to launch a campaign around ocean plastic. If he's not mindful, it could go something like this:

Jonathan: 'I'd like us to do something about ocean plastic. The way that we're polluting the ocean is unacceptable and our group really should be doing something about it. I'd like us to begin a campaign about it, declare the town plastic free and lobby the county council.'

Melissa: 'Wow, that's a big commitment. I agree it's bad, but we've already got a lot on with the other stuff we're doing, and everyone is a volunteer. I know it's important but I don't see it as a priority, I'm afraid.'

Jonathan: 'I think that saying we're busy is a bit of a cop out. This is a hugely topical issue right now. We have a responsibility to do something.'

Melissa: 'I think it's unfair to say we're shirking our responsibility when we're doing all of these other things.'

Jonathan:    'Well frankly, we could easily stop doing some things. That project that Toby is working on about the Amazon rainforest hasn't exactly taken the world by storm has it?!'

Notice that Jonathan has started the conversation not just with a sense of what he wants to work on (reducing ocean plastic) but also with a fixed roadmap of actions that they could take towards it. As he is so tightly attached, as things get heated Jonathan becomes sarcastic. Sarcasm is a manifestation of insecurity, and because Jonathan's plans are under threat he feels that way. When he doesn't feel that Melissa is on board, Jonathan ups the ante and accuses her of being irresponsible. This is only going to threaten Melissa more and now she is an opponent not a collaborator. Jonathan has put the issue ahead of his relationship, even though it is the relationship that he needs to influence. What if instead Jonathan leant back into a coaching approach and took this tack:

Jonathan:    'I'd like us to do something about ocean plastic. The way that we're polluting the ocean is unacceptable and we should be doing something about it.'

Melissa:    'I agree it's bad, but we've already got a lot on with the other stuff we're doing, and everyone is a volunteer. I know it's important but I don't see it as a priority, I'm afraid.'

Jonathan:    'That's a shame. What do you see as a priority?' [a big, open question]

Melissa:    'To be honest it's less that this isn't a priority and more that I'm really mindful that everything we say yes to requires more of the people we've got in the group, who are already doing a huge amount.'

Jonathan:    'I see what you mean. It's hard to balance [normalizing]. One of the things I appreciate most about you is that you're always making sure that we don't burn out [acknowledgement]. I know none of us has lots of free time, and I also know that you're not saying you don't care about ocean plastic [acknowledgement].'

Melissa:    'You know I do. But as you say, I also care about the group, and it's important that we don't put people off volunteering by overloading them.'

Jonathan:    'I agree, that would be a terrible outcome. I wonder what we might be able to do that wouldn't overload people, but would still feel like we're not sitting by while it's happening? [A question that opens up possibility] Because I know that like me, you also feel a sense of responsibility. [acknowledgement]'

Melissa:    'Yes I do. I saw a photo of a seal trapped in plastic and it haunted me. It just feels like we're fighting on so many fronts at the moment.'

Jonathan:    'Well if that's the case, then I wonder if it would help to do some planning? Maybe this isn't for right now, but we could work out what month we could do something? So we're not fighting on too many fronts at once.' [Customizing of a question to include Melissa's words]

Melissa:    'That's a great idea. You know the summer would be good because people naturally think about the beach then.'

Jonathan: 'That's a brilliant idea! [acknowledgement] Would you mind if I started to work up some plans now that we could use in the summer?' [goal setting]

Melissa: 'Absolutely. And maybe we can put this on the agenda for the next meeting? Joan and Antony often have good ideas.'

Jonathan: 'Thanks Melissa, I really appreciate you taking the time with me and opening this up to the group. I feel a bit better already.'

Melissa: 'You're welcome Jonathan, and thanks for being so flexible too.'

What did you see Jonathan doing here? Fundamentally, we hope that you saw Jonathan valuing his relationship with Melissa as much as he valued the issue or getting his own way. He also didn't abandon the issue in service of staying friends. Jonathan acknowledged Melissa's concerns and who she was as a person, rather than making her 'bad' for not initially supporting him. He showed that he knew that her resistance was not because she didn't care but was because she cared about something else. Fundamentally, he entered into the discussion as an exploration of the idea, without a fixed agenda to force through. He wanted to do something about plastic, but could be flexible about how and when that happened. At the end of the conversation, we saw that because they were able to agree a way forward, Jonathan actually felt better about the topic itself; and less attached and alone. When we are intentional, a coaching conversation can also help us to detrigger ourselves. When we are successful at influencing change we do not just get things done, we also leave with both sides feeling validated and valued.

## Becoming trusted advisors

We have all been guilty of hoping that we might change someone's mind and actions in just one chat or blog post. The old myth of information changing behaviour is hard to shake off, even when we know a lecture doesn't work, at least not on its own. Rather than hoping for an instant and Damascene conversion, we can aim to become a trusted advisor. This may not mean giving practical advice; the operative word here is trusted not advisor.[44] This means being a person with whom people can express fears and concerns, and who they trust to have their best interests at heart. Trusted advisors demonstrate care for the other person more than for their own 'angle' or agenda. In business, this could be the sustainability consultant that suggests a course of action that will benefit the client even though it cuts short their contract. In relationships, this means becoming the person that a friend can share their climate anxiety with alongside their ambivalence for change, all without fear of judgement. Over time, a trusted relationship has the power to incubate new and deeper ways of thinking.

This is important because while many of us find the threat of climate catastrophe initially motivating, the truth of the climate situation can just as easily deactivate us if we don't have the right support to sustain change. Two essential coaching skills to use here are normalizing and acknowledgement. People don't want to be shamed and blamed for their feelings about the climate crisis, they want to feel less alone. We can become a trusted advisor by actively asking someone how and if they want to talk about this subject, empowering them to drive the conversation, and by listening and acknowledging them as much as challenging their perspectives. In doing so, we create the conditions for them to be open to change, not turn away into fresh denial. Just as perhaps you once had someone to guide you through these choppy waters, the right support can encourage us to be open to new information or even seek it out on our own, knowing we have someone to turn to help us make sense of it.

Influencing others makes us feel valued and validates our own choices. Perhaps this is the reason that we can end up in an unhelpful tug of war when someone politely declines our suggestion, because while they are saying 'that idea isn't for me thanks' we hear 'I don't think your choice is correct', or worse 'you are wrong'. By contrast, when someone is open to our influence we feel seen and heard and form a deeper bond with them. Influence is a deeply affecting form of recognition of our own behaviour, not just an effective way to change that of other people. Put like that, we can see influence not just as a way of getting things done but as a way of connecting people together and changing the public discourse on climate change. When we approach influencing, we can hold multiple intentions beyond action that can help us to reduce our attachment to a specific outcome from a conversation. If all that we did was to make it safe for people to talk about and begin to have a better relationship with the crisis, we might change the collective mood music on this subject. We may not realize the ripples of impact that this kind of influencing can have, but just because we cannot immediately see it, does not mean that it didn't happen.

# 7 | Making it OK to talk about climate change

It can often feel like talking about climate change requires a lot of courage. When we broach the subject, people often look at their shoes and avoid eye contact. Yet we know there are many benefits of talking about this subject,[45] whether that communication is online, in our team meetings or the public commitments that we make as organizations, or in the way we speak to fellow parents at the school gates. There is a solidarity and a relief that comes from realizing that very many of us care and want to act but feel daunted or ill-equipped. In her book *Turning to One Another*, Margaret Wheatley says 'starting a conversation can take courage. But conversation also gives us courage. Thinking together, deciding what actions to take, more of us become bold. And we become wiser about where to use our courage.'[46] Simply communicating more compassionately about climate change can help us form new alliances and ideas, but so often it feels hard to do, whether online or face to face. Many people tell us that they happily talk about climate with people who 'get it' the way they do, but that they find it harder with those outside of these circles, even when those people are family or other friends and colleagues. Yet they also tell us that they want to share this important part of their life. When we are talking to a kindred spirit, there is an unspoken contract between us, which means that we don't have to ask permission for the conversation to take place. When we are speaking with someone with whom we don't have that understanding, we need to consciously create a spoken one. In this chapter we are going to show you how to do that using the tried and tested coaching skill of 'contracting', which is another word for 'agreeing the terms', in this case of communicating about climate change.

## Contracting without the lawyers

When many think of contracting as signing documents, when we talk about creating a contract as a coaching skill, it is instead about agreeing the parameters of our relationship, work or even just a conversation. In everyday life a contract tends to be legally binding and something we do when we get married or buy a house. The stakes are high and a contract ensures that both sides know where they stand and what they will do if something goes wrong. In a similar way, a verbal contract, such as with a new manager, establishes the working principles of a relationship (without the lawyers), so that both of you feel secure and sure-footed about how you will work together. The difference between the contracting that coaches and (good) managers do and the kind you do with a lawyer is that a legal contract is mostly limited to what activities you do together and does not usually include how you behave while you are doing it.

This is an important distinction when it comes to relationships rather than legal agreements because in relationships it is more often people's *behaviour* not their actions that causes the most conflict. Frequently it is not *what* we say about climate change but *how* we say it that matters. We can communicate the same information with compassion and lightness or with blame and frustration. For this reason, in our version of contracting we also look at behaviours and values, not just actions.

Contracting in relationships may seem alien in a world where we are encouraged to be individuals who broadcast, not communities that converse. Yet there is tremendous benefit to defining the rules by which we coexist. Through contracting we can ease the road of stepping into difficult territory together, giving both sides a chance to say how they would like to work together, what they will do if it is not working, and even how they want to have a simple conversation. Fundamentally, this is about transparency, allowing you both to say what you want, and express specific unspoken needs, so that both of you have full knowledge of what you are getting into. Because of the charge around the climate crisis, when our needs remain unexpressed, misunderstanding and conflict can easily arise because our actions can be perceived as thoughtless, heartless or careless. Articulating what is going on for us under the surface serves as a powerful reminder that our needs are in fact sensible and relatable, opening the way for more empathy and understanding. Here are some of the agreements that clients have asked of us, with the specific 'contract' in italics:

– 'I feel blamed, judged and guilty, and I'm worried that you are going to blame or judge me too. *I need you to be understanding and kind.*'
– 'People often bamboozle me with lots of information, and I don't feel that I can say I'm overwhelmed. *I need you to check that I'm doing OK, to make it easy for me to say stop.*'
– 'I'm fed up with people giving me half-baked solutions when I haven't asked for advice. *Don't tell me what to do!*'
– 'I do not want to become a vegan and *I don't want you to try to make me. I want you to help me work out how I can make a difference another way.*'
– 'People keep telling me that I'm worrying too much about climate change and sending me positive articles to "cheer me up". *I need some solidarity, and I want you to make me feel normal for being scared, not try to "make climate change better".*'
– 'I don't feel equal to the challenge because I don't know enough. *I need you to reassure me that I can still make a difference.*'
– '*Do not encourage me to set "stretch goals". I need you to help me find a vision for enjoying life.* I'm recovering from burnout and I want to rest, not push myself harder.'
– 'I feel so angry about all of this and I might raise my voice. Sometimes that frightens people, but I'm not being aggressive towards you and *I do not want you to be scared.*'
– 'I feel such deep sadness about this situation. I might cry, and if I do, *I don't want you to try to make it better or change the subject. I need to know that you'll be OK if I cry.*'

Have you ever felt any of those things and perhaps not been able to say them? All too frequently contracting only happens after a conflict, when people feel that there's nothing left to lose and are finally willing to be less guarded. Voluntary groups and organizational project teams tell us that the process of getting into action is relatively easy, but that there is rarely much discussion about how they will behave together, and when perceived poor behaviour causes conflict the group or team can splinter or disintegrate. Without any kind of common agreed behaviours it can be as easy to fall apart as it is to come together. Just because you agree that something needs to be done, you also have to decide how you will work together and how you will manage competing interests and beliefs. When people leave groups it is rarely because they stopped caring about the issue and far more often because they no longer felt welcome or had personality clashes they couldn't resolve. While a 'contract' cannot stop people from falling out, it can make it less likely by creating a protocol for what you will do when disagreement arises – helping conflict to remain contained and be more amicably resolved. This can help you to maintain positive relationships when you leave a group; something that we discussed in Chapter 4.

Many coaching clients who do this process with us express surprise at how useful it is, and how they wished they had done it at the start of their working (and often personal) relationships. The good news is that it is never too late to have this conversation. You can just as easily say 'We've known each other for 5 years and have often talked about this, but sometimes I feel uncomfortable afterwards and I'd like to work out how we can talk about these things more easily' as you can say 'We've only just met and already I'm telling you about climate change. Is that OK with you?' We don't always need a contract, but where we think there's the potential for storm clouds or discomfort because the context of the discussion or work is charged, an agreement about behaviour forms a solid foundation. Having a contract also allows you to call out transgressions of the agreement, such as 'We said we'd be non-judgemental when we talked about this, and I'm feeling a little judged' or 'We said that if I needed to talk about my concerns that you would make time, but today I felt you brushed me off.' Having established a set of rules for the working relationship, even those who avoid conflict can feel much more comfortable to voice dissent or challenge because there has been an agreement to which they can point back.

## Charged conversations require more permission

The mood that a topic generates can either attract us into conversation or repel us into avoidance. On an average day, we expect that you begin almost all of your conversations without asking permission to have them. In the majority of cases that is a perfectly reasonable approach. We doubt you greet your neighbour in the morning by asking if it is OK to enquire after their health; rather you just ask them. Likewise, you wouldn't spend much time in a work meeting asking if everyone is comfortable talking about an existing project (unless perhaps it was going wrong!). Those conversations have very little charge to them, and as such we can fairly safely assume that others will be OK to have them. When we talk with someone who relates differently to climate change than we do, it can be easy for the conversation to start to feel judgemental or guilt-inducing (even when we don't mean it that way) because this is such a loaded topic that many of us *do* actually feel guilty about. As Elizabeth Bechard's piece in Chapter 3 showed, a big part of what we do as coaches is to consciously make it feel safe just to be in a conversation about the climate so that our clients can then engage with it more easily.

Where a charge exists, we are wise to tread more cautiously. Much like a difficult feedback conversation where we are aware of the potential to hurt, anger or demotivate the other person, we approach the conversation more gently. During the conversation we look for signs that the other person is feeling safe: are they still communicating freely, do they maintain eye contact and is their body language still open? We may also notice how comfortable we ourselves feel, looking for any tension in our own bodies as a barometer of what is happening between us and them. These shifts in mood may be very subtle, but as humans we are adept at picking up on them. We are less good, however, at acting on them, often dismissing our intuitions, and hoping that if we just keep going, the situation will right itself. Often both sides leave a feedback conversation feeling incompetent, when if we had just voiced our intuition we could have rescued the conversation and built trust. Just as we can turn around a feedback conversation that is going badly, we can do the same to a discussion about climate change to ensure both sides feel good about it. More than that, the steps we take to course-correct a conversation can lead us to a much closer relationship with the other person, because we have shared our vulnerability by being honest. That can help them to see you as a trusted person to speak with in the future.

OVER TO YOU: Take a moment and recall a time when a conversation became charged in some way. What did you instinctively notice, and what did you do with those instincts? What could you have done?

## Types of contracting

So far we have discussed verbal contracting, but there are many other, more formal kinds that also signal our intention and can help us to engage people safely in the discussion. We call contracting that is overt or that you take your time over, 'macro-contracting', and contracting that is conversational, 'micro-contracting'. Another way of looking at this would be to see macro-contracting as creating the framework of the relationship and micro-contracting as addressing challenges to the health of the relationship as you go along. In the context of climate change, micro-contracting can be as important if not more so than a big standalone conversation. This is because in these conversations so many people feel out of control and hijacked by self-criticism. We have mapped this onto a grid (Fig. 7.1), comparing how coaches might do this, and showing how these behaviours also readily exist in non-coaching settings.

It is easier to use micro-contracting when you have earlier created a bigger set of macro-contracts for the relationship because you can refer back to what you initially agreed: 'Hey, this conversation is getting a bit uncomfortable. When we

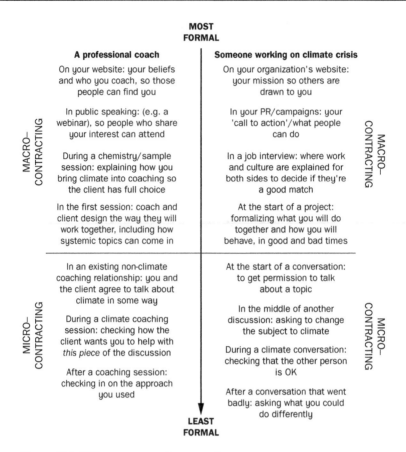

**MOST FORMAL**

**A professional coach**

On your website: your beliefs and who you coach, so those people can find you

In public speaking: (e.g. a webinar), so people who share your interest can attend

During a chemistry/sample session: explaining how you bring climate into coaching so the client has full choice

In the first session: coach and client design the way they will work together, including how systemic topics can come in

In an existing non-climate coaching relationship: you and the client agree to talk about climate in some way

During a climate coaching session: checking how the client wants you to help with *this piece* of the discussion

After a coaching session: checking in on the approach you used

**Someone working on climate crisis**

On your organization's website: your mission so others are drawn to you

In your PR/campaigns: your 'call to action'/what people can do

In a job interview: where work and culture are explained for both sides to decide if they're a good match

At the start of a project: formalizing what you will do together and how you will behave, in good and bad times

At the start of a conversation: to get permission to talk about a topic

In the middle of another discussion: asking to change the subject to climate

During a climate conversation: checking that the other person is OK

After a conversation that went badly: asking what you could do differently

MACRO–CONTRACTING

MICRO–CONTRACTING

MACRO–CONTRACTING

MICRO–CONTRACTING

**LEAST FORMAL**

**Figure 7.1** Different types of contracting

first started this project we agreed that we'd be honest if we ever found it hard, and I am sensing some bumps in the road. How are you finding this?' Because the relationship has been started in a powerful way, it is easy to repair minor punctures along the way; however, you can course-correct a relationship using micro-contracting at any time. It is far better to initiate a contracting conversation than to write off the conversation or even the relationship because we cannot summon the courage to check in.

When we seek to engage with others on climate change, there are multiple opportunities to begin the contracting process. Below we will describe when and how to contract in different ways, using examples of what a coach or an organization might do; however, these apply across the board. You may choose to use some of all of these. They are not intended as a checklist, but rather as suggestions.

**Macro-contracting**

1. WHEN: Communicating what you do publicly.
   WHY: At this level we are engaging people in an asynchronous or one-to-many conversation about climate change. As coaches and consultants, sharing our

beliefs helps people to understand what they will get from engaging with us. For organizations or campaigning groups, this also means that we are clear about what we want people to do in our 'call to action'.

HOW: Publicly expressing your mission and values through your website or social media, public speaking or writing, as an organization, group or coach. Just because everyone *can* visit our website or come to our lecture, doesn't mean they do. When we publicly declare our connection to climate and the planet, we can both attract people who want to engage in the subject and sensitize those who might not have considered that approach before.

2. WHEN: In the pre-relationship stage: the professional coach's chemistry or introductory session and the employer's job interview.

   WHY: It gives both coach and client the opportunity to check that they have the same expectations of the process. Similarly, in a job interview, the employer and the employee are doing the same sort of dance; finding out not just if there is a skills match, but if there is a culture match too.

   HOW: Coaches often ask contracting questions during a chemistry session, which are similar to those of job interviews:
   - What made you choose me/us as your potential coach/employer? What did you think you would be able to achieve here that you might not elsewhere?
   - What do you need to know about me/us, to work with me/us? What do I/we need to know about you?
   - What is the deal breaker here? What do you need to find out to know that we are the right coaching match/right company for you to work with?

3. WHEN: In the early stages of a relationship where the context is climate crisis; as a colleague at the start of a project/when joining a team; or as a professional coach in the first session.

   WHY: This provides a robust framework for exploring the goals you want to achieve together. As part of contracting you can discuss how you will balance ambition and uncertainty, to build in the permission to update goals if the external environment shifts.

   HOW: Build some of the following questions into your inquiry:
   - What is the main thing that we have to achieve together? How does this connect to the bigger picture of the climate crisis?
   - Think of someone/a team with whom you work really well. What is true of that and how can we recreate it here?
   - What makes you care about climate change and what impact do you want to have? What do you find hard about working on this subject?
   - What measurable goals do we need to create, and what do we want to agree to cut ourselves some slack if the external environment changes on us?
   - What is our intention for our relationship/this project, and how do we want to behave together?
   - How have other people treated you in relation to climate change? What of those experiences would you like to bring in/keep out of our relationship?
   - What would help you to trust me and what will break trust or cause a disagreement between us? How will we tell each other if that happens?
   - What do you need to know about my beliefs to feel comfortable sharing big fears and deep passions with me?

- How am I/are we going to hold you/us to account for the change you/we want to make?
- When will we check in on this agreement to update it and check it is working?
- Can we agree that it is OK for both of us to share when we feel stuck or uncomfortable? Perhaps we want to agree that we will pause to reflect?
- What should we agree to do on the days when the crisis is too much emotionally? Would it help to have permission to be upset or angry?
- We may also reassure someone that we are not attached to them acting in a specific way (or at all) in relation to climate change, rather we will help them to work out for themselves what they want to do. We may also tell them that we will believe in them, and in the planet, on the days when they do not believe in those things themselves.
- In meetings, just as in coaching sessions, it is also useful to ask what outcomes the client or team wants at the end. Many of us fear that we will be presented with a laundry list of impossible takeaways, but in practice this creates focus and almost always a reasonable set of objectives because people understand that they are as responsible for these outputs.

### Micro-contracting

1. WHEN: At the beginning of or during a conversation.
   WHY: To create the permission to broach the subject because you want to.
   HOW: It is easier to bring this topic into a conversation when you are open about it. Some useful approaches here can be:
   - Ask: 'Is it OK to talk about this?' 'I'd like to talk to you about …' 'I'd really like to hear your perspective on this.'
   - If in doubt, check: 'I'd really like to tell you about X, but I know it can make some people anxious. Are you OK if we talk about it?' If the answer is yes, you can add 'well if it starts to feel tricky will you tell me?'
   - Creating a micro agreement like the one above makes the other person feel more in control but still does not mean that they will let you know. Instead, use your intuition. Ask: 'As I started talking about this/as I said X, I sensed that it made you feel uncomfortable. I'm sorry if that's the case . . ?' and help them to tell you the impact it had.
   - Ask permission, especially with those who have experienced blame and unwanted advice from others. Even when a conversation is flowing smoothly, you may still do this: 'I have an idea, can I share it with you?'
   - It can help to explain our broader sense of purpose and what motivates us about the topic. 'I feel really excited by what we are doing in our community. Is it OK if I talk to you about climate change, and why we are doing it?'
   - When you first introduce the topic, avoid reaching for facts until asked for them, and avoid jargon and buzzwords. Instead ask questions. Using facts before they are requested often makes the other person feel ill-equipped to engage.
   - If 'climate change' is triggering, ask what term would feel easier; perhaps 'conservation' feels safer.

- When someone comments on a related issue or event, for example unusual weather, we can ask if they would be interested in talking about climate change. 'That seems to me like an example of climate change. Could we talk about it?'
- Focus on the relationship. While we want to start the conversation, long-term success comes from knowing when to *stop* and make sure that the other person is OK. By micro-contracting in the moment we give the other person control, and it is rare for someone to say they want to stop. If we don't ask, however, the other person allows it to continue because they want to stay connected, even though the conversation makes them feel bad. Checking in ensures that they don't have to trade their empowerment for your connection.
- Be at peace that their interest may be vague or non-committal. You are sowing a seed not forcing a plant to grow.

2. WHEN: After the conversation.
   WHY: While we want these conversations to go well, there are times when we leave wondering if we did more harm than good and worrying that the relationship needs repairing. We will discuss this in more detail shortly. For now, here are some micro-contracting phrases that can reset and deepen the relationship by demonstrating that we value our connection to the other person.
   HOW: Here are some approaches:
   - Keep it simple: 'Did I say something out of turn today? Did I upset you?'
   - Offer a rerun of the conversation, done their way: 'I felt worried after our conversation that perhaps I had made you anxious when I was talking about X. Is that true? Sometimes I can be so passionate about it that I don't notice if I'm overwhelming someone. I'd really like us to be able to talk about this again without it making either of us uneasy. What do you think?'
   - Explain your feelings, do not inadvertently make the other person wrong for their reaction. Instead of 'It looked like you were upset and you got defensive because I was talking about X' try 'I felt concerned/worried that I had upset you because you stopped talking.' Avoid telling them how they felt, instead say how *you* felt and what you *saw* in them. Avoid using the word 'why' because it has an interrogatory quality that raises defensiveness in others. Instead ask 'what made that happen?' or 'what made you do that?'
   - Find out how to avoid a repeat. 'I noticed that when we spoke about X you seemed to go quiet/withdraw/get angry. I'm concerned that I did something that wasn't helpful. Is that the case? What should we do differently next time?'
   - If this might feel too confronting, instead ask them to help you improve 'I'd really like to learn how to better talk about this subject. What could I have done differently?'

## An example of how a conversation can go wrong ... and how you can fix it

Let us consider an unsuccessful and fictitious conversation between Tim, who has been charged with embedding his company's sustainability strategy, and

Antonia, a colleague from Finance that he does not know very well. Tim has formed a great Green Team who regularly have energizing conversations and with whom he feels happy talking about the grim realities of climate change. Though it is difficult stuff, he comes away with a sense of solidarity from each discussion. Yet when he bumps into Antonia, the conversation leaves him doubting himself and the company's ability to change, and appears to make Antonia uncomfortable too:

Tim:       'How are things, Antonia?'
Antonia: 'Hey, Tim. Oh, you know, keep on keeping on! How are you doing?'
Tim:       'Pretty good actually, we've started to make real headway with the recycling initiative.'
Antonia: 'That's great. I wish I could say the same for our new product launch.'
Tim:       'Yeah, well it's a drop in the ocean really. We're behind the curve in doing anything when you look at the data. Just the other day there was a new report that said we're cruising towards environmental tipping points from which we can't come back.'
Antonia: 'Really? I don't know much about all that.' [Antonia is starting to look around her, and sound distracted.]
Tim:       'Yeah, and if you talk to young people they're really desperate. My niece says her school friends have eco-anxiety. I mean, can you imagine? Fifteen-year-olds worrying that the future is hopeless. When we were 15 we were drinking cider in the park!'
Antonia: 'Yeah. I spent my weekends buying junk from that hair accessories stall in the market.'
Tim:       'Yeah, most of which is probably floating around the ocean somewhere thanks to unscrupulous waste disposal companies!'
Antonia: 'Yeah I guess so … We didn't really know that then.'
Tim:       'No of course not, but the plastic manufacturers did. They've known for years!'
Antonia: 'Yes … Sorry Tim but I've got to dash off. I'll see you tomorrow, yeah?'
Tim:       'Oh, OK … See you then.'

Tim did what many of us do: he shared his passion. After all, this is an emergency. What Tim did not do, however, was ask permission or really listen. As he sensed her distraction, instead of learning more about Antonia and her response to what he was saying, Tim increased the charge in the conversation by talking about his niece's strong emotional reaction. This is really common because when we feel that the other side is distracted or uncomfortable, we often fill the void of lack of engagement with more talk, hoping that upping the ante will bring them back into the discussion. However, what we are not privy to is the internal processing that is going on for the other person, and by overloading them at this point we risk a greater disconnection. Too often we plough in without asking permission and then, as the conversation starts to run away from us or feel uncomfortable, we allow our inner critic to tell us how the other person is feeling, for example 'that they are careless'. If we had interviewed Antonia after this conversation, she may well have told us that she didn't feel knowledgeable enough and that as Tim was speaking she felt out of her depth and increasing despair. Rather than enrolling Antonia to his cause, Tim unwittingly scared her off. The good news is that it doesn't have to be hard to start on the right footing, and that even if we have had an encounter like Tim and Antonia's, we have not ruined things forever; we can repair it. It just takes being refreshingly honest, and being open to each other's needs.

OVER TO YOU: Look through the conversation between Tim and Antonia and write four open questions that Tim could have asked Antonia. How could you use these in your conversations?

Here is another version of that conversation in which Tim did several versions of contracting with Antonia. Again the contracts are in italics:

Tim:  'How are things, Antonia?'
Antonia:  'Hey, Tim. Oh, you know, keep on keeping on! How are you doing?'
Tim:  'Pretty good actually, we've started to make real headway with the recycling initiative. *Do you have a minute for me to tell you about it?*'
Antonia:  'Sure, though I know very little about our strategy.'
Tim:  'That's OK, we honestly didn't know much either when we started. It's been a huge learning curve. But we're getting there now. *I'd love to come and tell you about it properly next week. Do you have time?*'
Antonia:  'Yeah sure. It's great that you're making progress. I barely keep up with my home recycling, let alone manage it for the company! I honestly feel so confused by it.'
Tim:  'Yeah, I hear that a lot. I found a really simple website to make recycling easier at home. *Would you like me to send you the link, or do you think it wouldn't help?*'
Antonia:  'That would be really helpful, thanks. I'd like to do more.'
Tim:  'What kind of things would you like to do?'
Antonia:  'Oh, I don't know … it's all so complicated and overwhelming …'
Tim:  'I often feel that way. *Is it OK to talk about this, or would you rather not?*'
Antonia:  'No, it's fine, but thanks for asking. It's just hard to know what to do.'
Tim:  'Well *maybe when we meet next week we could talk about that a bit, and what might make it feel more manageable?*'
Antonia:  'Thanks, I'd actually appreciate that.'
Tim:  'No problem at all. It would be a pleasure.'

So what happened in 'take two'? It might be surprising to know that the two conversations are almost identical in length; the second contains 29 more words than the first and Antonia spoke for exactly 42 per cent of the time in both conversations. It probably didn't feel that way though. That is because Tim responded to Antonia instead of talking *at* her; the conversation was a back and forth, not a broadcast. Tim also asked Antonia's permission. She readily agreed, which might lead you to think that he didn't need to ask, but often we don't say how uncomfortable we feel, and requests like this can be reassuring and help to build trust. Tim was also paving the way for a much more charged conversation at their next meeting, in which he and Antonia might talk about her feeling of being overwhelmed. In the first example Tim asked just two questions – one was in his greeting and the second was rhetorical. In the second example he asked seven questions, five of which were some form of contracting. This really shows that it is not the action of talking or how long we do it for, but *how* we talk that changes the way people engage with us on climate.

OVER TO YOU: What do you do to make conversations about climate change more successful? What could you borrow from these examples?

## Improve your relationship ... admit you have made a mistake

So let us imagine that your conversation went really badly wrong. Maybe you feel confused, ashamed, guilty ... don't walk away. As C.S. Lewis said, 'You cannot go back and change the beginning, but you can start where you are and change the ending.' Believe it or not, you now have an opportunity to *improve* your relationship rather than just resetting it. Expressing concern regarding the health of your relationship and being open enough to recognize that you may have misstepped – 'Did I just say something to upset you?' – shares an authenticity and vulnerability that allows the other person to feel safe enough to be honest in return. 'To be honest, yes you did upset me a bit. I feel quite frightened about this subject.' It is noticeable that we often find it easier to share our deepest fears about situations *outside* of the relationship than inside it. For example, we may complain about our partner's behaviour to a friend rather than telling our partner themselves because we want to stay connected to our partner and fear any breakdown in that connection. However, when we are honest, we do the relationship more good than damage. If we take time to reset relationships (either in the moment with some mid-conversation recontracting) or after it (with a quiet apology and enquiry about how to do it better next time), we build trust and respect through our vulnerability and demonstration that it is the relationship that matters to us, as much as the subject. This is what the relationship psychologist, John Gottman calls a 'repair bid'.[47]

In all of these situations we are asking you to tune into not just what is being said but *how* it is being said. This means being attuned to what is happening in the atmosphere between you. This is the 'emotional field'; a force field between you, humming with all of the feelings either of your own making or that you are sparking from each other. As coaches we are looking to create a warm, safe and creative field between us and our clients, and we assume that you probably want to create something similar with your colleagues and friends. The first step is to notice the mood of the conversation. Many of us fail to notice this energetic space, but all of us can be trained to. We all know what it is like when you can 'cut the air with a knife' or when a team 'feels in a celebratory mood'. Once you can more readily notice the crackle or the comfort in the air, you can change the way that conversations run, by celebrating what is working or acknowledging what is not. For example, in a team that seems unwilling to talk, you can say, 'It feels like everyone's feeling unable to speak up, is that right? What would make it feel easier to speak?' Just as in a conversation you might ask 'I sense that this subject is making you feel unsettled, is that the case?' This can burst the dissonance and help people breathe out. Meanwhile, where there is resonance, positive contracting can build on it. With a team that is generating lots of ideas you might boost the emotional field with: 'You guys are on fire! What do we need to do to keep this happening?' In a one-to-one conversation that might sound like: 'That's such a brilliant idea! What would help you take it further?'

### How to contract so you can coach professionally about climate without 'bringing an agenda'

Many professional coaches want to impact this issue, and have a legitimate concern that doing so means wielding their own agenda within the sanctity of the coaching relationship. So here are some ways to contract this with clients while staying ethical:

#### Your climate change coaching practice in the world

Rather than asking what is allowed, ask: 'What do I want my impact to be?' When we wrestle with the idea of an agenda, we can lose sight of the fact that coaches are allowed to have a purpose too. Beware collapsing 'having an agenda' with 'having beliefs and values'. Although we strive to be neutral, we all hold a worldview, and in many cases our clients choose us because of our morals, values or beliefs. If we are open about this, we are ethical. You might rebrand yourself as a climate change coach or share stories about climate with current and future clients through your website, newsletter and administrative documents your clients receive. Alternatively, you might build products around it within a broader remit, such as one coach who facilitates seasonal walks in nature themed around climate change,[48] or Adam Lerner, below, who writes about it. Another way is to share your beliefs verbally in chemistry sessions. These are clear signals to clients of what they can expect from you as a coach, and your opportunity to share your purpose and vision. Your purpose will in turn help you decide *how* you want to use your coaching skills; whether in formal coaching or drawing on your coaching *skills* in activism. If you do want to coach people on this topic, consider who might really value this help, so that there is a clear shared agenda. Decide if the 'who' you coach is as important as the 'what' you coach them on. For instance, would it be fulfilling to coach environmentalists on non-climate change topics (such as work-life balance), or do you only want to talk with people about climate change itself?

#### Climate change coaching within sessions

What happens when a client brings a topic that is not related to climate change but as the coach we recognize a useful connection? It is possible to bring elements of the climate crisis into a conversation, providing the coach is willing to remain unattached, and offer rather than impose this lens on the client. Here are some tested approaches:

– Contract your interest with your client before the session so they can come to the session with any clarifying questions. Be clear that it is not essential – it is important that the client does not feel judged for *not* wanting to work on this.

- Introduce nature as a co-coach to reconnect clients to nature, rather than to talk about climate change as a topic (see Ruma Biswas' story on this in Chapter 8). Invite the client to reconnect to nature by noticing and reflecting on it during the session. They might simply observe their pot plant or you might alternatively coach outside.
- Notice and invite the client to step through their own open doors. If a client mentions feeling anxious about unusual weather or being overwhelmed by bad climate news, ask great questions to open the topic up. Of course, before you do so, check with the client that it is in service of the topic they brought to the session.
- Use climate change as a frame, more than a topic. If a client mentions systems that they want to change, ask them if they would like to add in climate or the environment as another dimension of the system. Similarly, if your client is doing long-term career or life planning, ask them what change they expect to happen socially, politically, economically and environmentally in that time, that might impact their plans.
- Understand that sometimes climate change has no place in the conversation. If the client wants to talk about a work conflict it could be a stretch to find a direct link to climate change. It is OK not to bring it in.

### Reframe your perspective on coaching and climate change

Examine the perspective you currently hold about 'the environment' or 'climate change'. The coaching industry mirrors the world, and initially many see this as a niche (just as sustainability used to be pigeon-holed into Corporate Social Responsibility). An alternative perspective is that climate change is the context in which all life is lived, which may change your definition of what 'climate coaching' means, broadening it from coaching activists to health coaching people with chronic diseases made worse by wildfires or leadership coaching leaders in communities suffering flash flooding. Many topics have been normalized in our industry (and societies) that once felt out of bounds. Talking about diversity, gender equality or mental health is now accepted territory for coaches and would not have been 30 years ago. Ask yourself how we might achieve the same switch about the climate crisis.

## In practice

### The pen can be as mighty as the coach

*Adam Lerner, coach, author of* The Understory *and founder of Solvable, dedicated to regenerative futures*

As coaches, we feel that almost everyone would benefit from professional coaching. We listen to leaders alienate and exacerbate disconnection within communities and with the natural world. We encounter debilitating anxiety and grief, *which manifest into climate distress. We witness the conflict of internal ambivalence like the two faces of Janus. We also know that the majority of people will never work with a professional coach. Sure, there are certainly misconceptions about the profession, but coaching also has equity, accessibility, and scaling issues. If we believe in the power of coaching to positively contribute to the world's most complex problems and yet most people will never work with a coach, shouldn't we ask how to expand our professional impact? I began to wonder if real-time conversation was the only medium for coaching, and whether there might be underutilized mediums that could create the same fruitful conditions.*

*After being arrested in 1963, Martin Luther King Jr wrote nearly 7,000 words of, what might best be defined as, profound coaching. King addressed the essay, 'Letter from Birmingham Jail', to eight moderate white clergymen because he felt they were 'men of good will' and sincerity. King named their fear, love, shame, and hope like any good coach would. It was his metaphor – the cup of endurance – that was perhaps his greatest masterstroke. Just as we would guide a client through visualization, King's cup helped the clergymen see the inevitability of the protestors actions: 'There comes a time when the cup of endurance runs over, and men are no longer willing to be plunged into the abyss of despair. I hope, sirs, you can understand our legitimate and unavoidable impatience.'*

*It was in part the power of King's Letter that inspired me to start writing* The Understory *with the strapline: 'to inspire leaders to act on the courage of their convictions in defence of the living planet and those who inhabit it.' From day one, I knew from decades in marketing that my chosen format would be fighting against the cultural tide. After all, even for someone as famous and brilliant as King, how many of us take the time to read all 7,000 of his words?*

*Our 24/7 culture reveres headlines over stories; reductivism over complexity; disruption over memory. I sought to create a publication that built a complex web of interrelationships by bringing forward voices who were disconnected by discipline, era, and nonconformity. After publishing a few issues of* The Understory *I realized it could also host community voices. I shifted the bi-weekly rhythm to publish a new issue in the first week and then community responses (entitled Reflections) the subsequent week. Each issue became the topsoil by which members of the community sprouted myriad forms of new life.*

*Now looking nearly back over thirty issues, I have found that writing fortified me with the courage to do inner work, publicly. I have modelled the behaviour of not knowing and even openly contradicted myself. In so doing I've helped to ease the transition of others into the climate movement by countering their own inhibitions, particularly their overwhelm. I recall one comment in particular that came to me privately in response to 'Issue Fourteen: Permission to Grieve'. The reader shared that until reading it she had neither realized that climate grief existed, nor that for years she had been silently grieving collective loss. She was now able to name her own grief.*

*Writing is an ideal medium for the gradual unsettling of ourselves to destabilize those limiting beliefs that prevent action. For those who feel called to pick up the pen, writing can be an additional medium to reach part of that large population unwilling to cross coaching's formal threshold. Whatever your chosen medium(s), go forth and make a difference in the week ahead.*

## Balancing power

In any relationship, our comfort is derived from how safe and in control we feel about what is happening to us. Climate change has the power to undermine that sense of safety, so it is more imperative that we give people the power to feel safe when we speak about it. While we might not feel it, we are to some degree powerful in these conversations and need to give some of that power to the other person. In his first conversation with Antonia, Tim was holding most of the power. He probably didn't feel that way (not least as his words failed to have their desired effect), but Tim *was* driving the discussion, while Antonia was on the back foot, unsure of where the conversation was headed. Every moment is an opportunity to help someone to regain their agency, not just to help them gain knowledge. A useful way to think about this is that it is about balancing power between the two of you.

As coaches we think about this a lot, because our role is to help clients to stand more firmly in their own sense of power. Feeling more grounded and secure ourselves also helps us to tune out our own self-judgement and be more present for the other person. Yet it is just as easy for the coach to take control as it is for the employer to do so during a job interview, and all of us must not be seduced by the feeling of having control. Really robust relationships are built on a sense of equal power. Many people we have coached complain that they feel like their power is constantly being taken away from them because the people they speak with tell them what to do or how to feel about the climate. This is where asking using 'micro-contracting' – e.g. 'So you feel stuck … How would you like me to help you with that?' – can help to redress the power imbalance. By continually offering the other person control of the conversation, we are under-lining the existence of their own resourcefulness by saying: 'I trust that you know what you need.' We are also standing firmly in our own agency because we are demonstrating our comfort in not controlling everything.

Sometimes we can start to feel unstable too, and that is when we might *ask* for some power back. An example of that might be, 'I'm not sure where this conver-sation is going, can you explain how this links to what you just said?' In everyday conversations both sides may be thinking these things and not saying them. As coaches, we see the tremendous value in being honest about what we are seeing in front of us and feeling within us because it deepens our relationships and enables the vulnerability that is part and parcel of change – a key principle of contracting is not being attached to being in sole control. During every interac-tion, we are looking to hand back power when we have inadvertently gained too much, or to reclaim power when we feel unstable. In this way we can model *rela-tional* not just individual empowerment (i.e. the security to *share* control).

Here are some examples of when we need to give away some of our power and when we might want to regain some:

### Hand back some of your power …

– When you have assumed rank over others; for example, you find yourself say-ing, 'I'm the/an expert / I've been working in this area since …'

– When you are asking question after question without explaining why or seeking permission, it feels like an interrogation.
– When the other person has put you on a pedestal and given you rank; for example, they say, 'You'll know better than me, you're the expert/you've seen this before.'
– When someone has lost faith in their own creativity they will often plead for a solution and ask you to solve their problem: 'I don't know what to do. I'm sure you do.' Instead explore their loss of faith and rebuild it 'What if you *did know*?'
– When you are driving a charged conversation without consent, either the content or the process. For example, Tim drove the content (he was in charge of *what* was said), but it is just as easy to slip into a process without explaining why it is beneficial; for example 'Let's try this' rather than 'would you like to try this idea …' This could also be talking without pausing to ask, 'What do you think?'

### Regain some of your power …

– When the other person is talking about something and you do not understand its relevance or are feeling unsure[49]; for example 'I'm not sure where this fits. Can you explain?' or 'Can you tell me where this fits in with the other things we've been talking about?'
– When you feel upset or destabilized by the content of the conversation; for example, 'as you told me how you are feeling I felt quite hijacked by the same feelings myself. Can I have a minute?'
– When you feel disrespected or dismissed; for example, 'I feel that my ideas aren't being taken as seriously as I had hoped they would be.'

At the start of this chapter we said that it took courage to talk about climate change with people, and perhaps that turned out to be not in the way that you might have expected. Courage is required in stepping away from 'telling' and moving into 'asking' and co-creating. When we do that, we also show people how to have equally safe conversations with other people and to feel more comfortable engaging with the topic itself.

# 8 When there's too much 'not enough'

## Dealing with scarcity

You may not know to call it 'scarcity' but the voice constantly telling us there is 'not enough' – time, money, resources, good people, political will – is nevertheless the most potent and unhelpful mindset in relation to the climate crisis. Scarcity talks us out of change on both an individual and a systemic level, leaving us feeling resourceless and defeated. In this chapter we will show you how feeling 'not enough' stalls us from taking action, and how managing scarcity can return us to empowered choice. We will also show you how it is not the facts of scarcity that matter, but the way that we relate to those facts. Whether you have been engaged in this work for a long time, or have only recently come to this, we are all susceptible to scarce thoughts.

### How scarcity manifests itself

Scarcity takes many forms, but it is usually accompanied by the word 'enough' or something related to it. The one articulation of scarcity that rings loudest about climate change is: 'there's not enough time!' In truth, the situation we face *is* scarce; we don't have lots of time to make big, systemic changes or yet have enough people actively involved. However, there is a difference between the *fact* of scarcity and the impact that it has on our mental state, because feeling the threat of scarcity changes the way that we respond for the worse. Feeling a sense of lack can reduce our imaginative capacity because we look for ways to protect not innovate, and because we dismiss ideas too quickly because there's 'not enough' resources to bring them to life. As Anna Cura explains in her story in this chapter, getting caught up in scarcity can get in the way of reimagining systems when we believe that we cannot change anything *until* we have resources.

As climate change coaches, we have worked with a lot of individuals who feel deeply alone in worrying about climate change, for whom scarcity is exacerbating their self-doubt and stopping them from moving forward because they 'have to make the right decision first time, because there's not enough time'. Many people find themselves being hurled over an emotional 'edge' from the way they thought the world was, to the way they now see it, and often they have gone there on their own – in many cases leaving their partners behind in the old world story of 'everything is OK'. Such a vulnerable, isolated emotional landscape makes them susceptible to the dual fears of 'can I do it?', and 'can humankind do it?' in which scarcity (and overwhelm) run rife.

Scarcity could be said to be socially and economically conditioned in us. Ironically, given that many of us live in a world of abundance (where any kind of food is available all year round), much advertising and marketing relies on scarcity to sell us things. It could be the sofa sale that 'must end this weekend' or the 'last chance to buy' clothes rack. These tactics activate our sense of threat that we will be missing out. If we engage our rational minds, however, we remember that we do not actually need any of these things. We also see scarcity baked into organizational business models, who embrace short production turnarounds like 'just in time' manufacturing and the hyper production of fast fashion that quite literally changes on a weekly basis. Each layer of the system has been designed to be opportunistic, moving at a pace that doesn't allow for thoughtful, conscious choice. We have all fallen victim to the lure of 'you aren't good enough now ... but you'll be a better person if' marketing. Companies do it because it works. This is a part of the current paradigm that we can all shift, by slowing the process down and noticing what happens to us when we experience scarcity. Consciously switching off the voice of 'not enough' and returning to our intrinsic values, reminds us what we care about and helps us respond differently.

When we feel we don't have enough physical stuff, many of us now try to bring a new perspective on what we already own – such as 'going shopping in your wardrobe first' – and we can apply the same principle to scarcity of knowledge or action. When we feel 'I'm not doing enough and I feel inadequate and ashamed' we can respond with 'I really care about this issue, because taking responsibility matters to me.' This is useful because when it comes to knowledge and action we often jostle with others. If we dial down the scarcity, we can stop competing and connect better with others. Connection to others who care also dials back the hesitation that scarcity can spark in us, leading us to act from a more secure, supported, less threatened state.

Common phrases that point to a scarcity mindset quantify amounts such as 'not enough', 'too few', or even 'too many', when 'the many' are doing the wrong thing (e.g. 'too many people buy flights on a whim'). More common, however, is for people to allude to scarcity indirectly. While some will say 'not enough governments are stepping up' far more will articulate it more subtly as 'It's all very well that our government is setting big targets, but what about China? What about India?' What they are not saying, but may feel is: 'We're just one country, too few are joining us/too many are doing the wrong thing.'

OVER TO YOU: Look at the sentences below and write an expression of scarcity below each. We have given you the first, and put potential answers for the others at the end of the chapter:

1. 'It's all very well banning petrol cars from 2030, but by then we will have reached tipping points from which we can't return. It'll be too late.'
   *Scarcity: There's not enough time*
2. 'When I go round my local supermarket I don't see anyone else agonizing over plastic like I do.'
3. 'If X party don't win the next election it's all over for the climate crisis.'
4. 'I'd need to do a PhD to be able to do anything useful about climate change!'
5. 'My kids want me to become a vegetarian, but it just feels so limiting'
6. 'Until the government makes us convert our fleet to electric, we'd be mad to do it. We'll completely lose our competitive advantage.'
7. 'I read a report about how bad our soils are, and how precarious our food supply is. I'm seriously thinking about stockpiling food in the garage.'

## How scarcity affects us physically and mentally

Scarcity acts on our bodies as well as our minds. To illustrate this, imagine that you are on a train on your way to the wedding of a dear friend. You have only 15 minutes between this and your connecting train, and all is going well when your train slows to a crawl. For what seems an eternity you move sluggishly along until an announcement explains that, due to an engine fault, the train will arrive 20 minutes late. Panic rises: you will miss your connection. Aghast, you realize that this will mean you will miss the wedding service, arriving after the meal has been eaten. This is your oldest friend, on their most important day. There is not enough time to catch your connection, no subsequent train to replace the missed one and not enough time to make the wedding. What might go through your mind, and what might happen to your body as you sit there, hostage to feelings of scarcity? Here are some common responses:

### Mental

- Self-criticism: 'Why did you think you had enough time to make this connection, you idiot! You are so thoughtless/so selfish for wanting that extra hour in bed; or so focused on your own wallet that you bought that cheaper, later train ticket. You have let your oldest friend down. How could you be so irresponsible?!'
- Trying to escape the feeling: Looking desperately for a way out, checking trains from other stations, bargaining with yourself about how much you would be prepared to spend on a taxi.
- Anger: Feeling thwarted and impotent, perhaps physically punching the arm of your seat, or digging your nails into your palms.
- Deep sadness and resignation as it sinks in that you have let your friend down, a feeling of the futility of everything.[50]

### Physical

- A feeling of dread in the pit of your stomach at being in the wrong and disappointing others.
- Clammy skin or sweating, a dry mouth.
- A racing heart and shortness of breath and/or short breathing in the top of your chest.
- Fidgeting: eyes searching around anxiously, legs jiggling and hands fiddling with things.
- Bold movements: perhaps standing up and striding to the corridor or pacing, to somehow outrun the feelings and clear your head.

Now think about the climate crisis and the last time that you watched something that brought home the scale and urgency of the problem. How many of the mental and physical responses above does climate change generate in you? The chances are quite a few, because our triggered minds cannot differentiate between the threat of global collapse and missing an event (with no offence to your fictitious friend and their imaginary wedding). This is why it is not the facts of the issue that we need to detrigger but our emotional response to them. When we are in the grips of a scarce mindset we are less able to access the prefrontal cortex, the part of our brain that governs logical thinking and rational decisions. Our job if we are to work with, coach and influence others about the climate crisis, is to help people move away from scarcity, back into their rational selves, so that they can make effective plans.

## In practice

### From anxious to activated – how simple coaching tools can move people

*Kelly Isabelle DeMarco, occupational therapist, National Board Certified Health & Wellness Coach, Climate Action Catalyst*

*In the company in which I work, I sit on our Green Team. That often means that people drop into my office to ask for advice. As a trained coach, I always try to also help them feel more empowered around the subject because often powerlessness is the main problem they are facing. A colleague arrived at my office looking worried and tense. She requested to meet because her attempts to quell her 9-year-old son's growing anxiety about climate change were not working. She fidgeted in her chair and let out a strained sigh. She had run out of answers to his growing list of questions. Her recent interaction with her son went something like this:*

Son:     *'Mom, what are we doing about climate change?'*
Mom:     *'Well honey, as we have talked about before, we are recycling and reducing our waste.'*
Son:     *'But Mom, what are we DOING about climate change?'*
Mom:     *'Well, we do have a garden and have started composting to make better use of our food waste.'*
Son:     *'No Mom. This is a HUGE problem! What ARE WE DOING about climate change?'*

*She looked at me imploringly and asked 'What can I do to help him? I'm not an expert and they didn't teach us this stuff in school. Where do I even begin?' After a brief pause, I asked if she was open to receiving a few questions from me. She nodded.*

*'It sounds like you think you need to have all the answers and be an expert to help. Is this accurate?' I asked. Nodding yes she said again, 'I never learned about climate change in school. It's all new to me. I'm at a loss for what to say. His questions make me anxious. And I really do not know if there's more we can do.'*

*I validated her feelings and reassured her that she was not alone. With new information coming out daily, it is hard to keep up. Her body relaxed a little and she settled more in her chair. Her facial expression softened. She looked at me uncertainly again and said, 'You are the expert on this. Can you help?!' I told her that we are all learning as we go and that goes for me too and that I don't have all the answers but that I might be able to coach her toward a solution of her own if she was open to it. She nodded.*

*I reflected back to her that it sounded like her son was the one driving the need for more information and answers and that he appears to be the one with the most energy and interest in the topic. She said that was true. I asked her to consider that she did not need to be the expert on this to help. Her eyes widened and she leaned in closer. I offered, 'What if you let your son take the lead? What if you explore opportunities for him to learn more and be the one to educate your family and share his own ideas for solutions?' Her energy visibly shifted and her facial expression brightened. I marvelled at how simple, powerful questions can really shift people. Her tone turned more optimistic as she said, 'Ooh, I like where this is going!' She then started to generate ideas of her own about library trips, educational shows and community engagement, and how these might spark more conversation and ideas for action. She said, 'Wow. A huge weight just lifted!' She acknowledged again that the hardest part for her was not having all the answers.*

*Allowing her son to take the lead and become the family expert would relieve her sense of overwhelm and her feelings of scarcity. Our conversation ended with a bit of role playing so that she could rehearse how she might respond to his questions. She left saying how remarkable it was to experience such a shift in only 30 minutes and to feel capable of empowering her son without needing to be an expert. I knew that I could have recommended books to read or videos to watch, or given her a list of actionable steps she could take. But it was clear to me that she was already overwhelmed. Additionally, I knew that doing so would reinforce her belief that she is not an expert. I knew that if I could help her shift her perspective, she could reclaim her sense of agency in this situation with her son.*

Kelly's client's dilemma is not uncommon. We want to support our loved ones, but we are in the grip of scarcity ourselves. The more we have at stake, the more threatening it feels. The client would not have felt anything like this much anxiety if her son was concerned that he 'didn't have enough time to write a big school report'. The type of scarcity we are talking about here is linked to existential threat, and while it may bring up feelings of individual scarcity ('I'm not knowledgeable enough') it springs from systemic scarcity. The key thing to remember is that this is a *mindset* that we can choose to adopt or not. We can make a conscious shift, and zoom out from the tight, closed perspective of 'lack' to a broader, calmer horizon. To do this we must put aside guilt and self-judgement, and embrace compassion – for ourselves and for every other part of this world of ours.

## Bringing us back to our senses – tried and tested methods for neutralizing scarcity

### Name scarcity – disconnect the feeling from the facts

Name the feeling early on to bypass judgement and criticism, avoid a lengthy story about facts and get straight to someone's desire for action. This allows more space for emotions to be processed, and develops someone's resilience because they will identify and resolve scarcity themselves next time. Phrases that work:

- 'It seems like you feel "not good enough"/there is not enough when it comes to this?'
- 'I notice you've said "not enough" several times now as you've described this – that sounds like a feeling of scarcity, is that right?'
- 'It sounds like you are feeling a sense of scarcity; like there's not enough of X, Y, Z?'

## Make it OK to feel this way

Normalizing the experience of scarcity and acknowledging the person's difficulty can in itself detrigger these feelings. Phrases that work:

- 'This is a really understandable response.'
- 'You are not alone. A lot of people feel like this/I've also felt like this.'
- 'This is a healthy human response when you see others not acting.'

## Acknowledge the person *and* the planet

Recognizing both people and the planet's positive qualities and achievements is an effective way to scaffold someone's sense of agency and help them believe that humankind can change. This acknowledgement soothes fears and helps to dismantle systemic scarcity. Phrases that work:

- 'I know you to be someone who … [point to their resourcefulness].'
- 'I wonder what other challenges the planet/this community/our company has overcome that once seemed insurmountable?' These do not have to be environmental, but signs of systems overcoming challenges, to prove the system's ability to change.
- 'Humans have changed systems for the better before … [insert example, e.g. the end of slavery, the success of the civil rights movement].'

## 'Show me the money!' – ask for the proof

Often our fears lead us to conclusions, without much fact-checking. Asking someone, gently or light-heartedly, whether they have any proof to back up their fears can help unpick assumptions from truths. If they *have* really done the research, be ready to accept they are right, but more likely your question will send them on a fact-finding mission that is in itself a form of action, jumpstarting their process of change. Phrases that work:

- 'Have you talked to anyone who could help you find out if that's always the case?'
- 'You know, sometimes we can buy an idea that our mind sells us without really inspecting it. I wonder how much you've inspected this thought? Is it really true?'
- 'I wonder if that's always the case? What research have you done?'

## The bigger picture

While it may feel that the world is in a downward spiral of not enough, as Matisse said, 'There are always flowers for those who want to see them.' Helping someone zoom out from the tight, scarce view to see the beauty and kindness around them can often loosen a scarce mindset. Phrases that work:

– When someone says 'I feel like I'm the only one who cares about plastic waste' you could respond with 'What about if we looked up some of the organizations that are working on it? Are there any you are inspired by?'
– When they say 'We've only got X years!' – find ways to show the expansiveness of time and how much can be done in a short amount of it. 'What were you doing X years ago, and what did you get done between then and now? What did you get done in the last year or month?' 'What has the world accomplished in this time?' 'What new ideas and technologies could we develop, with the same time again?'
– When you hear 'I don't see any other companies making big commitments like this' point to movements or industry leaders that are. 'I get it. Did you see that X Company just made a big announcement?' 'What companies do you know of that *are* making big commitments?'
– Map the system: A more practical approach for someone who believes that there is 'not enough' is to map the system out to see who else is out there and what they are doing. This helps to find collaborators and spot any gaps that they could usefully fill.

### Be contrary … as a thought experiment

This is useful when you have a really good relationship with someone based on care and respect. Where that is not the case, you can first contract it by asking, 'Is it OK if I ask you a question, simply as a thought experiment?' Phrases that work:

– 'What if there *was* enough time? What if you *did* know enough?'
– 'What *is* good here? What *is* there enough of?'
– You can also introduce humour here if you get on well. 'You are right, you must be the *only* person that cares about this … in the whole world!' Essentially, you are joking at the expense of their inner critic, and so this takes some hamming up on your part to demonstrate that you are not criticizing the person themselves.
– Agree with them! 'You know what, you are right, there is nothing that you can do/nothing that can be done. You should just leave it to others.' The fact that they are talking about it shows their willingness to act. They need help to overcome their fears.

### Offer a metaphor

Metaphor can help someone connect to the emotional truth of the situation, by showing them how they feel right now. It can paint a less charged picture than the dire scenario that someone is gripped by. These do not need to be elaborate, often those we use in common speech are most effective. Phrases that work:

– Suggest a metaphor: e.g. 'You sound like a kid in a sweet shop!'
– Pick up a metaphor they used: 'You said you feel like a bear in a cage?'
  Directly ask: 'What's a metaphor that would help you think about this differently?'

**What is good about scarcity?**

There is a gift in everything, even scarcity. We also only feel threatened by the loss of something we care about. Seeking the gift here could mean recognizing how much we value nature, or our children. Phrases that work:

– 'Putting aside what's hard about feeling this way, what is good about it?'
– 'I can hear how much you care, so it seems there is something good here too …?'
– 'What are you grateful for about having these feelings/thoughts?'

**What are the beliefs or values that scarcity steps on?**

Our fears also sting because they trample on our values. Helping someone to understand which values are threatened can help them regain confidence. Phrases that work:

– 'What's important to you/what matters to you about this?'
– 'It sounds like this is stepping on a value for you. Do you know which one?'
– Appeal to a universal value that most people share, such as responsibility. Appealing to someone's natural sense of responsibility might look like 'I know that you take being responsible very seriously' or 'I know that you are a responsible person' or even 'I know it drives you crazy when you see people being irresponsible.'

**The present moment**

While climate scarcity is mostly a condition of the future (that can also find its roots in events in the past that taught us negatively about loss or lack) it is not, however, in the 'right now'. Experiencing the present is a powerful way to draw us out of this mindset and deescalate the feelings. Simple grounding activities and noticing our physicality can help. While that may sound like a stretch for an everyday conversation, in trusted relationships or where you see real distress in the other person, it can be transformative. Embodiment can bring us back to 'who' someone is not what they are thinking. A focus on the present and what the other person can hear, see, smell or touch right now can coax them back to themselves. Phrases that work:

– Where you don't know the person very well or are in public, point out something in your direct environment. 'Can you hear that bird?' 'Feel that sun on your skin?' 'Listen to that rain!' 'What a deep blue sky!' Then pause and let them drop into the experience of using their senses, so that they disconnect temporarily from their thoughts. This may need you to be silent or to comment on your own sensations. 'I can feel my shoulders relaxing in this sun' to hold them in this moment.
– 'It sounds like there are a lot of unhelpful thoughts here. We'll pick them back up, but just for a moment, can we take a couple of deep breaths and let the thoughts go?'
– 'Something that helps when you feel this way is to notice what's going on right now. What can you feel, smell, taste, touch or hear around you where you are?'

### Connect with nature

This is an extension of the present moment idea, but deserves a mention of its own. Connecting to nature is transformative. It puts us in a relationship with the world that we seek to serve, and to breathe out, and allows nature to guide us. You do not need to be in nature to do this, a pot plant or spider's web in a room offers a path to connect. Even in the midst of a city we can find some aspects of nature. Phrases that work:

- 'If you look at that (pot plant/scene outside) what draws your eye? What would it tell you about this/tell you to do?'
- 'Find something in nature that you are drawn to and simply look at it. Notice how it manages to exist in this complicated, difficult world, in which it has no control.'
- Invite the other person to 'spot the good': 'What can you notice that you feel grateful for?' This brings in a sense of abundance, scarcity's opposite.

## In practice

### Nature as co-coach

*Ruma Biswas, coach and team member, Climate Change Coaches*

*I first discovered nature as a co-coach on the banks of the River Thames in England, on a coaching walk with a client. Nature also partnered with me as a co-coach on the shores of the Arabian sea in Mumbai and then more recently in Hong Kong in the mountains. Every single time nature has stunned me with her agility, her flexibility and her ability to reveal the truth. It is the last especially that draws me to coach in nature because it is revealing the truth that is the hardest and also the most transformational part of any coaching relationship.*

*My coachee, Sam, was the managing director of a retail company when he came to me for coaching in order to make some key decisions and set a new vision for his company. Sam is a visionary but he often has trouble articulating his vision. Although he likes the idea of being disruptive, his tendency is to be cautious and sceptical of new ideas. During Covid-19, Sam's entire being underwent a shift that he found extremely uncomfortable: his forecasts did not work any more and operations that worked like clockwork fell by the wayside as new procedures were called for every single day. His business sold essential items, and it witnessed a surge during Covid and Sam had to respond to rapidly changing government regulations.*

*It was a cloudy day with the wind picking up as Sam and I trekked up the side of a mountain outside Hong Kong. On the way up, I suggested that Sam notice, pause and record four things that inspired him. By the time we met at the top, dark clouds were hanging heavy as if about to split open at any moment. Sam's mood was exactly the same – dark and brooding. Instead of being inspired, he was frustrated with his lack of fitness and his shortness of breath. Nevertheless, I asked him what he had noticed, and he said they were all things he would never otherwise notice: an intricately woven spider web, that he likened to his*

*eye for details and for building a home and security; a brightly coloured butter-
fly representing expression; a beautiful red leaf among green leaves, which
spoke to him of diversity; and the ever-changing cloud patterns that he felt were
not permanent.*

*Sam started describing the areas where he was stuck at work, and the things
that kept him second guessing himself. As we progressed through our conversation, I witnessed Sam speaking almost in tune with what was happening to the
weather. As Sam expressed his frustration and how immobilized he felt, the
atmosphere became heavier and heavier with the wind almost disappearing. As
Sam shifted through various perspectives, there were moments of rumble and
wind picking up, even a light drizzle and finally as Sam explored possibilities,
a small portal opened up in the clouds allowing the sun's rays to stream through
into the sea below the mountain. As Sam witnessed that, really seeing and not
just looking; he stopped speaking and I held the space in silence as he processed
the shift that seemed to have hit him. I will never forget what he said next:*

> *Just because I can't see the sun behind the clouds, doesn't mean it isn't there.
> I always considered nature as dangerous, as if one has to be alert all the time
> when in nature and that's how I have lived my life – alert all the time! How I
> relate to others is also with that same level of alertness and wariness. I just
> realized that sometimes all I have to do is observe and portals of opportunity
> will open up. The clouds are not permanent – they are always moving. But if
> I am always looking down and never up, I don't see that.*

*This realization hit home for Sam in a way that we could not have created in a
closed room. The rest was easy. Sam said he was inspired to metaphorically look
up more and like the spider to keep rebuilding security again and again despite
it being destroyed again and again. He committed to listening to diverse views,
like the red leaf, and yet to show his own colours like the butterfly. Finally, he
said that when he feels stuck, he will wait for the clouds to part and the sun to
shine instead of struggling on. When I next saw Sam, he was more relaxed. He
told me he had still had moments of struggle but he had our session encoded in
his brain to fall back upon. Nature holds very many answers for us. Sometimes
it just needs eyes to really see and not just look.*

## When there is an inherited history of 'not enough'

As we write this book, sitting here in the Global North in the 2020s, it is easy to
lose sight of the fact there are many people around the world (and also within a
mile of our homes), for whom personal scarcity is a very real state of their current lives. As Anna Cura alludes to in her piece, for many there is not enough food
to feed the adults as well as the children, enough money to heat or light homes or
enough protection from violence. These cannot be ignored when we are talking
to people about the climate crisis, not least because the most financially insecure
and socially vulnerable people will be (and already are) the worst affected by it.
Scarcity may also be a past lived experience that plays into the way the person
relates to climate change. Not having enough as a child may make someone seek

to consume more now as a form of reassurance that those times are behind them. It may mean buying more than they need or it could be a desire for abundant opportunity that leads them to do too much.

Just as our upbringing and past experiences shape our sense of our own and others' resourcefulness, so it shapes our experience of scarcity. Almost everyone has a past experience of 'not enough', but some have it to a more extreme degree than others, and for some it is handed down through families as a part of previous trauma. From slavery and civil wars, to the Holocaust, the Second World War, and the Cuban Missile Crisis, if your forebears fled war or persecution, or have lived through global unrest that directly threatened their lives, it is likely that their experience of scarcity shaped the way that they raised their children and their grandchildren in turn. Just as the climate crisis is not the first existential threat for many people, it is not the first experience of profound scarcity either. When we work with this, we need to be aware of these things and to contract our conversations with compassion and ask how (if they do) they would like to relate differently to their scarcity.

## In Practice

### Tackling feelings of scarcity in the food system

*Anna Cura, senior researcher, Food, Farming & Countryside Commission*

*Scarcity underpins so many of my daily conversations around hunger and hardship. As the whole issue is about lack of food access or financial means, inevitably the sense that 'there is not enough' drives both the way that we talk about these issues and our proposed solutions. At the individual level, I often see that some people do not have enough food or money to access it and that 'if only they had more money, they would be able to buy good food like everybody else'. Money would give people freedom of choice and dignity in participating in our current food systems. But are these food systems the ones we want to participate in and maintain in the future? And is money really the only answer to solving hunger and redesigning our food systems? At the system level, there are stark social inequalities, with some of us living in abundance and some of us living in genuine scarcity.*

*As a practitioner, I am always walking a tightrope between acknowledging the pain, hardship, hunger and injustice, while also wanting to encourage positive change. As someone privileged enough not to have experienced hunger or hardship, I always need to check-in with myself whenever I get frustrated about conversations around scarcity. It is much easier for me to ignore scarcity because I have not been traumatized by it in the same way, so this belief that 'there is always a way' comes easier to me. But this also means I know the benefits of an abundance mindset. What frustrates me most is that focusing too much on scarcity leads to solutions that call for more resources, stalling action now while we wait for those resources. Scarcity tells us that we cannot act until we have more, but very often those resources are outside of our control. The whole approach feels disempowering and demoralizing, and misses opportunities for action.*

*Is that to say that we should ignore scarcity? Absolutely not. But there is a difference between needing resources to create change and needing resources to accelerate change. I prefer to talk with people about the latter, because it implies that we already have options and that action can be taken now, under current conditions. In order to get there though, there is a lot of emotional work that needs to happen, both within myself and with others.*

*The first stage in any conversation will* always *be acknowledging someone else's experience. I hear out the frustrations and empathize with them by naming the feelings I see expressed. I may say things like: 'This sounds terribly frustrating. I can see how annoyed/exhausted/stuck you feel about this. You clearly care so much about your community.' I also show genuine curiosity for their experience, with: 'This is so important. Tell me more. How is that evolving?' I have to walk a careful line between acknowledging someone else's experience of scarcity, while also acknowledging and being open about not having experienced this particularly. One way I do this is to think of other contexts in which I have experienced it, and ask myself: 'How would I feel if someone said this to me? Would it invalidate my feelings?' I have found this really helpful when speaking to someone who has directly experienced hunger or who is running an organization working tirelessly to provide food for their community. This is where it is an advantage to not have experienced these situations because I may not get triggered when talking about this and can remain calm, open and loving. I next ask how they are currently coping, pointing out what they are already doing and celebrating that together. I find that this can only happen if we have first acknowledged their emotions around scarcity, in order to clear space in their minds and hearts. I ask them about what they are hoping to do next ('How are you hoping to tackle this problem at the moment?') and what resources they already have at their disposal to achieve that ('Who is helping you with this? What do you already have in place to make this happen?'). Slowly, they remember that they are excellent problem solvers who can always find a way. They remember what it is that they do have and slowly move towards a sense of feeling resourceful, even in the face of constraint.*

*When my approach does not work, it is because my own scarcity gets triggered, not around food but about time. I remember talking to someone about what opportunities they may have to move forward and being met with a barrage of what did not/could not/would not work. I totally understood where the sentiment came from, but I was unsure I was able to dedicate the time needed to walk through it with them. Would it take 5 minutes or 3 hours? Could I afford this amount of time as part of my role? I often wish I felt able to support my peers emotionally as part of doing 'the work'. Individually, we need so much more compassion and acknowledgement to begin the healing process and not let ourselves be tied down by scarcity. Organizationally, we can map out current skills and assets that we already have access to, and carve our path of action from there. At a systems level, we need to acknowledge the role of emotional support in healing our society and invest as much as possible in nurturing our relationships. Our souls are in pain and need healing if we want to have the energy and courage to tackle the challenges of today.*

## Scarcity of human resourcefulness

There is another form of scarcity, bigger than the climate context, which is scarcity of human resourcefulness; in other words, that other people are not resourceful enough to solve their own problems. Many of us are conditioned to problem solve, and when someone shares a challenge or a vulnerability, we race in with advice. Sometimes this is because we have a good idea, but more often it is because we believe the other person when they tell us they 'don't know what to do'. If you can pause for a moment and refuse to believe in that story of scarcity of imagination, you will transform the way you interact. That can be difficult because when we are in a conversation we are not completely neutral. Often our own sense of scarcity is there too, so make a point of noticing how you feel when someone voices their scarcity.

A useful trick here is to consciously recall the other person's capability. What have they already overcome and achieved in your recollection? This simple mental exercise will remind you that they are, in fact, extraordinary and don't need rescuing. To go a step further, you can ask them to tell you how brilliant they are, as a way of defeating not only your voice of scarcity but theirs too. 'Sounds like you are really finding this hard. I wonder when you've encountered a difficulty like this before and what you've done about it?' All of this helps you to ground both of you in a belief that they are not hopeless, but rather they do not know what to do *right now*. Believing in someone's brilliance changes the way that you ask questions and even your body language. For example, if Kelly's client said, 'I just don't know how to make my son feel less anxious about climate change' Kelly could ask how she calmed down his anxiety previously (not what she's previously done about the climate crisis). Solutionizing about the climate will only spark more scarcity, but there are few parents who have not got experience of calming down a distressed child. These questions stop people in their tracks as they realize they are more competent than they feel.

OVER TO YOU: Consider the last person that you were tempted to rescue or give unsolicited advice. What happened in the conversation to make you do that? What could you do instead?

In this chapter we have shown you that scarcity can take many forms, sometimes manifesting itself as not enough, and sometimes as too much (of the wrong thing). We have explained that it is not the facts of 'not enough' but the way those facts make us feel and react that we can first attend to if we want to help others to get into action. We looked at how coaches manage this and how policy practitioners use recognition of scarcity to improve conditions for innovation. We have given you tools and shown you how believing in the resourcefulness of each other can deepen our relationships and empower others to believe in their capability again. Next we explore scarcity's relative, overwhelm, and how 'not knowing where to start' can similarly paralyse us.

Suggested answers to the earlier scarcity sentences:

2. Not enough people care.
3. There are not enough politicians/political parties engaged on this/there is not enough time.
4. I don't know enough/I'm not good/clever enough.
5. There won't be enough choice.
6. We won't make enough profit/not enough of our customers care about it.
7. There won't be enough resources/we won't have enough food.

# 9 Overcoming overwhelm

Agatha Christie wrote that 'imagination is a good servant and a bad master',[51] and our imaginations certainly run wild about climate change. Just as it can generate feelings of 'not enough', it can also paradoxically create 'too much'. Climate change presents us with a dizzying vision of a very different future for which we are not currently equipped. Pull on the threads of this subject and it unravels like a ball of wool, with every elegant solution quickly becoming a poor compromise. Working out how to make an impact is overwhelming when there is a long list of problems, and the skills required seem highly specialized.

## Identifying overwhelm, systemically and individually

Just like scarcity, modern life also frequently creates in us a sense of overwhelm. Where once the news was broadcast twice daily, now the 24-hour news cycle provides a rolling depiction of how desperate our world is and how dreadful its people are. We used to eat fresh produce only in season, while now many of us can access whatever we want, whenever we want it (even in the middle of the night). We have more clutter in our lives than ever before, to the extent that some of us rent storage facilities to hold the overflow of our unnecessary belongings.[52] Social-media inspired FOMO (fear of missing out) now has an official dictionary definition,[53] and compels us to constantly take part. As with people, so with the planet, and our over-consumption and wasteful use of resources have led us to degrade global soils, fill the atmosphere with greenhouse gases, cut down valuable forests, pollute rivers and seas, reduce our biodiversity to perilous levels and acidify the oceans …

Breathe. It is easy to read that list and feel overwhelmed. Complex systems challenges are hard. They cannot be solved by one person, community or organization but require a multi-pronged, long-term approach that naturally challenges our fondness for simplicity and short-termism and creates uncertainty, which we are wired to resist. In the face of such monumental change and sense of loss, our minds can halt at the point of choosing action, feeding us the phrases from Figure 9.1 to make us doubt not just our ability to *do* the right thing but to *choose* what the right thing is. If we stay in a holding pattern and do not land the plane, we keep our minds safe from the risk of choosing, but we also experience great dissonance.

Some signs that we are overwhelmed are:

– Failing to make decisions, putting things off, missing or forgetting deadlines.
– Finding reasons to dismiss new ideas that we might normally embrace.
– Talking about how much we have on and how we don't know how we will do it all in time.

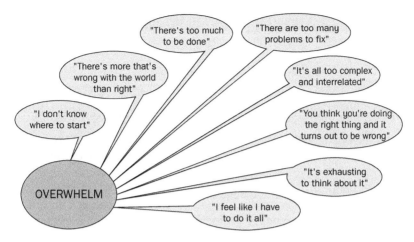

**Figure 9.1** Articulations of overwhelm

- Being preoccupied, distracted or not present and using our phone/social media as a distraction.
- Being unable to stay with a topic or to handle any new information.
- Being snappy, short-tempered and rude.

When faced with all the possible actions that we can take, it is easy to believe that anything new will be time-consuming or hard to learn, especially in the context of a life lived with little 'slack' available for new things. This is where scarcity comes into play.

## Tools for handling overwhelm in others

Coaching tools can help people to 'see the wood for the trees', calm the panic, zoom out and get a new perspective. Here are some tried and tested tools:

### Name and normalize the feeling

Naming and normalizing the experience of overwhelm can reduce panic and help someone to separate their emotional response from the facts. Phrases that work:

- 'It sounds like this is pretty overwhelming?' (Naming)
- 'Feeling overwhelmed is a really normal response to a situation like this.' (Normalizing)
- 'It's hard to do this and all of us can be overwhelmed sometimes.' ('Sometimes' reassures that this is a temporary state.)

### Acknowledge someone's strengths and values

Remind the other person of their ability to be decisive, capable and act within their values. Phrases that work:

- 'I can hear how much thought you've put into this and how much you care about it.'
- 'Even though you're overwhelmed you sound really committed to seeing it through.'
- 'I know you as someone who always finds a way/makes a good decision.'

### Ask for specific historic examples of them dealing with complexity

Asking someone to tell you about a past occurrence of handling complexity can return them to that ability now. Equally, when someone's overwhelm is driven by systemic disbelief (e.g. 'I see all these countries doing nothing. What hope is there for our government?') invite them to recall times in which the system has triumphed in the face of adversity. In both cases, the examples do not have to be environmental. We are simply looking for evidence of managing complexity. Phrases that work:

- 'When else have you handled so much information before? What helped you then?'
- 'How have you successfully managed this feeling of overwhelm in the past? What could you use from that experience to help you here?'
- 'What past examples do you have of us (the system) tackling complex problems well?'
- Consider giving an example of a complex problem that the world has solved, such as the abolition of slavery, nuclear non-proliferation treaties or changes made to tackle CFCs.

### Working with black and white thinking

In our desire for simple climate solutions we can mistakenly expect a singular fix to a multi-faceted problem. We can often discount perfectly good ideas because they won't work for *everyone, everywhere,* or *all of the time.* For example: 'The trouble with all of these so-called sustainable choices is that they're so expensive. I don't know how they think people on a budget can afford them.' An alternative perspective is that those on higher incomes can be the early adopters who support products to reach economies of scale (when the price lowers for everyone else, more on this in Chapter 18). We can also show the bigger system of environmental initiatives, pointing to money-*saving* schemes, such as swapping groups and libraries of things.

### Thinking through priorities

It is common to lose our ability to plan well when we are overwhelmed. You may hear something like: 'If I can just clear everything off my desk, then I'll be able to think straight.' The unfortunate truth is that our commitments to mitigate the climate crisis will keep growing, not diminish. Instead we can help someone to develop a new relationship with their to-do list, and accept that this is the way it is (for now at least) and help them prioritize and sequence tasks. Phrases that work:

- 'If you had to put these things in order of priority, what would go first?'
- 'What if you could do this in steps? What would the steps be?' Here you can even walk the steps along the floor with the other person to give them an embodied sense of progress.
- 'How does nature do lots of things at once? What or how could you borrow from that approach?'

### Return to what matters and refocus from there

Returning to purpose and values can clear the path back to clarity. When we feel it is up to us to change the world, we can take on a lot of unpalatable activities. If we see ourselves as a leaf on a huge tree of change, however, we can play our part with more joy. Individuals can also define a theory of change that says *how* they make an impact. This can help cut through the noise of 'all the things I should do' to 'what I want' and 'would have the most impact'. Phrases that work:

- 'What matters to you about doing something about this?'
- 'Which of these ideas/solutions/problems is the best match for your beliefs and values?'
- 'Let's list the ideas, circling those you feel you "should" do, and those you would love to do.'
- 'What are you here to do? What's your purpose? And which of these ideas best allows you to bring that to life?'
- 'What do you want your legacy to be? What does that mean you should do?'
- 'What do people say is your real impact on them? How could you use that super power here?'

### Change the perspective, with a metaphor or visualization

Sometimes feeling overwhelmed is a bit like walking into a beautiful house and only noticing that it needs a good dust. We can easily focus on the negative and fail to see the rest. Zooming out can help to put things into perspective, where the person feels at ease and can approach the situation with fresh, more capable eyes. Phrases that work:

- 'What's a totally different way of looking at this?'
- 'When did you last feel a sense that nothing needed to be done? Tell me about that, what was it like? If we applied that perspective here, what would you do?'
- 'Imagine that you are in a hot air balloon that is steadily rising above the houses. Can you see the world getting smaller? Can you see where you used to be, down there? What do your ideas look like from up here?'
- 'What's a metaphor for how overwhelmed you feel? Can we adjust that metaphor to make it feel more manageable?' An example here might be: 'I feel like I'm standing on a landfill site, surrounded by acres of rubbish that I'll never get through … If I stand on the top of the pile I can see green fields all around the site.'

## Recognize when it is also scarcity and put down perfectionism

Overwhelm is a cousin of scarcity, because its underlying causes can often be an expression of 'not enough'. This shows up as the need to make perfect choices. If we ask: 'Why not just start anywhere?' and get an emphatic 'I can't afford to waste time getting this wrong!' then we are hearing the voice of scarcity whispering that there is not enough time or resources with which to afford failure. Others may have created scarcity and self-criticism about their own ability to overcome overwhelm itself: 'I should be able to work this out by myself! I'm useless!' Gently reminding them that isolation can slow down climate action is a compassionate response. Worrying about making a mistake and having to backtrack can lead to analysis paralysis. We can feel we have to work on something until it is 'perfect' to avoid any flaws that could lead to criticism. Offer the option of contracting with others that 'it is a pilot/prototype' to overcome this. Kindly reassure someone that a multitude of imperfect actions (not a limited group of perfect people) are needed, not least because action boosts the collective mood and reduces the charge around this subject. We may also remind them that there is no one agreed-upon 'perfect' set of actions or centralized authority judging whose are best. Phrases that work:

- 'If you could only do one thing, what would it be?'
- 'If the first thing you did didn't matter, what would you do?'
- 'If no one was looking, what would you want to do?'
- 'What would it be like to do something small, perhaps even unrelated, just to get moving?'
- For someone who is feeling beaten by all they 'should do': 'What do you love doing, that you could do to shift your mood, so that you can come back to this fresh?'
- For someone who feels that they are not being useful to the world because they are stuck: 'How could you be useful in your company/community *right now*, perhaps in a completely unconnected way? This is not the "big idea", but something to connect you to others.'

## Get out of the road

Just as with scarcity, it is easy to end up in a mental tussle trying to convince someone that their actions are worthwhile. If someone is feeling overwhelmed it is usually because they care but are frightened. If we don't engage with the fear, that care can take over. Instead of resisting when they tell you they cannot, momentarily agree that there is indeed nothing they can do and that it is OK to walk away. This often releases people from their mental trap, though it is important that you do this where you have a good relationship so as not to seem callous. Phrases that work:

- 'I think I'm inadvertently trying to convince you to do this. It's OK to not do anything. Would you rather walk away from it?'
- 'Maybe you're right. Maybe this is too hard/maybe you are too busy. Maybe you shouldn't do anything. What do you think?'

- With some humour, where the relationship can take it: 'You've convinced me! You're just not the right person for this. The right person would have to be ...' [and then describe someone impossibly perfect that could never exist].

### Help them to find a tribe

Isolation is one of our worst enemies of action. Helping people to find supportive others dramatically increases their chances of sustained success. You can also be a vital support. This is where the act of contracting can also be supportive, not just clarifying. Phrases that work:

- 'How can I help you feel less alone in this? How can I avoid inadvertently making you feel crazy for wanting to do this, as you've told me your family does?'
- 'What can I do to help you feel normal/secure?'
- 'Who do you know who also cares about this (even if you don't currently know them well)?'
- 'What groups might you be able to connect with?' You may need to reassure them that going to one meeting of Friends of the Earth does not mean becoming a lifelong member, but that they can test out a few groups and then choose.

## *In practice*

### *Overwhelming information*

*Maia C. Rossi, climate change, biodiversity and sustainability advisor*

*As a sustainability professional, I focus on climate change, biodiversity strategies and sustainable and climate finance. In my career I have worked with many different sectors including government agencies, intergovernmental organizations, financial services, development institutions, mining, energy and construction companies across the world and in very diverse country contexts such as the UK, US, EU, the Middle East, Central Asia, China and in most of the African countries. I'm used to handling lots of complex technical data and am currently writing a PhD at the Business School of the University of Bath in the UK.*

*On a very warm summer's day, I was preparing to kick off a project as the carbon strategies technical lead. Usually I take a lot of time during the proposal stage to understand the specificity of the challenges that the organizations I am helping are facing. This time, however, for many different reasons, the contract had started, and I knew pretty much nothing about it. I knew the manager was personally engaged in the climate change challenge and wanted to position the organization he works for as a leader in the market; however, I did not know what his or the organization's actual strategy was. When I finally met him, it was clear there was no strategy, that he was very nervous and kept repeating that he wanted to make a difference, that the future looked gloomy and that he wanted to change it. This manager wanted to act fast and bring results faster; however, there was more to it for him than shining brightly to top management or achieving a business-related goal. He wanted to do the right thing and be done with feeling guilty about not taking action.*

*As is usually the case, there was a lot of confusion between us about language, legislation and processes. Environmental, social and governance risk management frameworks and climate-related standards, reporting and legislation are fairly new, and it can easily be overwhelming to understand the context and decide where to start. Some of the language can sound like gibberish, so it is not uncommon for managers to pick up some words and frantically repeat them in the hope of communicating their willingness to act and their feeling of being lost in translation. The problem could have been solved easily by me taking control, focusing on specific tasks, preparing the carbon assessment and doing the reporting. However, this would have solved the short-term goal of providing the organization with a carbon inventory and not the long-term one of showing them how to take action in a complex system, without feeling paralysed by both fear of being left behind and fear of the unknown.*

*Climate change action is not an individual act: it is a long-term strategy made of many different decisions. So we designed the project with a large training and internal stakeholder engagement phase. That had the double purpose of understanding the organization more and training the key actors, including our manager, in how to extricate meaning from the jungle of terminology and possibilities. We agreed on a framework where they did the work under my guidance, going at their pace. It took double the time it would have taken me to interview key personnel and prepare the work; however, the learning was deep. At the end of the project, carbon-related terminology was not something scary and unknown any more; instead, the team had mastered the process and was able to propose a carbon strategy that was achievable, tailor-made to the organization and, more than anything else, owned by them.*

*I believe that the role of the expert or consultant in the sustainability and climate change space really is to bridge the organization's world with a technical, scientific world so that the basics of climate science are understood and people are empowered through learning. That newly acquired knowledge and critical approach are useful tools in the hands of employees and employers who then can take informed actions. At the end of the day, they know their organization better than we do, so they must be the ones who design the strategy. I found that not knowing the scientific basis, fundamental legislation frameworks and basic language is what creates confusion, paralysis and inaction. I can see this slower approach applying any time a team needs to solve a challenge or take a decision. The problem with climate change is that it is part of everyone's life, and to be part of the change, we must understand its dynamics. Adding this learning phase could seem time consuming; however, its cascade effect of positive outcomes is definitely worth it.*

## Using coaching tools – an 'in practice' conversation

Let us put together two imaginary characters: Owen who has started career coaching with Jonathan, the professional coach that we met in Chapter 6. Owen has had a traditional career but recently became a father and feels compelled to make his career climate focused. As this is all new to him, he feels overwhelmed about where to start. Jonathan has just asked him what he is drawn to …

Owen:      'I want to do something about carbon capture, but I just don't know what.'

Jonathan:  'OK, what interests you within that subject?'

Owen:      'Well I care about forests and I thought about conservation work, but then I thought what about the oceans? I read that underwater forests are much more important than trees on land. And then I heard that actually farmland can be a huge piece of this and thought maybe I should work on lobbying farmers. And then I went to my local garden centre and was incensed to see so much peat compost on sale.'

Jonathan:  'That sounds like a lot of good ideas. And that they're also a bit over-whelming?'

Owen:      'They are. I don't know where to start, and when I research I fall down a rabbit hole.'

Jonathan:  'What would it be like to pick just one or two of them to research?'

Owen:      'I could, but what if I pick the wrong one and waste valuable time?'

Jonathan:  'Ah, so it sounds like there's not just some overwhelm here but also some scarcity. Like you've got to get it right the first time, because there's not enough time?'

Owen:      'Yes, that is exactly how it feels!'

Jonathan:  'OK, well, as a thought experiment, how much time would you think is an OK amount of time to waste? Two weeks? Two days? Two hours?'

Owen:      'Ha! Put like that, I guess I can afford two days.'

Jonathan:  'Great, so if you had two days to research, which is your favourite out of that list to take a closer look at?'

Owen:      'I think it's the first one, forests, because I really love spending time in them, and working with forests wouldn't even feel like work.'

Jonathan:  'Sounds like you have a real connection to the forest itself, not just its potential for carbon capture. What is it about forests that you love?'

Owen:      'I go to a local forest near me quite a lot when I feel stressed and it always calms me down. It's so peaceful.'

Jonathan:  'So you have a great way of managing stress then. I wonder if the for-est also helps when you feel overwhelmed?'

Owen:      'Yes, it really does, it helps me put everything into perspective.'

Jonathan:  'I'm curious, if you could ask the forest for some wisdom, right now, about how overwhelmed you feel, what would it say?'

Owen:      [pausing to think] 'Probably that there's enough time and not to expect everything to happen at once. I'm always struck by how slow and steady everything feels there, and yet how much is going on. All of the pointless activity that humans have created is totally absent, it's just the essentials that get done in the forest.'

Jonathan:  'I can really picture that. How could that idea be useful to you as you start to research?'

Owen:      'I guess I should stop letting myself get distracted, but focus on the goal.'

Jonathan:  'Would it help to define a goal for the research or will that make it feel too tight?'

Owen:      'I think I just need to say that I'm going to spend two days finding out how forests contribute to carbon capture and not other reasons that forests are useful or other ways of capturing it that aren't in forests. And maybe I can start to define some criteria for me doing something about it all, as I'm researching.'

Jonathan:  'Sounds great. When will you do it, and do you need anyone to create accountability around it?'

Owen:      'I've got some time next weekend, and I'll let you know in our next session.'

Jonathan:  'I look forward to hearing how you get on.'

Notice how Jonathan asked open questions and trusted Owen to have the answers, but that he also didn't allow Owen to wallow in disempowerment but helped him to find a new perspective from within Owen's own thoughts. Given that we know that Jonathan is also in a voluntary environmental group, he also did not get caught up in Owen's story or bring any of his own knowledge to the session. Managing our own responses is a huge part of the conversation. When we steer the discussion away from the other person and onto our own thoughts and feelings we can think of it as moving the spotlight away from them and onto us. When we ourselves feel overwhelmed, we can do this as a way of answering our own questions and salving our own anxieties, but beyond feeling some solidarity, it does not help the other person move out of their overwhelmed state.

## In practice

### The presence required to coach climate change topics

*Liz Hall, author of* Mindful Coaching *and editor of* Coaching at Work *magazine*

*One challenge inherent in climate coaching is that we are in this mess together. Our client's overarching issue of wanting to do more and/or better to tackle the climate crisis is our issue too. As coaches, we easily risk tipping into overwhelm and empathic distress ourselves, which means no longer being in service of our clients or taking care of ourselves. I've been there myself, especially a few years ago when I awoke more deeply to the extreme nature of the climate crisis. Personal actions such as making more environmentally conscious travel choices and seeking out contributors on the topic of climate-crisis related coaching for the magazine I edit no longer felt enough. I felt distressed, anxious, overwhelmed. I drew on the strength of community and dialogue, heeding Kelly McGonigal's research that community is one of the antidotes to empathic distress, and I benefited from some powerful coaching conversations. I went on climate protests, and I launched Climate Coaching Action Day through Coaching at Work, which sees coaches from all over the world raising awareness of the climate issue and coaching's potential contribution. I also explored stepping more ably into climate coaching.*

*In coaching, many clients echoed my own feelings of fear, anxiety, ennui and languishing. Hearing their experiences triggered an intense emotional response, and maintaining my long-standing mindfulness and compassion development practice has been vital to me staying present and resourceful for climate coaching clients. Before each session, I get grounded through a 5- to 10-minute mindfulness practice, and I draw on mindfulness mid-session if I'm triggered. I notice my emotions and how they echo in my body, and anchor myself in the present moment through becoming aware of my breath or my feet on the floor, and settling my nervous system so I can be present again. Tell-tale signs that I'm lost in my own process include my body shape shifting from sitting back relaxed to hunching my shoulders, feeling contracted, with a knot in my stomach.*

*Managing myself mindfully in this way means I can tune into what is happening for the client, and between us, and extend this to include what else might be emerging. With one client, for example, we explored their sense of not being good enough. I asked what the impact of this might be generally and on our coaching relationship. They said it was inhibiting them from being their authentic self (and this played out elsewhere too). I then guided them through a centring practice to help them attune to what the wider system, including Mother Earth, was asking of them. This galvanized them to take action without fearing others' judgement.*

*I also offer clients a simple mindfulness practice to begin the coaching session. I also suggest practices they can use outside of the session, including awareness of breath meditation to reduce anxiety or self-compassion practices (such as Kristin Neff's Self-Compassion Break). Clients who care deeply about the Earth yet feel they are falling short can be very self-critical, which can be very depleting. Building their self-compassion helps them be more resilient, able to set boundaries, and to forgive themselves for not saving the planet single-handedly. 'Difficult' emotions such as grief, anger and anxiety commonly arise in climate coaching and drawing on mindfulness, bodily wisdom and compassion supports me and my clients to work with 'strong' emotions.*

*Getting more nuanced in understanding empathy and compassion has helped me appreciate that a compassionate response requires attuning with a coachee without blurring so I experience and understand what they feel without confusion with what I feel. If I feel there's blurring, I'll take a few deep breaths, consciously adjust my posture and maybe self-soothe with a gesture such as placing my hand on my belly. And I'll remind myself that the coaching is not about me. I've discovered that we have far greater capacity for compassion (for self, for others and to be able to receive compassion) than we may realize. It takes a mindset of committing to developing our ability to extend compassion to ourselves and others, as well as to receive compassion. As a coach, this often does not require me to 'do' very much. Sometimes the most empowering gift I can offer a client, particularly if they are grappling with how to address climate issues, is a safe container, deep listening, my presence and my love. Emotions are messengers. If we ignore their messages and their echo in the body, we– and our clients – can become paralyzed and overwhelmed in the face of the enormity of the climate emergency.*

## The end game – cultivating a mindset of acceptance

We often notice how much the desire to control features in people's change journeys. The loss of control – either to allow our creativity to take over or serendipity to intervene – is destabilizing. We feel unsure to start the journey if we cannot draw a detailed roadmap and know the steps along the way. Yet paradoxically we must give up our certainty if we are to stay the course. There is a paper-thin line between being highly motivated and highly attached, and attachment can put a lot of pressure on us. Acceptance – that no one person can solve this and that we cannot do everything – can lead us to true calm, and a place from which we can act more intentionally. This is where defining our unique contribution can help, as can developing a long-range sense of purpose, and a strong identity and sense of values, and we cover this in Chapter 20.

We may not always be fully conscious of our attachment in our minds, but our bodies can tell us if it is happening. Think about when you have felt pressure to succeed in a tightly defined way, and how it changed your outlook. What happened to you physically? Perhaps your body's response mirrored that tightness with shallow and rapid breathing, a dry throat, tense shoulders or a sore back. In this contracted state, did the world seem expansive or smaller? Could you take the long view or did the timelines seem short and urgent? Did you sleep well or lie awake ruminating? Often in this state our worlds feel smaller, more urgent, and all consuming. Fortunately, we can de-escalate attachment using some tried and tested methods:

### Connect to the present moment

Because attachment is most commonly about the past or the future, consciously anchoring in the present can prompt a healthier perspective where we can take stock. Right now notice: what can you see? What can you hear? What can you touch? What can you smell or what can you taste? This is mindfulness in the moment, and can be done in any situation, without anyone else knowing. Mindfulness is being mindful of what is going on and most especially of our thoughts. This can mean catching ourselves mid-thought: 'I notice I'm worrying a lot about this issue' and finding a way to either put the thought down or connect to our senses, perhaps by noticing the pen in our hand.

### Take deep breaths to reduce anxiety

Physiologically, anxiety rises up the body, making our breath short and high in our chests and our shoulders tense. Mental anxiety signals to our bodies that there is trouble and our bodies then respond with tension and shortness of breath. We can reduce it by consciously breathing long and slow and down into our feet, sending our breath lower into our stomach. This technique allows our bodies to signal to our minds that we are calm.

**Ask yourself 'What is really at stake here?'**

Before you answer 'The planet, stupid!' check that you have not inadvertently become the hero with the silver bullet. What is really at stake and how important is it really? While we all want to feel useful and do meaningful work, we have many opportunities to do so.

**Ask yourself 'What will really happen if this change does not happen?'**

We often catastrophize around failure, without getting clear on what the actual consequences. It is exceptionally rare that a life is at stake, yet our attachment can trigger us to feel that.

**Ask yourself 'What is really important to me here?'**

This question will lead you to the heart of the matter. If you like, this is the baby in the bathwater. There may be a lot of bathwater that you will happily throw out.

## Becoming intentional

Think back to the way that we described the imaging of climate change in Chapter 1 and how disempowering those images can make us feel. They remind us how distant and difficult the problem is, and in order to feel less alone with those feelings, we share them with friends on social media. Sharing it does not really make us feel better though. Worse, our friends then feel that same sense of overwhelm, so they share them too, to feel that they have done something … and on it goes. We can instead choose the way we engage with the world to reduce our sense of overwhelm and that of others. We can make more conscious choices as consumers; buying quality over quantity so that we are surrounded with possessions we cherish, not overflowing cupboards that make us stressed. We can make time to do nothing, or at least very little, and not feel guilty for the weekend spent on the sofa, or for making time for our morning meditation. We can reform our work diaries to create breaks between appointments, so that our time together is less flustered and more intentional with time to prepare properly so that we enter meetings feeling present. Finally, we can be more conscious of the way that we talk about climate change with others, online and in person, so that we are not overwhelming them with facts, but rather saying enough to spark their interest.

OVER TO YOU: Think about how your life contributes to your own overwhelm and how your actions might contribute to the overwhelm of others. What would you like to do to change that?

# 10 Understanding intrinsic values to overcome resistance to action

Throughout this book we have explained how the beliefs and principles we hold most dear can dial back our fears and reaffirm our confidence in what is possible. Values sit below everything that inspires and fires us to take action and can also be a potent tool when we encounter resistance. When we come up against someone else's resistance to change, it can be easy to characterize them as stubborn and pig-headed or write them off as an enemy or a roadblock. But their resistance holds important information for us and by ignoring their concerns, we miss a useful opportunity to gain insight and collaborate. The poet Rumi wrote in 'Beyond', 'Out beyond ideas of wrongdoing and rightdoing, there is a field. I'll meet you there.' If we can venture beyond our judgement, we open up a space for greater understanding. There, we can see what legitimate values and concerns lie beneath someone's resistance. In this chapter we consider how learning what motivates someone is the key to overcoming their blocks and reticence. We also look at how, when values are heard and honoured, we strengthen our chance for long-term change and find better ways to work together.

## What are values and how do they differ from behaviours?

Values are principles that guide our behaviour and underpin our actions. They are a representation of the beliefs that we aspire to live by. Values drive our reactions to life events when they are both present and absent. Russ Harris describes values as:

> our heart's deepest desires for how we want to behave, how we want to treat ourselves, other people and the world around us. They describe what we want to stand for in life, how we want to act and the sort of person we want to be.[54]

For example, we may have a strong value of kindness and make a point of treating a waiter well or of variety and never visit the same restaurant twice. You can think of values as the sunlight that feeds the photosynthesis, essential for helping any plant to thrive. When we live closely aligned with our values, we have more fulfilling experiences, and if we engage in activities that are not associated, or even clash with our values, we can find ourselves demotivated or conflicted. As a result of the insight and energy associated with values, exploring them can be a highly effective coaching tool.

Professor Shalom Schwartz has mapped universal values commonly found across all major cultures (see Fig. 10.1). His theory of basic human values groups 58 values under 10 principal headings shown in Figure 10.1: universalism,

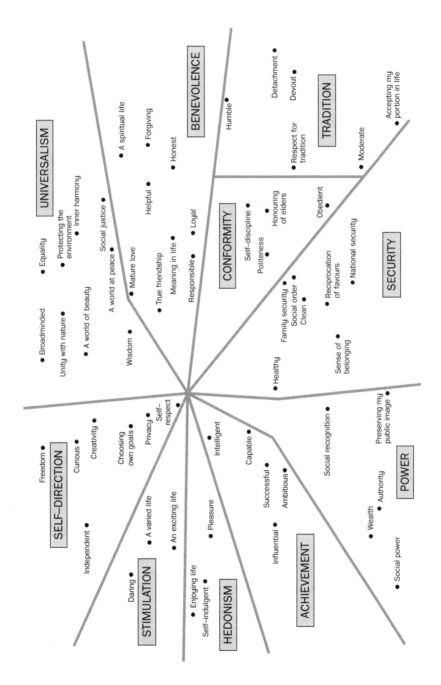

**Figure 10.1** Schwartz Values Map[55]

benevolence, tradition, conformity, security, power, achievement, hedonism, stimulation and self-direction.

Schwartz's theory then arranges these values into four organizing principles, broken into opposites:

1. Self-transcendence – universalism and benevolence – and its opposite:
2. Self-enhancement – achievement, power and hedonism (which it shares with openness to change)
3. Openness to change – self-direction, stimulation and hedonism – and its opposite:
4. Conservation – security, conformity and tradition

You can immediately see how, depending on our values, we can clash over what we believe to be the right course of action on climate change, depending on whether we want to conserve traditions or transform societies, or prioritize climate justice or work within existing power structures. However, the values we hold are not always congruent within us ourselves. Two opposing values held at the same time can cause us inner conflict and ambivalence, blocking us from moving forward unless we can prioritize one over the other. A strong value of protecting the environment may lead someone to contemplate reducing their meat consumption, but a value of freedom (to choose what they eat) and choice might hamstring someone from committing to such a dietary change.

Values are in the eye of the beholder; a single value can be instantiated in different ways by different people. This makes it very important that we understand not just what someone's values are, but how they express them. To test this out, imagine asking three friends if they value honesty and what honesty means to them in practical terms. The first may say that honesty is about telling the truth when you do something wrong, while the second says it is about challenging someone when you do not agree and the third could say it is about being clear about what you expect of people. All three are an expression of 'being honest', yet all hold different meanings. They could easily come into conflict with each other if they assumed that 'honesty' should always be instantiated in their own way.

Values differ from behaviours in that values are why we do something and behaviours are how we do it. Sometimes there can seem to be little difference between a value and a behaviour, but while values can motivate our behaviour, what they cannot do is dictate one single way in which they are brought to life. The same value can be manifested in a multitude of different behaviours. It is simply that for most of us we have never considered alternative ways to bring our values to life; we stick with what we know. Take punctuality, for example. It may be that being on time is really important to you and it drives you crazy that your colleague is late to meetings. It would be easy to assume that this is because you value *punctuality* and the only way to improve this situation is for your colleague to be on time. However, punctuality is a behaviour not a value. We can tell that because punctuality can only be lived in one way: by being on time. If you reflect on what it is about punctuality that matters to you, you might notice that you feel *disrespected* when someone is not punctual. With your newfound reframe that this is actually an issue of *respect*, you may recognize that your colleague is very respectful once in the meeting, and you may be more able to either let their lateness go, or tell them how punctuality for you is a sign of respect. When we stay at the level of the behaviour, we can usually only see one path ahead of us, but when we drop down to the value that we are trying to honour with that behaviour, we can find multiple ways to do so. Teasing apart values from behaviours can deescalate a trigger around a specific behaviour and broadens our ability to honour our values.

When values and behaviours fuse, it can be harder to collaborate and easier to excuse our resistance to change. This can show up as conflating values and behaviours as a cover for poor behaviour, for example, 'I'm disorganized because I value spontaneity' (when plenty of spontaneous people remember birthdays or plan their workload well). Equally, some organizations have rightly been criticized for crafting impressive-sounding values that do not translate into behaviours at all, which can act as a cover for greenwashing as well as law-breaking.[56] Management author Patrick Lencioni framed this as a fad that: 'swept through corporate America like chicken pox through a kindergarten class. Today, 80 percent of the Fortune 100 tout … values that too often stand for nothing.'[57] Too often the staff within organizations are operating within a different, unspoken set of values that are shaped by the norms and processes of the company. This fusing and confusing is particularly problematic in the context of big systems change because we can also fuse our values and behaviours together as societies. We might fuse adventure and air travel together to assert that 'the world values adventure too much to give up flying' when adventure can be achieved in multiple ways. Governments may claim to value inclusion and then create policies that mitigate against refugees. Using values as a cover for unhelpful behaviour or failing to recognize the difference between them only undermines the true usefulness of a values-led approach.

OVER TO YOU: Choose your four most important values from the universal values diagram in Figure 10.1 and write a sentence for what each means to you.

## Extrinsic and intrinsic values

It is important to distinguish between intrinsic and extrinsic values because they are directly correlated to the kinds of behaviours we engage in. Like intrinsic motivation, intrinsic values are those that we find inherently rewarding, such as caring for others or a connection to nature. Extrinsic values, on the other hand, are based on rewards that remain external to us, such as wealth, status or the approval of others. According to Common Cause Foundation, intrinsic values are positively associated with political engagement, concern about social justice, environmentally friendly behaviours and lower levels of prejudice. Extrinsic values, on the other hand, are drivers of higher levels of prejudice; less concern about the environment and corresponding behaviours; weak (or absent) concern about human rights; more manipulative behaviour and less helpfulness.[58] Intrinsic values are often bigger than us, and tapping into them can be a powerful way to return us not just to ourselves but also to more self-transcending principles. Intrinsic values are a particularly useful lever for anyone who is demotivated or questioning themselves and their abilities because it allows them to shift the focus from them to taking a stand for something much bigger than themselves. However, as Tom Crompton from Common Cause Foundation writes below, it is not always easy for people to act (or talk about acting) from their intrinsic values.

## In practice

### The values perception gap

*Tom Crompton, co-founder of Common Cause Foundation*

*Across almost all countries (and we have data from around 90) most people place more importance on intrinsic values (such as equality, creativity, social justice, community, environmental protection or friendship) than they do on extrinsic values (such as social status, public image or financial success). This does not apparently arise as a result of bias in the way that people report their values (for example, because they want to project themselves in what they believe is a more positive light). This is good news. To the extent that a person places greater importance on intrinsic values they are also likely to report higher well-being, be more civically engaged, feel closer connection to their community, report higher levels of support for social and environmental action and have a lower ecological footprint.*

*Unfortunately, however, it seems that people are often held back from acting in line with their intrinsic values. One reason for this is that people worry that to be seen by others to act in line with these values could leave them looking a bit odd. Sometimes such action is directly attacked – people who act in line with their intrinsic values may be labelled as 'do-gooders' or are perhaps accused of 'virtue-signalling'. But it seems that we are held back from acting in line with our intrinsic values in more subtle ways, too. This has been called the 'norm of self-interest' and is the perception that social norms require us to act in our own self-interest. For example, the reasons that people provide for donating to charity rarely acknowledge an altruistic motive. They are more likely to explain their action in terms of self-interest ('I got a free gift', 'Who knows, maybe my donation will lead to the medical breakthrough that saves my life down the road').*

*The pressure for us to behave in ways that can be seen to align with our own self-interest, or to explain altruistic behaviour in self-interested ways, has a tragic consequence. By acting in this way we then project a public image of someone who cares less for others than is actually the case. Put another way, we tend to behave in ways that exaggerate the importance we place on extrinsic values and downplay the importance that we place on intrinsic values.*

*This probably helps to explain why the majority of us (almost four in five people in the UK) tend to underestimate the extent to which typical fellow citizens value intrinsic values and overestimate the importance that they place on extrinsic values. We too often believe that people are motivated primarily by things like fame or wealth, when, in fact, they are more likely to prioritize fairness or kindness. The way that people are portrayed in the mainstream media adds to this. It is easy to see that there's the probability of a vicious circle arising here. Our observation of others' behaviour is likely to inform our assessment of their values (leading us to underestimate the importance that others place on intrinsic values). But at the same time, our tendency to underestimate the importance that others place on intrinsic values will reinforce the social norms of self-interest and undermine our own commitment to act in line with the intrinsic values that most of us prioritize. If others care less than us, then why should we go out on a limb to help?*

*This is a tragic situation. But there's hope here too. These insights suggest that public commitment to new levels of social and environmental action could be enabled simply by conveying a more accurate perception of what most of our fellow citizens value. Not by changing people but by seeing people for who they are. Our surveys in the city region of Greater Manchester, UK, found, for example, that respondents who underestimated the importance that fellow Greater Mancunians place on intrinsic values also expressed lower support for action on climate change – including less support for use of public money to insulate homes, less opposition to airport expansion and less support for local government action on climate change. On the other hand, we have worked with a wide range of arts and cultural organizations to develop some approaches to achieving a more accurate perception of others' values.*

*For example, we have examined the assumptions about others' values that underlie some of these organization's public communications. Often there is a tacit assumption that people will be motivated to help because their own interests are promoted. Volunteership may be promoted on the grounds that participants will receive a boost to their CVs. However, volunteers report being primarily motivated by meeting new people or the camaraderie of fellow volunteers. We celebrated these motivations through posters in public spaces that conveyed this in volunteers' own words. In another initiative, we facilitated strangers (in this case visitors to a museum) to meet and explore with one another what really matters to them in life, to help people recognize this values perception gap.*

*There's a role for diverse organizations and networks to support efforts to convey the simple insight that most of our fellow citizens care more for one another and the wider world than we imagine. Today a great deal of energy is exerted in trying to persuade people to care about climate change because it is in their immediate short-term interest to do so – 'climate change will impact on food costs', for example, or highlighting how action to improve our domestic energy efficiency will have immediate and personal economic benefits. Though perhaps successful at the time, viewed in a broader context such appeals may be counter-productive: they risk reinforcing the impression that people are primarily motivated by their own self-interest – a belief that is inimical to ambitious and sustained action on climate.*

## Why it is important to help someone articulate their values

While all actions are in some way driven by our values, it is often the case that most of us could not name them. We form opinions and make decisions every day that 'feel right' without ever needing to articulate why. That makes a lot of sense – imagine if we had to stop and reflect before every decision, it would make food shopping impossible for a start! The result, however, is that most of us are not practised in naming our values and struggle to articulate why something bothers us or makes us resist change. Instead we stick our heels in because it 'feels wrong' without really knowing why. When we are tussling over an action, it may feel counterintuitive to help people explain why they disagree with you. Yet when we are arguing, our fundamental human desire to agree and be in the warmth of community is at odds with our need to stand up for what we hold dear, and it can often result in both sides feeling disempowered and at fault. Rather than escalating the disagreement (or damaging the relationship), finding out the legitimate reasons for resisting can bring us back into mutual respect. When we realize that someone is not being difficult but holds meaningful concerns, we no longer see them as a roadblock.

The power in a values-based approach is that once someone can name what is behind their reticence, it opens the door to exploring how that value can still be honoured in other ways. We know from thousands of hours of coaching that when we help people to surface their values, they feel more confident and less reticent about action, and at the same time, they also understand what has been holding them back. This also partly explains why people do not often take advice. It is not that people don't listen to your ideas, but that those ideas are more likely to be adopted when they chime with their own deeply held beliefs. If you have a colleague who values humility, no amount of encouraging them to 'be a bit more pushy if you want to get on in this company' will convince them to do so. Until they can see a way to be more assertive *and* maintain humility, they will feel inauthentic doing it and resist. Sometimes simply asking what their value is will get you an answer, but more often we have to use our intuition and then offer a suggestion.

## Methods for surfacing values

As many people cannot name their values without help, in many cases, we have to dig someone's values out of a story that they tell us. It can be useful to imagine

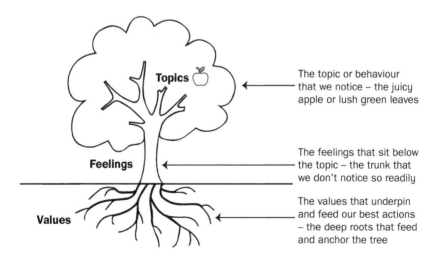

**Figure 10.2** A tree of values

values as the roots of a tree, anchoring and grounding it, and that we have to travel down from the leaves (the story), and through the trunk (their feelings) to get to them (Fig. 10.2).

This requires us to listen below the story with its ins and outs, and below the feelings of frustration, excitement or curiosity to hear the values that are being expressed underneath. As we said in Chapter 6, person-focused and attuned listening means not comparing stories but focusing our attention on the other person. If we can do that, we can hear the beliefs or values that someone holds dear.

### Method one – listening for values

Values are everywhere, so we have lots of opportunities to listen out for them. When we hear something, we can offer it to the other person. We can only guess at, not dictate values however, and coaches describe this as 'throwing pasta at the wall to see if it's cooked'. If the value sticks (the person agrees in a way that sounds convincing), it is right. If it falls off, we need to get cooking again. What we do not want to do is get attached and insist that we know best. If we are wrong, we drop the suggestion and get back to cooking up new ones. Only the values-holder can confirm or correct your guess. If you are wrong it could be your own values that you are sensing in their story. Try to name the potential values that are expressed in the below statements. We have suggested example values in the first statement. Refer to the Schwartz model of basic human values in Figure 10.1 if you need some inspiration. More suggested answers are at the end of the chapter.

- Statement one: 'I'd like us to be practical here.'
  *Potential values*: Pragmatism; preparation; thoughtfulness.
- Statement two: 'I watched a bumble bee on a flower the other day and was so struck with its concentration and how it had just one job to do.'
  *Potential values*:
- Statement three: 'I get really angry when people tell me that sustainable products are too expensive. They should try harder. We've all created this mess and we all need to do something about it.'
  *Potential values*:
- Statement four: 'I like managers who trust me to make the right decisions and allow me to solve problems myself.'
  *Potential values*:

### Method two – quick access questions for values

In addition to great listening, there are two superb questions that can help you to instantly unearth values and that we encourage you to use liberally. When we ask these questions, we shortcut the story and get straight to the value. These are useful if you don't feel comfortable suggesting values and would rather the person articulate it themselves:

1. 'What matters to you about that?'
2. 'What's important to you about that?'

Let us go back to our imaginary manager, Tim, who is having a conversation with his equally fictitious facilities manager, Greg, about putting solar panels on the roof of their building. If Tim were to specifically probe for values, the conversation could look like this:

Greg:   'I really like the idea of solar panels on our office roof, but I do think we need to research it properly, not rush in.'
Tim:    'OK. What matters to you about not rushing in?'
Greg:   'It's about making sure that we don't make a mistake. I've seen that happen too often.'
Tim:    'I know what you mean. What would you say is important about not making a mistake with this?'
Greg:   'Well it's about efficiency, isn't it? I don't like to waste time. And I don't like things that aren't thought through properly.'
Tim:    'So is it also about preparation as well as efficiency?'
Greg:   'Yep, that's true. I like things to be well planned.'
Tim:    'OK, so it sounds like you hold some values around preparation and efficiency.'
Greg:   'Yes, I suppose I do!'
Tim:    'So what do we need to do to make sure that we're well prepared and efficient when it comes to solar panelling the roof?'

You may have felt that Tim asking Greg those two very similar questions was repetitious, yet the surprising thing is that Greg will almost certainly not have noticed; very few people ever do. Instead, the questions surfaced Greg's values and gave them both powerful information to work with. As we will show you later in this chapter, this is how we can zero in on not only the reason for resistance, but the solution to it also.

### Method three – hearing values in stories of doubt

While not everyone proclaims their values loudly, we do share our doubts and it is just as possible to extract values from those. As an example, let us consider Rita, the imaginary global VP for sustainability at an investment bank. She is doubtful of her ability to convince the CEO of her new strategy and tells her friend Clive.

Rita:   'I'm such an idiot. I went to see the Executive Committee and totally confused them all with too much detail. I could see them glazing over as I was talking. I've blown it.'
Clive:  'That sounds really tough. How do you feel now?'
Rita:   'Like I've let myself down. I did so much preparation but I just feel like I missed an opportunity.'
Clive:  'Sounds like there's something important to you about not missing opportunities?'
Rita:   'Yes, it's about not wasting time. We don't have much of it!'
Clive:  'So what matters about that? Is it a sense of progress?'
Rita:   Yes, yes it really is. I want to move this forward.'
Clive:  'So what can you do to move it forward, from where you are now?'

Rita:   'I probably need to go and see the COO and ask what she thought of my presentation and who else I need to speak to, to check their understanding of what I said and move it on. Thinking about it, I would have wanted to do that anyway. It might not be such a mess after all.'

Look at how simple and fast it was for Clive to help Rita understand why she was beating herself up and what she could do about it, and consider how much longer the conversation could have been if Clive asked her about the presentation. That would be Clive coaching the 'what' (the presentation) not the 'who' (Rita and what she believed in). While we all need a 'Clive', we can also listen more closely ourselves to our own inner critic, and ask 'What am I *really* afraid of here?' The answer will be close to or directly a value, which can liberate us from doubt and put us back on the right path.

## In practice

### Connecting to values to design actions

*May Bartlett, regenerative futures and coaching lead at Solvable*

*'Whenever you are ready, you can open your eyes.' The burnt-out young woman sitting across from me slowly returns to the room as her eyes open. 'How was that?' I ask her, sensing new alignment in her body. She begins to transport me to the water's edge as she describes a moment standing on the beach, taken by the vastness of the ocean. Sand in her toes. Sea breeze in her hair. Sun on her cheeks. Salt in her nostrils. As she relives this experience, I can hear the values pouring out of her. Presence. Belonging. Connection. Nature. Love. Solitude. Quiet. Spaciousness. Beauty. I reflect these values back to her and see life in her eyes. 'I have not felt this seen in years,' she says.*

*This moment with Natalie, a past client of mine, reflects a visualization exercise I have done hundreds of times with my coaching clients. I guide them to a happy memory from the past so they can connect with their core values. Natalie came to me for coaching for the same reason many of my clients do: she was feeling unfulfilled and burnt-out by work and wanted to do something to help make the world a better place. She was a graphic designer in San Francisco working for an agency with a culture steeped in modern western values. Efficiency. Growth. Profit. Competition. Greed. All of which were counter to the ones we had just uncovered together in session.*

*In this moment of connecting with her core values, she surfaced the inner conflict she was experiencing. She had been trying to succeed in a value system she did not believe in. She moved to the US from Sweden after high school and quickly fell into the American rat race; working long hours, competing with her peers for the better job, defining her worthiness by her career and buying the belief that busier meant more success. Now, she found herself caught between two worlds, wanting to live a meaningful life helping the planet and feeling trapped by a system that was extracting her energy similarly to the way it extracts Earth's resources.*

*For Natalie, our work became a process of freeing her from old beliefs about success and worthiness and aligning her life with her values. This resulted in her leaving her design job and starting a business making sustainable meditation cushions. She was able to use her design skills to offer a product that was aligned with her values of presence, spaciousness, quiet, solitude and connection, while honouring the health of our planet.*

*Values drive our actions and are at the root of our climate crisis. Societies that cause the most harm are built on extractive values of greed, growth, profit, luxury, convenience, efficiency, etc. Unfortunately, Natalie's experience is not unique. Nearly every time I do this exercise with people, similar values emerge: connection, belonging, care, connection, presence, contribution, peace, etc. These conflicting values make climate change work challenging. Because mainstream society is out of alignment with our fundamental human values, we must go against the grain to honour them. If we want to address the root of climate change, work is needed to realign society with our core values to fundamentally shift our relationship with ourselves, other beings and the planet. Viewed this way, climate change activism could be the willingness to act in alignment with our true values and reject systems of harm.*

### How can values be used to help people overcome resistance to collaboration or action?

Now you can see how values can be a powerful motor; there is almost nowhere that they are not useful. Values really come into their own, however, when we meet with resistance. If we do not take the time to surface the values below someone's resistance – and instead reiterate all the reasons that change is advantageous – we can entrench someone more firmly into their position, from where they may also enrol others to join them, so that they don't feel an outlier.

Below is an example of a real coaching demonstration that we held in front of a room of local authority leaders, in which we asked someone to come to the front and share a real issue that they were currently resisting at work. A woman stepped up and as she explained her position, and the values beneath it, her colleagues watched slack-jawed as they realized that all along their team-mate had not been difficult, but, in fact, had been standing up for some important principles, many of which they also shared. We have mapped this conversation onto the Flynn model of sustainable change (Fig. 10.3), so that you can see how the conversation moved the person from dissonance to resonance as their values surfaced. Pay attention to the way in which the questions helped this person to become open to influence around a topic that she was strongly against just 5 minutes earlier. Start at the bottom with statement one, and work your way up.

This conversation really did take a matter of minutes and ended with the member of staff realizing that she was not averse to eco-housing at all, provided that some of it was allocated in advance to low-income families. She did not abandon her scepticism, but she was now open to discussing it with others. Having come into the conversation looking defensive, her body language at the end seemed calmer and more open, as she sensed that the coach was not trying to sell her on

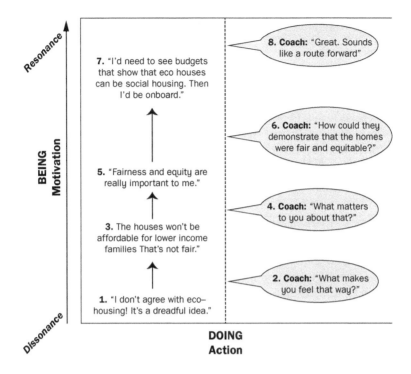

**Figure 10.3** Building resonance and clarity to overcome resistance

the idea but just understand her resistance better. As she sat down her manager raised a hand and said, 'In the two years that we've been talking about this issue, I never realized she felt that way.' Another leader later told us that you could hear a pin drop during this exercise. All too often we all get stuck at the level of the issue and fail to understand the big, legitimate principles beneath.

Rather than tell the other person why they should care, why they don't have a choice or why they are wrong, we can help them to instead understand why they are resisting. Contrary to what you might expect, when someone articulates their feelings and values, their relationship to the topic changes, and rather than digging in deeper, as they become more conscious of the reasons for their resistance they are able to better articulate (and contract) what they need to stop resisting. Few of us enjoy being an outsider or holding an unpopular viewpoint on our own. This kind of 'mining' for values allows us to test our assumptions about someone's resistance. To do so requires a coaching mindset of curiosity, positive regard and non-judgement to by-pass our impulse to 'other' those who disagree with us, so that we can move through resistance to collaboration.

## Overcoming 'cannot and will not!' – a practical example

Let us turn here to our fictitious sustainability consultant, Nisha. She has been doing consultancy work with an equally imaginary paper manufacturer. The CEO is keen to reduce their carbon footprint and is worried that the writing is on the wall for his company if he doesn't find a way to do so. Having found a sustainable source of wood, he is most concerned about the carbon cost of transporting paper, due to its weight.

### Part one – the story begins …

Nisha assumed that tackling transport would be straightforward because the company is eligible for a government grant for electric trucks. But when she spoke with Gary, the haulage manager, she hit a brick wall. 'This electric vehicle stuff is just a massive waste of money,' he said.

> They won't cope with the long distances … The charging points are too few and far between. This is a bad solution to the problem and if we wait a bit longer a better solution will come along. Look at other areas of the business first.

Nisha felt dismissed and deflated. Explaining the grants was pointless because Gary objected to the very idea of electric trucks. Defeated, she fell back on the excuse that the CEO wanted it to happen. 'Well then he can come down here and tell my boys that it doesn't matter if they're late for a delivery because the van ran out of juice miles from a charging point!' Gary stormed. Nisha retreated into herself. Could she work around Gary and get it done without him? Not likely, and bargaining felt like a dead end. Gary seemed fundamentally against the new strategy. Seeing the dejected look on Nisha's face, Gary spoke up:

> Look, I'm sorry. I'm not trying to make your job difficult. I agree with what you are trying to do, just not how you are trying to do it. This company won't survive unless we do something radical; I do get that. But electric vehicles are just creating another set of problems. I read the other day that there aren't enough precious metals to make the number we need globally, so it is an impossible dream anyway, whatever that bloke at Tesla would like us to believe. Messing about with my trucks is not going to make us fit for the future.

### Part two – finding each other right

How can they make this work when Gary is against and Nisha in favour of change? If you doubt that they can ever move forward, remember that fundamental desire to be in community. Gary is waiting to be enrolled as a supporter, he just needs the change to fit with his values and needs. First, Nisha needs to be able to hear not Gary's objections but the values underneath them. Second, she needs to treat Gary as 'good and right' and get curious. If Nisha sees Gary as justified and reasonable, this will in turn encourage Gary to behave that way. Feeling attacked or maligned just makes people retreat further. Remember that our threat response has a hard time differentiating between a real threat to life and a new company strategy.

Nisha's initial reaction to Gary was to assume he was against the whole strategy, when in fact he was in favour. That led her to consider side-stepping him or buying him off. It is easy to depict those who disagree with us as enemies to be dealt with, but that makes everything they say confirm that bias, and we can subtly transmit that negativity through our tone, body language and facial expressions. The more we imagine the other person as difficult and wrong, the more that attitude affects our micro-expressions, even while we may be telling them how much we value their ideas. People are hugely perceptive of such non-verbal cues. We know when we are liked and when we are being disrespected, regardless of the words that are spoken. Whether Gary will be able to fully articulate it or not, if Nisha is not approaching the conversation by giving him the benefit of the doubt, he may struggle to trust her. This can lead to an intractable situation in which Nisha resorts to command and control: 'The boss said do it, so just DO it!' While we know logically that we are far more successful when we take people willingly with us, we can often play the warrior when someone stands in our path.

Does this mean pandering to everyone's ideas and making decisions by committee? No, it means allowing everyone to feel that their values and beliefs have been satisfied. Remember that values can be manifested in a myriad of behaviours. If you are anxious about 'death by committee' decisions, perhaps that is because you have a value of efficiency, or progress, or decisiveness? You may therefore not mind if everyone is consulted about a decision, as long as it is clear from the start that not all ideas will go forward and that the process will not drag on beyond a certain amount of time, at the end of which a decision will be made and stuck to. You can still instantiate your value of efficiency *and* consult widely.

### Part three – finding common ground in shared values

The key to fruitful collaboration is respect. Even when we do not share a person's interests or understand the way they live their life, we can still respect their values and beliefs. Maybe Nisha spends her free time watching her local hockey team while Gary prefers solitary fishing. Gary can still understand that Nisha finds connection in a match-day crowd, just as she can understand the peace that Gary feels on the riverbank. While their values and interests are different, they can find points of respect. If we translate this over to the climate crisis, we can understand why people push back so hard when we suggest that aspects of the world need to change, because for them, those very things could be what bring their values to life and give their life meaning. In our story of the electric vehicles, Nisha and Gary can both get clear about what matters to them about the electric vehicles.

Nisha:   'I liked the idea of electric trucks because there are government grants for them. It's cost-effective and quick to do. That feels efficient to me. But if you have equally efficient ideas, I'd be up for hearing about them.'

Gary:   'I don't like electric trucks because it feels irresponsible towards my staff who will struggle to charge them. It will make their jobs harder and slower, so in fact it seems much less efficient in terms of operations, even if it's efficient in terms of buying them.'

Nisha:   'Hmm, I can see the difference between the point of purchase and the point of use now. Nevertheless, we need to reduce the company's footprint and the trucks are a big part of that. What can we do?'

Gary:   'Well you know I've been thinking for a while that we're not very efficient at route planning. Sometimes we have Jim driving 200 miles when Andrei was round the corner on another delivery. That's really inefficient and costly. I wonder if we could get some kind of logistics software that could streamline this for us, so I'm not poring over a map working it out manually. That would reduce overtime costs as well as fuel use.'

Nisha:  'That's brilliant, Gary! Could you investigate what software is out there? I'd be happy to look into whether we can get a subsidy or grant for it. It's definitely easy to make the case for how it will reduce emissions.'

Gary:  'Yep, I can do that. And you know, don't hold me to this now, but I did also just think that if we had a more predictable delivery system we could look at electric trucks … one day … because we'd be able to plan around charging points as well as delivery points.'

Nisha:  'Gary you just made my day! Would you be able to incorporate stopping at charging points into your research for the right software? I know that we're not going electric yet, but it seems efficient to be ready for that day. If nothing else, in 5 years we won't be able to buy new petrol trucks.'

Gary:  'Yep, that makes sense to me too. I know there are logistics companies that have software that shows them where petrol stations are. It can't be hard to swap out petrol stations for electric charging points.'

What happened here? Nisha let go of being attached to a specific activity and instead returned to the principles that she was trying to achieve. She and Gary both shared a very strong value of efficiency, which they then leveraged together. If they had not shared this however, it could have worked just as well, because by expressing their values, they would have created a sense of mutual respect. It is easy to disrespect someone who just says 'cannot and will not', but much harder when they say 'this is about fairness' or 'I really care about honesty'. When we return to these underlying values, we can become much more collaborative. Nisha and Gary solved the problem together, and even realized that they could have electric trucks one day. As we said in Chapter 6, by letting go of the road-map (electric trucks) but keeping the destination (reducing the carbon of haul-age), Nisha was able to see why Gary was upset, and to recognize that what felt like a solution to her would actually create a problem for him.

OVER TO YOU: Note down your own emotional reaction to this story. This points to your own values and beliefs, and can also help you to develop empathy for someone else's resistance. What will you do differently next time you encounter resistance?

## When it isn't values

It is not always the case that someone is resisting change because they feel their values are being stepped on. There are other types of resistance to listen for. For example when someone seems blindsided by the conversation or unwilling to have it, you could fall back on contracting. Similarly, if they tell you that there is not 'enough' or seem overawed by the enormity or complexity of the situation, you might consider using the tools we gave you for scarcity and overwhelm. You may also feel that someone's lack of a clear sense of purpose may be holding them back from caring in a broader sense, and that helping them connect to the bigger change agenda will help them find a foothold with which to climb on board. Finally, as we shall explain in Chapter 13, it may be that someone is angry or frustrated and cannot engage productively in the conversation, in which case we may have to step away or help them return to better behaviour before we can have this conversation at all.

## Moving from competition to collaboration

Adopting a more collaborative approach can not only help us in relationships, but help us to also model the behaviours that the world needs. The current paradigm tells a story of competition out of kilter with collaboration. While competition is not bad per se, its excess has had negative consequences. Companies compete for resources at the expense of the planet; politicians compete for short-term partisan votes at the expense of cross-party policies that support our long-term survival. Rather than fighting about the challenges, we ourselves can help more people across the world into a relationship with the climate crisis, if we first seek to better understand them. We do not need to agree on everything or see the world in exactly the same way to make these connections; we need to find the good in each other, not the bad. The ripple of positive change that comes from this is only possible when we put down our own feelings of threat, and create empowered, equal relationships that enable others to find their own security and empowerment. And on it ripples …

## Potential values from method one – listening for values

Statement two potential values: Focus; dedication; simplicity
Statement three potential values: Responsibility; commitment, tenacity
Statement four potential values: Autonomy; freedom; creativity

# 11 Defining a dream to run towards not a nightmare to run from

What if instead of continually feeling threatened by the spectre of climate collapse, we could find a deep sense of fulfilment in playing our part in changing the paradigm? What fresh ideas would we have if rather than being gripped by scarcity, we could feel enlivened with meaning? How would our destination change if we followed a map towards a dream instead of running from a nightmare? Welcome to the world of purpose. So far in this part of the book we have helped you to dial down the forces against change. Now it is time to dial up the forces in favour of it. There is no better way to do this than to tap into our deepest sense of purpose and meaning. In this chapter we show you practical ways to define a sense of purpose for yourself and others.

But before we start, we want to acknowledge the purpose-driven elephant in the room. In the last decade, so much has been written on and spoken about purpose, that people often feel a tremendous pressure to have a well-crafted, multidimensional and immutable sense of it that they can use as an elevator pitch. Many feel rightly unequal to this task and become either blocked or sceptical. So let us be clear. We are not talking about a statement of intent or beautiful words that you put in a frame in your kitchen but rather the deep *feeling* of purpose, fulfilment and meaning in your life. Some call it being in flow, others call it contribution. Whatever name you give it, you will recognize the feeling of it from the moments that you have felt a real sense of being in the right place, doing something that feels made for you and feeling a level of motivation that propels you forward through difficulty.

You can experience purpose anywhere, from the Saturday morning kids' football team that you coach to a high-level policy meeting. Purpose is also not hierarchical: the office cleaner may feel a deep sense of meaning while the chief technical officer feels their job is pointless. Seniority may guarantee more remuneration, but it does not guarantee more purpose. Instead, meaning comes from joining the dots between what we do and a wider impact. For the cleaner, this could be noticing how much their colleagues appreciate the immaculate communal kitchen when they are stressed. For many purposeful people, this is unspoken and unconscious. If you enjoy your work or life, you probably do not need to spend time sweating this. When you are standing on the cusp of change however, purpose can be a wayfinder and a way of uniting an external problem with our internal desire to address it, bringing together who we are with what we do.

Finding a need that will satisfy us is a great way of engaging with change; almost like a secret doorway into change that is unique for each individual. So helping others to find a sense of meaning in climate action, can fuel them through the inevitable difficulties that change throws at them. Nietzsche famously said,

'One who has a why, can bear almost any how.' Because a sense of meaning provides us with intrinsic motivation to act it is the cornerstone of our resilience. In the context of the climate crisis, helping someone to find a sense of meaning within their anger, overwhelm and fear can sustain them in ways that those raw, threatened emotions cannot. More than that, in the midst of a complex, global threat, the long-term quality of purpose can help us all to weather the short-term shocks, shoring up our determination and ability to handle uncertainty. Like a north star or a compass in someone's hand, that big picture context of why they are acting helps keep them on course when the external world changes, and they suddenly have to change tack at short notice. It can also help decide which piece of the jigsaw puzzle someone should hold, and what their contribution should be; a powerful tool when it comes to climate action, where there are so many simultaneously pressing problems to solve.

Feeling purpose*less* can also have a powerful effect on our behaviour, decision-making, ability to relate to others and mental well-being. When we feel no purpose and that our efforts feel meaningless, our response is often to disconnect from our emotional selves because it is just too hard to be with the feeling of wasting our lives. For those whose sense of meaning has waned, even simple tasks can drag and it can feel hard to love their role, paid or unpaid, even if they are still making what others would consider an impact. As Gillian Benjamin says in her piece in Chapter 20, once extinguished, we feel the loss of meaning much more keenly than if we hadn't had it at all.

It is important to recognize, however, that not everyone seeks out purpose in the same way. Some people may volunteer to fulfil that sense of meaning and go to work just to pay the bills. While introducing a sustainability agenda can inject motivation and a new sense of pride in your team or organization, evangelizing a purpose-driven career can backfire if those on the receiving end feel alienated rather than inspired. If you are influencing within an organization (either as a consultant, coach or as a team member) go lightly and notice if you are holding a perspective that 'everyone should do meaningful work' or that 'meaningful work is the only worthwhile work' – purpose looks different for everyone.

## In practice

### Reviving purpose

*Alison Maitland, coach with a specialism in leadership, sustainability and inclusion*

*It is easy to fall out of purpose with our career, even if what we do looks meaningful to the outside world. As a marine environment adviser, Katie was working on biodiversity projects in the fossil fuel sector. But the sense of purpose that had attracted her to conservation had waned. She felt stuck, under pressure, and dragged in different directions. She wanted to create 'a new system' instead of fighting within the existing one, but she was also fearful of stepping out of the familiar into the unknown.*

*Katie contacted me after reading an article I had written about women's leadership and seeing that I had integrated nature and climate into my coaching. An early breakthrough in our sessions came when I invited her to do some deep breathing, which helped her to slow down and recognize that she needed to make space in her life to allow for new possibilities to emerge. Asking her the question 'What makes you angry?' released emotional energy. She spoke about her anger at the damage that plastic waste is doing to the oceans. Underlying this anger were her love and compassion for species such as turtles and whale sharks, and her longing to do something impactful to protect them. As we explored these and other core values, Katie was able to articulate what she stood for with clarity: respecting people, places and planet. I acknowledged how confident and persuasive she was when she expressed her purpose, and I asked what it was like to say it out loud. This reinforced for her how meaningful it was.*

*I invited her to visualize herself in a place she loved or felt inspired by, where she could access her inner leader. At first, she found herself in a familiar marine environment and expected to stay there, but then the scene changed into a woodland. The leader she had initially pictured as male morphed into a female. As she talked, I noticed a shift in her energy and posture. I told her what I had seen and asked what she was feeling. It was an emotional moment, as she realized she had been resisting thinking of herself as a leader at all. Now she saw that she could apply her strengths and purpose to a new environment and lead in it.*

*Reconnecting to her deeply held purpose helped Katie to push through the discomfort she experienced about making a big career change. After identifying what had been blocking her about leadership and moving past it, she set about investigating her next steps. We revisited and elaborated on her purpose in subsequent sessions, adding 'sustainability' – something I articulated that was so innate to Katie that she had not even seen it as a value before. She said this helped her to prioritize options according to whether they were sustainable and respectful of people, places and planet. She zeroed in on ways to reduce plastic waste and pollution. She was able to let go of peripheral activities and apply her energy to what was most important.*

*Noticing her easy connection with metaphors from the natural world, I continued to bring these into our sessions. Another 'lightbulb moment' came after working with her to create a picture of who she might be in the future, seeing herself and the environment around her from different perspectives, for example from ground level and from high above the Earth. This time, she envisaged herself up in the mountains, confident, relaxed, happy and fulfilled. I find that this kind of visualization work helps to switch off our 'internal chatter' and open our creative minds to the possibilities that we are yearning for. The mountains evoked Katie's love of adventure, climbing and taking on tough physical challenges.*

*The next morning, Katie knew that the business she wanted to start would do more than just reduce the impact of plastics. It would disrupt demand for plastics at its source. She had set her sights on a major consumer market and was on course to innovate, disrupt and live out her purpose.*

OVER TO YOU: What brings you a sense of aliveness and what leaves you feeling the opposite? What do you know about your own sense of purpose?

## Employing our own purpose to enrol others

If purpose is individually sustaining, in relationships it can be deeply enrolling. Even the most bashful among us come alive and are more compelling when we speak from a place of purpose because we are authentically sharing *why* we are doing what we do, not just the pedestrian components of *what* we are doing. Consider these sentences:

'I drive an electric car because the government offered free road tax.'
'I drive an electric car because I want to be part of a clean air revolution.'

Which would you find more inspiring?

Too often instead of explaining why we are motivated to act, we talk about 'what' we are doing. Perhaps to be practically instructive to others ('I do this, you can too'), or because as Tom Crompton said in his piece in the last chapter, we assume that while we are intrinsically motivated, others are not. Marianne Williamson wrote that we shrink 'so that other people will not feel insecure' around us,[59] and perhaps there is a feeling that speaking about meaning is arrogant or boastful, or even that we will be considered gullible. The problem with focusing on the practicalities of what we do, is that this territory is ripe for disagreements, criticism and even shaming, and we can widen the gap between those we are trying to influence when we stay at the level of what we do, not why we do it. When we share the bigger picture, we invite them to envision themselves there too and we don't dictate how they join us.

As we said in Chapter 2, it is tempting to try to enrol people by 'selling' a nightmare vision that vividly describes how bad things are. People do not buy nightmares however, they buy dreams. That does not mean hiding the truth or giving false promises, but it does mean creating a compelling alternative future to act on behalf of and a sense of what Joanna Macy and Chris Johnstone call 'active hope'.[60] If we can speak about the sense of meaning we derive from our environmental choices and actions, we naturally characterize them positively, which makes them attractive to others. On the other hand, when we share the terrifying statistics about the crisis *without* a sense of why they are personally meaningful to us, we can understand why someone might not want to 'buy into' learning more. This does not mean hiding the truth but making sure that we share the connection we feel to the planet too, so that the information we then share falls into that context of our sense of bigger meaning.

## How to help others to identify their purpose

Often when we share what brings us meaning, we find that the other person starts to describe their own. So what if we could do that directly? Let us now look at how to connect someone to what is meaningful for them. Here are some tried and tested ways that you can use, one to one, or one to many, to explore individual or shared purpose. The first is an all-purpose coaching method and the next three are designed specifically to support purpose in the context of systems and climate change.

**Method one: looking back for red threads**

Scan backwards over someone's life and identify meaning from previous experiences. Everyone has felt a sense of purpose, even if only fleetingly. It does not have to be related to climate change – it is not the *what* that we are looking for in this process, but the *why*.

Here are some approaches for this method:

1. Ask: 'Tell me about three key moments in your life where you have felt real meaning.' When the person describes them, move away from what they did to how they felt while they did it. 'So it sounds like in all of those roles, you felt a strong sense of autonomy? What is it about autonomy in those experiences that has real meaning for you?'
2. 'Let's look at times that you have felt really lacking in meaning/purpose/ fulfilment. What was missing?' You may have to dig under the 'what' to get to a sense of why and how, for example when someone says 'I felt totally micro-managed. I just wanted to make decisions by myself' you might ask 'what mattered to you about making decisions?' rather than 'what did you want to make decisions about?'
3. 'Can you give me an example of something in your life that has filled you with pride?' This may be small, but like an iceberg it will have depths of purpose below the water.
4. Plot a timeline of their life. This is something that psychologists in the UK National Health Service also use to help patients make sense of the past, and its simplicity belies its dramatic effect. Draw a line across a piece of paper (landscape format), and mark decades evenly along it. Ask the other person to first write down along the line what they did in their life. They choose where to start – it can be at 2 or 20 years old – you want to capture events or roles that felt significant, either because they were purposeful or it's opposite, dispiriting. Once you have a line of activities, ask them to draw a curve along it, with peaks above the line to show high points and troughs below the line to show crucible moments or lows. Then ask about the highest peaks and the lowest lows: 'What was the common feeling when you were in the peaks?' 'What was consistently missing in those lows?' Eventually, you will see a pattern emerging. 'I derive real meaning from solving problems, and in my low points I was micro-managed and made to use someone else's process. Being creative gives me meaning.'

**Method two: your stake and the need in the world**

All of us realistically have a stake in the climate crisis, but exploring our stake is a way of finding the meaning that the crisis holds for us personally, and that we can use as a motivating starting point for our climate action. Like purpose, everyone's stake is different. We should not assume that parents' only stake is their children, or that those without children are not motivated by a responsibility to the next generation. We have to find out. Perhaps your stake is a love of nature or animals, while for others it may be about equality or justice.

Questions that can help you unearth someone's stake are:

1. 'What is the need in the world that you feel drawn to meet?'
2. 'What do you want to create in the world?'
3. 'What is the legacy that you want to leave the world?'
4. 'What speaks to you about this challenge and what about it matters to you personally?'
5. 'What is important about this issue to you personally?'
6. 'What is something in the environmental sphere/climate crisis/biodiversity loss that most interests or matters to you?'

## Method three: inverting and utilizing righteous anger

You can almost feel the blast of wrath and power that comes off the phrase 'righteous anger', in the printed words alone. Righteous anger is our sense of fury at a situation, which unlike blame springs not from our insecurities but from our values. A woman can feel righteous anger when someone makes a sexist comment not just because she herself is offended but because she is offended on behalf of women more broadly. To unearth the purpose in someone's anger, go beyond *what* they are angry about (which can lead to a disempowering conversation), to fully understand *why* it makes them angry. This will lead you right to the heart of what is important to them: 'Because this is about equality!' or 'Because everyone should be treated fairly.' By looking beneath our anger we are able to find out what is precious to us, our dream for the world and the dearly held values that are being compromised. With this rich material we can create a vision for the world we want to build and goals that are compelling enough to motivate us into action. For example, 'I'm motivated by creating an equal world.'

Questions to help you unearth purpose beneath someone's righteous anger:

1. 'What do you find unbearable or intolerable?' 'What matters about it?'
2. 'What about this is so angering for you specifically, and why?'
3. 'What would the world look like if the opposite were true?'
4. 'What role would you play in creating a world that would feel meaningful to you?'
5. 'How would you behave in that new world, beyond what you did, who would you be?'

## Method four: seeking purpose in the natural world

We need only look to nature for a moment to see purpose writ large. Every aspect of our natural environment is engaged in deeply meaningful endeavours. You will not find your average oxeye daisy killing time on social media. The singularity of focus and symbiotic relationship with its ecosystem, makes nature an inspiring place to reflect on our own purpose, free from the extrinsic clutter of 'what people will think'. Reflecting on the purposefulness and collaborative approach[61] of the natural world can dial down our competitive tendencies, and dial up our understanding of collaboration. Nature moves in step with its bigger reason for existing, regardless of the obstacles. A stream carves a path through a landscape, using the simplest, easiest route forwards, changing direction if it reaches solid rock, but sticking to its purpose of constantly flowing. Plants find ways to grow on the most barren, windswept hillsides, to fulfil their imperative to survive and propagate. It would be easy to dismiss these activities as meaningless attempts at survival, but that would be to assume that our human lives are not also ultimately about survival too. The flower that blooms facing the morning sun is reaching for the best conditions to thrive. Similarly when we pursue meaningful work, we are seeking that same source of nourishment. Using nature as a lens can declutter someone's mind and help to bring a new perspective to what is purposeful to them. Someone might watch a bug industriously traverse a patch of earth and realize that they derive real meaning from solitary achievement. Contemplating a forest canopy, imagining the vista from a mountain range or picturing ourselves by the sea can connect us to why we choose to act and revive our belief that there is something worth fighting for. Partnering with nature we realize we are not fighting alone but are in this together.

Questions to help you identify someone's purpose with the co-coach of the natural world:

1. 'What is meaningful to you about what you see in nature?'
2. 'How is your purpose alive in nature?'
3. 'Can you see something in nature from where you are sitting right now? It could be as simple as a pot plant. Take a few moments to just observe it. What is *its* purpose? If it could talk, what would it tell you yours was?'
4. 'Observe a really small animal, ideally a bug, and notice how it manages to live in a world that is so inherently precarious to its survival. What is *its* purpose? What are you noticing about your own purpose from observing it?'
5. Go for a walk in nature and invite them to notice the environment around them. Then ask, 'What draws your attention here? What about that really speaks to you?' And from this question unpack into subsequent questions, such as 'What meaning do you derive from that?' 'Is there a sense of fulfilment for you here? What is it?'

You may well start with one method and end up in another as you explore the subject. For example, someone may begin with a sense of their stake in the crisis and then tap into their righteous anger around it. In which case, just follow their lead. As you do so, make sure to regularly acknowledge the person for their ideas and notice out loud how you are seeing their sense of purpose manifesting in their body language or facial expressions. For example you might say: 'Your whole face came alive as you said that' or 'It is really inspiring to hear you talk about equality this way.' Your positive feedback will grease the wheels of further exploration, and lead to a deeper understanding of their motivation to act. Do not be distracted by crafting the perfect purpose statement. Remember that it is the *feeling* of it that we want people to connect to. We have coached many people who made a deep connection to purpose by just saying one word.

## In practice

### Finding a shared sense of purpose

Jess Scott, volunteer campaigner, London Citizens

*A few years ago, London Citizens organized a meeting attended by institutions across the city. The purpose of the meeting was to work out what London Citizens, as an alliance of institutions, would ask mayoral candidates to commit to in the run up to the upcoming election. I was invited as a member of my church, itself a signed up member of London Citizens. I found myself, on a rainy evening, in a room full of people talking, with ambitious ideas being shared and vast topics being discussed. Climate change was one of the topics. What kind of pledges could we ask of the candidates? When should London achieve net zero? What do we need to demand for this to be viable?*

*Those of us who attended from my church met up after the event, energized by the heady visions we had been invited to imagine. Between us, it was the question of climate change to which we kept returning. I cannot remember the moment we decided we'd do something as a group, or quite when it was that we named this as our focus. I suppose that in itself speaks of the nature of purpose – it can arise without prompt or planning, finding you more than you find it. What we recognized, in those early meetings, was a resonance: 'what I cared about, what I was troubled by, so were you.'*

*We asked at church if anybody who cared about climate action would like to join us after the service for lunch. A handful of us gathered in a kitchen and shared food as we took turns explaining what had drawn each of us to be there. We started wondering together what it might look like to take some action in our own community, working out where our own context of agency was within the broad vision we'd heard about at that first London Citizens meeting. Each of us shared ideas that we have since worked on together: running a fair energy campaign to encourage people to switch to green energy; leading climate themed services; and undertaking an 'eco audit' of all the work happening within our church.*

*Finding our shared sense of purpose took several steps. In turning up at that first event we became visible to each other as people who cared about the same thing. Having become visible, we could share some more – our worries, our reasons and our hopes. In doing that, we came to realize the resonances between our experiences, recognizing that what might have felt solitary was, in fact, shared. In finding our shared experience, and in sharing a context, it became possible to imagine action and change at a workable scale.*

*The magnitude of the climate crisis can make us feel individually ineffectual. Uncovering a sense of purpose with others made action start to seem possible. The complexity of what this crisis demands can easily overwhelm us. Finding purpose with others enabled us to set manageable parameters, to particularize the problem, and to give ourselves some hold on it. The threat of the climate crisis can leave us feeling desolate. Finding purpose together has offered us great consolation, and taking action alongside each other has brought us joy.*

## From purpose to goals – landing the plane, not just painting the sky

For a sense of purpose to be useful, the plane has to land, not just circle the airport writing words in the sky. Or as Henry David Thoreau put it 'If you have built castles in the air, your work need not be lost; that is where they should be. Now put the foundations under them.'[62] We have shown you how to help someone to identify and articulate what is meaningful, which can be used as a powerful lever to convert dreams into tangible plans, actions and behaviours. At this point we ask *how* someone will bring their purpose to the world, translating it from a *feeling* into tangible behaviours and activities. Too often we can generate the former without connecting to the latter, which can leave us feeling dissatisfied and adrift.

To connect the two is to some extent a strategic process; however, dreaming is also important at this stage. We want to go beyond purely cognitive solutions that reproduce ideas from our current paradigm and instead step into the realm of new possibility. To bridge someone's castle in the air with foundations on the ground, we can start the conversation by asking them to list all of the things they love or enjoy doing. This will identify the activities that are intrinsically meaningful for them, in which they experience 'flow'. There is no need to stick within the confines of paid work – coaching that Saturday morning football team might yield a great list of activities through which their purpose could be brought to life, such as teamwork that demonstrates collaboration or training regularly to cultivate excellence. Here you are looking to listen, ask questions and paraphrase what the person tells you back to them, not diagnose a list of activities on their behalf.

This leads on to determining their unique talents and how they can be applied to tackle a problem in a way that is inherently fulfilling and energizing. What we love and what we are good at creates a virtuous feedback cycle whereby we do it more willingly because we love it, and the more we do it, the better at it we get. It is not always easy to subjectively identify our talents though and we often need other people to tell us what they see in us and think we are good at, in order to recognize it. Seeing the magnificence of another person is a key coaching skill because it not only uncovers blindspots, but also offers a deep acknowledgement of how they show up in the world. Similarly, believing that each individual has a unique contribution to offer the world is important when it comes to climate change, where soon very many types of work could be perceived to be climate work.

Questions to help someone recognize their unique talents and abilities:

1. What are you naturally good at?
2. What do your close family/friends/colleagues say you are great at? (If this is hard to answer, ask if it would be helpful to actually ask some.)
3. What is it that people admire about you? Does that link to times when you have felt really purposeful?
4. What skills do you have that you are most proud of?

Now that you have defined someone's interests, skills and talents, you can map these onto the issues in the world that they are most drawn to working on. Good questions to facilitate this process are:

1. 'What problems in the world are you drawn to or find yourself thinking a lot about?'
2. 'What issues do you commonly share news about in your social media?' A person may even want to do a quick check of their last month to see what has particularly inspired or frustrated them.
3. 'What issues ignite your righteous anger?'
4. 'When it comes to other people's problems, what are you good at solving? What does this tell you about how you could do this more widely?'
5. 'In what way can your purpose allow you to bring to life your stake in the world or the future?'
6. 'How can what you are great at and your purpose be part of systems change?'

To bring all of this together, we can circle back and ask:

1. 'How does that connect to your sense of meaning?'
2. 'How can we overlay your purpose onto that?'
3. 'In what way does your purpose enable you to become the person you know yourself capable of being?'
4. 'In what way does your purpose support you to create the world you want to live in?'
5. 'Now that you know this, how can you connect with others from a place of purpose, and what will you be connecting with them for?'

## Helping someone to identify purpose – a fictitious conversation

Let us look at how these coaching tools create a more empowering conversation between fictitious friends Monica (the youth activist that we met earlier) and Paula, who is not sure about whether to get into activism. They have met for coffee in a shopping centre:

Paula:   'I know you've got your group, but it's not for me. I feel like nothing I do will make any difference. I look around here, at all these people shopping, buying stuff they don't need, and I just think it's too big a change.' [Here, Monica could try to convince Paula to join her group, but look what happens when she does not.]

Monica:   'I really get that feeling. [Normalizing] But what matters to you about us changing?'

Paula:   'Well it's obvious isn't it? To make sure we have a planet to live on!'

Monica:   'Well yes, but all of us also have a very personal reason for taking action. For instance, I've realized that for me it's about *showing people that there's another way to do things* [Monica's sense of purpose]. I've always been like that. Even when I was a kid I was trying to get my mum to try new things that I saw on the TV.'

Paula:   'But that had nothing to do with climate change.'

Monica:   'Yep, quite right, but it motivates me, it's what I enjoy doing. That's important if I want to stick with things. I realized that the stuff I gave up on didn't have that in it. So I've tried to find ways to do that in my activism. For instance, I write the group's blog that's read by lots of people outside of the group because it instinctively feels like a way to help those people see things differently, but I just turned down writing the group's handbook because that's preaching to the choir, so it's not as exciting to me.' [Monica showing how she lands her purpose into action]

Paula:      'Wow. It's fantastic that you know that about yourself. I have no idea what motivates me.'

Monica:   'I bet you do if you think about it. Want me to help you work it out?' [contracting]

Paula:      'Yes. I'd love that!'

Monica:   'OK, so let's start with where you are right now. You mentioned looking around here and feeling defeated. What exactly do you feel about that?'

Paula:      'Well, I feel frustrated that these people don't get it. And really angry with all these companies and our government for pushing this "buy, buy, buy" message when they know it's not just bad for the planet but for us too. I bet most of these people are using credit they can't afford.'

Monica:   'I think I hear that you care about more than the planet?'

Paula:      'Yes! This is so damaging for ordinary people. And they'll take the brunt of this. Not the fat cats running these companies.'

Monica:   'Sounds like this is an issue of … what … justice … fairness?'

Paula:      'Yes! Both! I actually don't feel angry with these guys shopping. They've been conditioned to do it.'

Monica:   'Great, we're really getting somewhere! [acknowledgement] So if you could remake the world, what would it look like?'

Paula:      'Oh I don't know …' [empowerment dipping]

Monica:   'Go on, I think you do know.' [acknowledgement of Paula's resourcefulness]

Paula:      'I would help people to find ways to enjoy themselves without shopping. Look how miserable they all look! I want people to find happiness by being together, and doing things that don't endanger them or the planet.'

Monica:   'Wow! You help people to find happiness by being together, and doing things that don't endanger them or the planet?' [repeating Paula's words back to help her hear them]

Paula:      'Well I don't know about that … I work as an admin.' [empowerment dipping]

Monica:   'Sure, but that's still what motivates you [acknowledgement]. Can you think of a time when you've helped people be happy together?' [pauses to allow Paula to think]

Paula:      'I'm not sure it counts, but every Christmas I'm the one who buys a jigsaw in the charity shop because everyone loves doing it, and it's the one time on Christmas day when there's a kind of simplicity and peace in our house.'

Monica:   'There you go, you already do it. I wonder what you'd do if you could help more people to find happiness without endangering anything?'

Paula:      'I'll have to think about that … maybe it's as simple as showing people a happier, more connected way to spend their time? You know, don't take this the wrong way, but the reason I haven't joined your group is that I don't like the idea of protesting. It feels angry. I want to put the opposite into the world.'

Monica:   'I totally get that. [acknowledgement] What could you do to bring more happiness?'

Paula:   'I'm going to research initiatives that are focused on happiness and joy. Maybe I can even do something at the office.'

Monica:   'Brilliant. So how do you feel about these shoppers now?'

Paula:   'Ha! Well, it's still a massive mountain to climb right? But I don't feel so angry. And I feel like I've got more of a sense of what I can do. Thanks Mon.'

Notice what Monica did here. She acknowledged Paula, particularly when she sensed her self-belief dipping, and gently held a space open for Paula to explore. She also did not dismiss Paula's example of the jigsaw at Christmas or allow Paula to do so. Purpose can be lived in the way we speak to the man in the corner shop, it does not only show up in grand initiatives. Monica helped Paula to identify what gave her meaning and helped her to hold onto it when Paula started to doubt that it could be *systemically* useful. It did not suspend Paula's sense of the challenge of climate change, but it did help her to feel more empowered to act.

Purpose is a great way not of reducing the scale of the challenge, but of *growing* the size of the person to meet it. This is what it truly means to focus on the 'who' someone is, not the 'what' they do. Using this process, we can start to flesh out how someone can create their definition of a meaningful life, and take it from an aspiration to a potential set of new behaviours and activities. In the next chapter we will take this a step further and look at how to create goals from these strategic choices and how we can hold those goals lightly enough to create flexibility when the rapidly changing context in which we work changes.

# 12 | Setting goals and getting into action

This is where the rubber hits the road, when we help someone decide what they will do, by when and with whom. This is where the plans come together into something we can measure. First we have to work out what we are aiming to achieve, and then we need to break it down into steps that we can track. You may call them goals, targets, strategies or even dreams. Whatever label you chose, you are identifying the specific areas on which you will focus and by definition those on which you will not. The steps here are simple and familiar, but we will also prepare you to avoid a couple of bear traps created by the systemic and complex nature of climate change.

## Combining the head and the heart – designing motivating goals

For many of us goal setting will evoke memories of a tedious performance review process or at best an emotionally empty planning exercise. If we want to create goals that motivate us not just in the calm of the planning but later in the heat of implementation, we need to engage both our thoughts and feelings. While our logical selves can see the world as chess pieces to be moved back and forward, our feeling selves recognize the need to use our emotional intelligence to cultivate relationships. Our emotional selves also improve the process of goal-setting. Professor Richard Boyatzis argues that the word 'goal' reduces our cognitive capacity and encourages us to play safe. In his research,[63] he placed participants into functional MRI scanners and observed changes in their brains depending on whether researchers asked them about 'goals' or 'dreams', and found that 'dreaming' resulted in a bolder set of ideas than 'goal setting'. He concluded that even when we have to get things done and be held accountable, our solutions are greatly improved when we begin the thinking process by framing the bigger dream within which our goals sit.

Richard Boyatzis suggests engaging in long-range dreaming by asking someone where they want to be in 10 to 15 years' time. It is also acceptable to ask about their dream for a specific initiative or for their organization, and this may be more of the focus for those that struggle to visualize a positive future or are triggered by fears of an apocalyptic one. The language of dreaming takes us into our feeling selves and taps into our sense of longing and meaning. While this process may be effortless for people whose work already feels purposeful, research shows many people are not motivated in their jobs. If you are coaching or trying to influence someone who is demotivated by their work, you may need to spend longer exploring the dream, unpacking their scepticism and gently opening the door to an idea they may have assumed was closed to them. If that is the case, simply make it OK for them not to have immediate answers, and keep reassuring them that everyone is allowed a dream.

A simple way to flick this switch from goals to dreams is to ask questions that appeal to our values or what we yearn for and enjoy doing. Here are some examples:

1. What is your dream for 10 years' time?
2. What do you yearn to create in the world?
3. What will make you feel proud?
4. What would give you true meaning in your work?
5. What matters to you about this goal? How can you put that front and centre in the goal?
6. What inspires you about this goal?
7. What excites you about working on this?
8. When you think of doing this, what do you dream of achieving?
9. What do you intuitively know about the other people you'll be working with. What do you appreciate about them? How could you adopt their mindsets to make this goal a reality?

## Behavioural goals alongside action goals during systemic change

Goal setting can often be a straightforward, cognitive process of determining the direction of travel and the milestones along the way. We want to get a new job, move house, start a new project, even fall in love. And just like falling in love, we focus on the 'happily ever after' end result moment – the door closes on the happy couple and the credits roll. We know, of course, that in reality this is just another beginning, and that goals evolve rather than ever finishing. The other problem with focusing on a notional end point is that we forget about the behaviours that we adopt to get us there. As Maya Angelou so famously said: 'People will forget what you said, people will forget what you did, but people will never forget how you made them feel.'[64] Because in the climate crisis we are playing a very long game indeed, the way that we behave on the road towards our goals can be as if not more important than the attainment of the goal itself, as Jill Bruce's story in Chapter 6 demonstrates. Even if our goal is to lobby for specific policy change, if we ride roughshod over others in the single-minded pursuit of the end goal, and fail to collaborate skilfully, we can damage our long-term ability to influence by alienating people.

In the rapidly changing context of climate change, focusing as much on behaviours or processes as on outcomes enables us to take into account the shifting external environment. This is also important when considering the problem from a complex adaptive systems perspective in which the whole is greater than the sum of its parts. As old systems die away and new ones emerge (as we are currently seeing with the fossil fuel economy) the environment and the actors within it are in a complex and continual dance of reshaping each other. Thus, if we have fixed, linear goals that we assume operate in a vacuum, we will be constantly blindsided when the real world context kicks in and derails our goals or makes our targets irrelevant. We explore this further through the lens of emergence in Chapter 18. When we recognize that the world of 'doing' is constantly changing, then developing our 'being' or behaviours as we move into action is all the more important.

Here is an example of how building in behavioural alongside action goals, as our fictitious character Rita, head of global sustainability for an investment bank, helps one of her imaginary staff in her Asia regional team determine goals that will support their carbon zero agenda:

Rita: 'So you've seen the new strategy. What do you think it means for your territory?'

Vikas: 'It means a lot of change. If we can get four of our offices on board by the end of next quarter that would be a strong start.'

Rita: 'Great. What will signal to you that you've got them "on board"?'

Vikas: 'Knowing them, it'll be that they are independently taking action, without me having to keep checking on them.'

Rita: 'Sounds good. So it sounds like that behaviour itself is a measurement of success. If we framed that as something we could measure, would it be "staff taking the initiative"?'

Vikas: 'Exactly. Because some of these sites can be quite "command and control" oriented, and the staff sometimes hang back. I want them to feel able to come to me with ideas.'

Rita: 'It sounds like you've got a real instinct for the culture of those sites?'

Vikas: 'Yes. I can't speak for them all, but I started out in the Mumbai office before I moved here to Singapore. I know that there are a lot of good people there who don't feel comfortable raising their voices. I want them to feel happy to do that with me.'

Rita: 'This is great insight Vikas. It sounds like there are some behaviours for them to develop around taking initiative, and that there are also some for you to employ to make it feel safe for them to speak up.'

Vikas: 'I'd not thought of it like that, but yes, I think I could do more.'

Rita: 'So what do *you* need to do to help them feel comfortable?'

From here Rita and Vikas can design the behaviours that Vikas will consciously use and how he will use them. For example, he might make a point of always finding time to speak to someone in those teams one-to-one when they ask, no matter how full his diary. Focusing on the behaviour first automatically leads to *how* we can bring desired behaviours to life, which then takes us to a set of 'doing' actions, which lead us to an eventual 'doing' goal. It also puts us in a relational mindset, recognizing that we can have a real impact on others with just our behaviour alone. By contrast, if Rita and Vikas took a 'doing' approach to 'getting four of our offices on board by the end of next quarter' they could find themselves considering a more dictatorial approach. If Rita has asked Vikas 'what *action* will you take to get the teams on board?' he would be more likely to feel that he had to 'make them'. A behavioural lens also enabled Vikas to reflect on the 'who', of the staff, not the 'what' of the strategy. In this case, that helped him realize that there were good staff who lacked confidence. This allowed him to start his goal-setting process from a place of respect and positivity. The other benefit of focusing on behaviours first is that this broadens the impact beyond Rita and Vikas' specific sustainability agenda. In helping staff to be more proactive, Vikas was not just helping the organization with its new strategy but also encouraging behaviour that would be widely useful. Knowing this, Vikas may be able to engage the management of his sites more effectively in the new strategy, because he will be able to demonstrate this wider benefit beyond their sustainability targets.

While coaches help people to set independent goals, for those making change within organizations or communities, goals are usually co-dependent and shared. Whether for an overall organizational strategy, or for a team or a group embarking on a particular project, setting goals together can make sure that everyone is similarly aligned and motivated from a shared sense of purpose. This takes people to the high-performance zone because purpose is enlivening. Contracting is also our friend here because we can agree on joint intentions both in terms of what we will do and how we will behave. It is useful to additionally discuss the consequences of missed goals in a spirit of compassion and team building that designs out blame, resentment and anxiety. We look at this in much more detail in Chapter 18.

## Mental traps when goal setting in a climate context

The current climate context sets up a few mental traps when goal setting that can threaten to derail our best intentions. Here are some common pitfalls that we have encountered as coaches and some solutions that can make goals more attainable:

### Scarcity

Good goals take time to set. If the voice of scarcity tells the person you are working with that 'we don't have time for this', explain that rushing this stage leads to poor problem analysis and misdirection of efforts – a waste of time for everyone all round. Instead we can remember the hare, who seems certain to reach the finish line first and the tortoise whose slow and steady approach gets him there sooner. Rather than seeing goal-setting as a means to dictate action, encourage its use as a process to get really clear about why someone is doing something, with whom and how. It may be that this process initially poses more questions than it answers, which can prompt further research. Not all of the answers may be at hand when you start, so setting expectations about how long it will take is a useful way to manage scarcity. Scarcity can also show up as a feeling that we must continually have something to show for our efforts. We can learn from the way nature works and take heart on the days that our efforts are not visible or worthy of social media, that we have been putting down roots or spreading mycelium underneath the forest floor. One day our work will burst forth in blooms for everyone to see. In the meantime, we can take comfort that our work is still worthwhile.

### Big enough to matter and small enough to manage

A frequent problem with goal setting is that we get carried away and create unmanageable unspecific goals of which we cannot track the success (or failure). Equally, when we lack confidence we can also set goals that are too timid, and do not present enough of a challenge or create enough change. A great rule of thumb is to ask 'is this big enough to matter and yet small enough to manage?' In other words, will this solve a problem that I and others care about, or have an impact that really matters? *And* can it be broken down into steps and actions that are within my or our capability to manage and execute? When we are looking to make broad-scale social change, as we are in this context, this formula can save us a lot of heartache (when the goal is too big to manage) or a wasted effort (when the issue does not matter enough). Instead we can set goals that are tangible enough to put our hands around while still being inspiring.

### Protect against overwhelm by identifying unique contribution and timing

An alternative way of looking at 'small enough to manage' is to define what someone's specific talents and skills are best suited to doing. As we explained in Chapter 9, the climate crisis creates a feeling that we have to do everything, right now. Help someone to consider where their particular goals fit into the wider systemic change that they are trying to make, among the many other actors who are also effecting change. This allows for more spacious planning that can include rest and reflection. While our inner critics may want us to do the really hard things, we are much more likely to go the distance when we experience some degree of ease. Focusing on our talents and skills increases our overall chances of success. That is not to say that we should not challenge ourselves, but we may want to avoid setting goals that require us to radically change in multiple dimensions at once.

### Help someone personalize their goals

It is an obvious point but one that we all frequently forget when we are trying to change the behaviour of others. A good goal is something that we can affect ourselves, but as coaches we hear goals like this all the time: 'The CEO signs up to X agreement', which can only be a goal for the CEO themselves. Instead, a goal can be 'Effectively influencing the CEO by doing X and Y by Z date'. This means getting clear on the real problem we want to solve (in this case the need to influence). In helping someone set *their* goals try to avoid setting what psychologist Russ Harris jokingly calls 'a dead person's goals';[65] goals that a 'corpse would do better than you'.[66] An example of this is 'drive less' for which a more 'alive' goal would be 'cycle more'. We are often guilty of setting this type of goal because we are more familiar with what we don't want, than what we do. This process requires a simple flipping over of these to find a 'towards' goal underneath an 'away' goal.

### Recognize the rapidly changing context and help someone see the system

The operating environment and public mood about climate change are changing at pace. Accepting, and helping others to accept those shifts as positive signs of disruption (even when they feel like retrograde steps), means cultivating self-compassion that can allow us to return with renewed energy to a new round of planning. Good goal-setting recognizes that our actions do not operate in a vacuum. We need to be sure to map everything we are already doing so as not to fall foul of what Daniel Kahneman and Amos Tversky call the 'planning fallacy' in which we employ an optimism bias to planning and underestimate how long tasks will actually take. This is very important in a context where so much needs to be achieved, and in which people frequently burn out. It is also important that people connect to others in their field as they plan, strengthening their chances of success by working in concert and sharing processes. Mapping the system can help to set goals that are genuinely needed and identify existing resources and collaborators.

## In practice

### A SMART way to set climate goals

*Charlotte Lin, coach*

*Goal-setting is difficult, but climate action goal-setting is tremendously difficult. Systemic climate goals often seem unachievable, while personal climate goals can seem insignificant. As a coach, I have found that when clients describe goals vaguely (as desires or aspirations), they rarely succeed. An example of a vague goal regarding climate change could be 'to go green this year'. However well-intentioned it may be, this goal just is not SMART. The SMART framework is excellent for defining vague goals, in which SMART stands for Specific, Measurable, Achievable, Relevant and Time-sensitive.*

*Jessica, a professor of International Studies, spends her waking moments reading, talking and worrying about the climate crisis. She is also introverted and equates climate leadership with 'being pushy and overbearing'. In a goal-setting session, we looked at how she was struggling to maintain momentum in a climate action group she started with her colleagues. The initiative had seemed so promising, but a few months in it felt like pulling teeth to get anyone to participate. Jessica felt alone, confused, discouraged and even embarrassed.*

*Jessica wanted the group to meet monthly to discuss a climate topic and potential actions, but while the group resonated with climate action, they lacked the guidance, motivation and experience to conduct self-directed actions. The group stagnated due to not having clear actionable plans, while Jessica's own dissonance around their inactivity grew. That clashed directly with Jessica's disempowerment around climate action and leadership. Jessica needed an actionable plan to overcome her triggers in order to help her group become more self-directed, and successful.*

*We agreed to use SMART goal-setting to explore how she could turn her group around. Examples of SMART framework coaching questions that I used included:*

- *Specific: What would be the 'who, what, where, when, why and how', of the meeting?*
- *Measurable: What kind of measurable details could you add to this plan?*

- *Achievable:* How achievable do you feel your plan is right now? Is anything missing?
- *Relevant:* What would be an exciting outcome for you if you achieve this goal?
- *Time-sensitive:* When exactly will you implement this plan? When will you see results?

*The first turning point came immediately after helping Jessica get more specific with her goal, using the* **Specific** *coaching question above. With details like, 'the first meeting could be discussing what people's interests are, and making a list of the ideas and then also making a schedule of who wants to lead future meetings based on their interests', Jessica felt more confident because her plan made more sense than before. Having group members take turns to lead was also significant for Jessica – because they could develop experience and autonomy for self-directed climate actions, instead of relying on Jessica all the time.*

*A second turning point came when she remembered her own power as an experienced educator, capable of designing engaging learning plans suitable for the participants. Jessica said 'because they are my colleagues, I assumed we are at the same level for everything … but it's true that I've been much more immersed in climate'. She realized she was operating at a level that exceeded most group members' level of climate activism – a cause for the group's drop in participation. With this realization, we were able to add the appropriate measurable, achievable, relevant and time-sensitive details to her plan. She eventually decided to propose to the group that every month, members would take turns designing and leading an hour-long interactive workshop or presentation around a climate topic of their interest. She felt much more hopeful and was able to reflect on her negative emotions around her leadership. She said, 'I have a much bigger idea of what I want to achieve in the long run. And that could be a big reason why I am so concerned about my success at the moment.' She also realized that her expectations around success were creating pressure that made her endeavours seem more difficult.*

*What stood out to me was how swiftly the SMART goal-setting process alleviated Jessica's fears. The clarity and simplicity of her resulting plan gave her a renewed sense of confidence to approach her group with more conviction. Jessica subsequently told me that she is no longer triggered by the engagement level of the group, and the group has been productively meeting for months since she implemented her plan. Jessica's situation highlights how, for ambitious climate leaders, the stresses of climate change can bring on self-induced pressure that make things feel harder. When trapped in a storm of negative emotions and inaction, SMART goal-setting can provide a clear way forward.*

## An example conversation – coaching skills to make goals more measurable and manageable

When we set targets for ourselves or with others, we want them to be measurable and manageable, so they stand a chance of success. It is not always easy to define measurement criteria for our goals but coaching skills can help us to tease out ambiguities and ground our targets in reality. Here's an example involving imaginary Monica from earlier in the book with her coach, Adrienne:

Monica:     'My main goal is to set up a successful beach-cleaning group.'
Adrienne:  'Great, how will you know that you've done that?'
Monica:     'We will meet regularly and the beach will be largely clean of litter, most of the time.'
Adrienne:  'Can we make that a bit more measurable? For example, how do you define "regularly", "largely" and "most of the time"?'
Monica:     'Good thought! OK, so "regularly" would mean meeting to do a litter pick every fortnight, and "largely clean most of the time" would mean something like a 70 per cent reduction in litter.'
Adrienne:  'How would you measure that 70 per cent?'
Monica:     'I guess looking at 10 metres squared of beach and see that it's changed.'
Adrienne:  'Great, so you'll be a group that meets fortnightly and sees the difference it's made because a representative bit of beach will have 3 pieces of litter where once it had 10?'
Monica:     'I guess … though that's making me feel pretty exhausted just thinking about it. I'm not sure measuring bits of the beach is why I want to do this.'
Adrienne:  'That's good to realize. What do you dream of creating here?'
Monica:     'I really want to change people's behaviours about littering the beach. I want people to stop and think before they drop litter and take it home, and I also want them to pick up other people's litter out of a sense of ownership for their beach.'
Adrienne:  'What would help you achieve those things?'
Monica:     'There's probably a goal before any of these goals, which is about forming a group and asking them what they think we should do; designing it together, so that ownership is there from the start. I probably need a couple of goals just about the group itself, and another about researching what similar initiatives have done that have worked and failed.'
Adrienne:  'That sounds like a way to make sure that when you come to it, the action you take is worth the effort. Shall we work on those new goals then and drill down to make them measurable too?'

Here Adrienne stayed separate enough from Monica to be able to notice that her early measurement criteria was vague. She did not judge Monica for this but helped her work methodically through the implications of her plans, so that she could quickly see that her initial ideas were aiming in a direction that didn't satisfy her. While most coaching models appear to offer a straight line from challenge or desire at one end to goals and action at the other, in reality people move backwards and forwards, redefining what they want, before landing on final goals. Adrienne helped Monica think more clearly, and ultimately saved her months of mistakes. Monica's next step would be to define some timelines with Adrienne, but if Adrienne were a colleague not her coach, they would also clarify how Monica's goals will impact on the colleague's plans and put more emphasis on accountability.

## Creating accountability that feels comfortable

When goal-setting is done well, accountability naturally flows from that process, because the person responsible for the goals has driven their design, had time to become really comfortable with them and got clear about what they involve. When these things are imposed on people who don't feel able to push back, the goals are more likely to fail. Accountability is an important part of achieving our goals and for this we can turn to the skill of contracting. Contracting can help us to get clear on accountability, so that the person who owns the goals feels safe from blame. It can also help us to get clear not just on what we will do but what will happen if goals or deadlines are missed. This is especially important when we are working in a complex, shifting context, in which our goals can become obsolete. You could ask: 'We both hope that these goals are going to get done as we've designed them, but real life often does not go as planned. If you don't manage to get these done in this particular way or on time, how would you like us to talk about it? What will make that conversation easy?' This creates permission, ahead of time, to have that otherwise difficult conversation.

Some people struggle with the idea of goal-setting because accountability can feel pressurizing. Leaders within organizations often resist climate goals, targets or accountability because they either fear censure or have had bad experiences of being publicly attacked. A solution here is to frame goals as intentions, thinking of it as a journey that is being embarked upon, not a proclamation of perfection, and we cover this in more detail in Chapter 15. We may see pushback on accountability simply because someone hasn't finished thinking about their goals and is not ready to declare them yet. Pushing Monica to set goals about counting litter would have got a measurable but meaningless goal. Taking time here, offering reassurance and probing to ensure that goals are fulfilling and flexible can create a stronger appetite for accountability.

## In practice

### Supporting leaders to set climate goals

*Ellie Austin, sustainability consultant and co-founder of Twelve*

*As a climate change consultant, I work with company founders and senior leaders to help them establish their company's response to the climate and ecological crisis. I mostly work with consumer brands who are making the products we use and love every day in our homes – from furniture to fashion, to food and drinks.*

*For one of our clients, we ran a two-day sprint workshop to develop a climate strategy for their business. The company founder, Sam, was so enthusiastic and open to thinking big to tackle the climate emergency. He pulled in 15 of his team members for the workshop and we had a brilliant two days together. We helped them connect emotionally with the climate emergency, clarified the areas where they can take meaningful action, and came up with some big goals – including a target to reach net zero emissions by 2030. I felt their team's excitement and energy around this headline climate goal. After the workshop they said they could not wait to tell the world about it.*

*And then they went quiet. Really quiet. For weeks it was a struggle to get them to engage with the plans we had created together. We suddenly could not get agreement on the next steps to take or even the next meeting agenda. One day I called the company founder, Sam, to try and understand what was going on. We had a tough conversation – he sounded flat and quiet on the phone – the opposite of his usual buoyant self. He told me he could not commit to the net zero goal.*

*It became clear that Sam had significant doubts and fears around the goal. While he had no problem with setting a net zero goal itself – the problem was telling the world about it. He was scared of being called out for greenwashing and he was envisaging the worst case scenario of a journalist or a competitor asking him exactly how he meant to get net zero, backed up by detailed data from day one. I asked him to be OK with that uncertainty – to be OK with not knowing. He said he thought others would ridicule him for announcing something that he didn't know how to do. I could tell that he was feeling overwhelmed and vulnerable, and that he was carrying too much personal responsibility for 'fixing' the climate crisis.*

*I responded by being vulnerable too. I told him I didn't know exactly how we would get there either. Sam laughed and breathed a sigh of relief – it sounded like some of the pressure he was putting himself under was being lifted. I asked him if he thought much bigger companies, with far greater resources and even bigger climate targets, knew exactly how they would achieve them. He sounded thoughtful as he named a couple of leading sustainable brands that had not published detailed plans yet. He started to reflect that we are facing an urgent crisis, and he reiterated out loud that he must try to do his best as a business leader, even if that meant admitting he did not have the answers.*

*Without realizing, I was using coaching skills to acknowledge his fears and say that I shared them. I could hear in Sam's voice that he sounded stronger and more confident, and that his resolve was returning. We ended our conversation by agreeing to get his team together again to create a more detailed internal action plan, and I am really proud that Sam's company has since published their net zero goal – and been applauded by their customers for it.*

## Our bigger goal – rolling with uncertainty

Whether as coaches or changemakers, one of our most important goals in relation to this crisis can be to model the ability to tolerate uncertainty. This means not getting too tightly attached to the specifics of our short-term goals and always keeping in mind the bigger frame of the paradigm we want to create. We still need to hold people to account, but we also need to show compassion for others when it is hard to imagine a new world or act for it and to challenge self-judgement when we hear it. We all want to be a safe pair of hands for the people we work with or coach, but we can mistakenly assume that in order to do this we somehow need to know the future or be several steps ahead of them. Remember that it is impossible to know everything about what is unfolding before us, just as it is impossible for someone to achieve 100 per cent of their goals. People do not need us to have a crystal ball, they need a partner on their journey into unknown territory, who will believe in them, help them to navigate the uncharted and encourage them to forgive themselves and carry on when their best laid plans do not work.

In this chapter we have shown you how to set behavioural as well as action goals. Goals that also have flexibility baked into them, allow us to handle uncertainty and change, and make us resilient through our failures. When we model the ability to accept that we do not know the future, we show that we trust our inner resourcefulness to adapt us to new, challenging circumstances. It can be soothing to notice how the natural world adapts when all around is changing. Notice the weeds that against all the odds push through the pavement. We can set powerful goals and intentions, provided we also create compassion that allows those goals to change as the world does. We are not in total control of the environment in which we set goals. They are a destination on a map but the journey towards them is uncharted and exciting. Who knows what bird will pick up our seeds and where they will spread them?

# 13 Rage against the machine

## Turning anger and blame into forward energy

There are many reasons to feel angry about climate change. We get angry on behalf of our children, for those who will (or are) feeling the worst effects but have done least to cause them; we can even direct anger towards our previously unaware selves for our own complicity. For some this anger lives on the outside, like an active volcano, for others it simmers below the surface rarely bursting forth. Harnessed well, anger can be a form of energy, spurring us into action. If left unacknowledged and unaddressed, anger can manifest in a range of unskilful behaviours – such as aggression, criticism, blame and judgement – that eat us up and damage the relationships that support us or which we want to influence. What matters is not whether we are angry, but what we do with that anger.

In this chapter we will look at three ways of being with anger:

1. Avoiding unskilful behaviour – understanding anger so we can respond better.
2. Helping others to be with their anger – so that they can reclaim their agency.
3. Using anger as a resource that can be channelled into motivation and action (considering the gifts in anger and how we can put this angry energy to work).

We will give you the tools to work through anger and cover the core coaching skill of challenging, which allows us to call out poor behaviour while treating someone with care and respect. We will look at how climate anger shows up in relationships and how to handle it skilfully.

## Step one: understanding anger better

### Anger as a fact of life, not a category of individual

Before we begin, it is important to recognize that anger visits all of us. As Alexsandr Solzhenitsyn wrote:

> If only there were evil people … committing evil deeds and it was only necessary to separate them from the rest of us and destroy them. But the line dividing good and evil cuts through the heart of every human … who … is willing to destroy a piece of their own heart?[67]

It is very easy to feel threatened by someone's anger, and in response label them as bad, wrong, and an 'angry person'. The problem with labels like this is that they stick, becoming an identity that we can find hard to escape. Some labels do not affect us too much. 'Izzy is passionate' may not damage her relationships, but where the depiction is more charged, such as 'Tom is an angry man', the effects can be much more damaging. As climate change coaches, we like to imagine people as being like Dorothy in *The Wizard of Oz* in the aftermath of a tornado, with her home shattered around her. Given that terrifying experience, we could forgive Dorothy for acting out or having a tantrum. We will not collude with her anger or allow her to hurt others, but we can separate out Dorothy from her circumstances and see that she is not an aggressive little brat but a frightened child in need of love and help. Because when the storm has subsided, Dorothy will still be a little girl, she will not be a tornado. So too, the most angry of us is still deserving of love and can be separated from the circumstances that make us act out.

### Externalizing anger unskilfully – misplaced anger and its destructive effect on relationships

While anger can be harnessed to give us momentum, when untamed it can negatively impact us and our relationships. In the climate space, we see anger frequently manifesting externally as blame, of either the system or of people or organizations within it. This kind of systems-level anger can still trigger another person to respond angrily too, if the anger is directed at something they hold dear or with which they identify. Someone may be attacking the farming community, the urban wealthy, meat eaters or frequent flyers, and we might in turn feel under attack because their anger seems to give them the power to destroy something we care about. This might involve us being defensive, or blaming right back. Sherman and Cohen[68] write that when we feel that our identity is threatened, all of us try to demonstrate that we are essentially good people (the good guys). When someone directs their anger at things that we consider to be a part of our identity (for example criticizing flying when we see ourselves as globetrotters), this threatens our sense of being on the right side of the argument.

Similarly, when we blame others for climate change, it is sometimes a displacement for the self-blame we feel because our own positive self-perception is threatened by the idea that as humans we have damaged our planet. When we direct our blame at the system, we are often expressing our guilt. Many of us have not come to terms with the fact that we have been, if not driving then at least along for the ride, during the destruction of our world. The guilt that this knowledge engenders in us is easily converted into anger at those who were driving. In the short term, that saves us from having to work through guilt but not forever. We have coached people who have answered the question: 'who are you really angry with?' with a quieter answer: 'myself'. This is another reason for self-compassion and compassion for the system. Anger used badly can be polarizing and ruin not just our relationships but the action that is possible within them. Just as easily as environmental groups can form, they can splinter and disintegrate when anger, contempt and blame create toxicity. This can frustrate those who joined looking for a sense of hope and agency but who leave meetings feeling *more* hopeless. We find similar things happening in green teams within organizations, when too much time is given to venting anger unskilfully rather than harnessing that energy.

Sometimes our anger can get displaced onto those around us. We may be feeling stressed about a bad environmental news story when a loved one tells us that we are out of milk. Or we may be feeling thwarted about our bigger impact, when a colleague tells us that a project is going wrong. We can fly off the handle at the person in front of us, when really we are angry with ourselves, the system or simply the global situation. Whatever the trigger, when we get angry at someone else, we often combine the physical signals (such as shouting or slamming doors) with verbal criticism because we want to make our anger their fault. Anger that manifests as blame, criticism and judgement is often a sign of our insecurity or sense of powerlessness, and these tactics are an unskilful way to distance ourselves from the things that make us feel that way. Similarly, that insecurity can also show up as sarcasm, aggressive defensiveness or angry, silent brooding. Consider how many of us feel angry with politicians for not acting decisively on climate policy. Contemplating this reminds us that they hold some power over our futures, which can make us feel trapped and powerless. It can feel soothing (at least in the short term) to externalize these feelings and rage and blame the 'self-serving political class' instead of being with our powerlessness.

It is also the case that we can sometimes experience systemic anger directed at us for being the person who holds the space for that systemic conversation. You might hear:

To a consultant: 'All this strategizing won't stop the glaciers melting! You're wasting my time!'

To a coach: 'Our navel-gazing isn't doing anything about those bush fires blazing!'

To a manager: 'Go ahead, restructure my department! It's just rearranging the deckchairs on the *Titanic*, we are still going down!'

To a partner: 'I cannot believe you thought it was a good idea to buy a petrol car because it was cheaper! You don't care about anyone but your wallet!'

---

### When aggression is unsafe: leave

If someone is hostile or outright aggressive towards you, you are in no way obliged to tolerate it. While here we show you ways to work with anger, we are not suggesting that you have to, and especially not if you feel in any danger. While it is sometimes the case that we say something that inadvertently triggers something in another person, this does not make their anger our fault and not our responsibility to resolve. It is easy to find yourself in the firing line when someone feels defeated and helpless and does not know what to do with those feelings. In most cases, you will be perfectly safe, and their anger is only an expression of frustration, not a threat to your well-being. If, however, you do feel physically or emotionally threatened: leave. Do not feel a pressure to stay and support someone who could hurt you. Instead report the incident so that more appropriate, trained professionals can help. This in itself can be a powerful signal to someone who has become used to using their anger unchecked. Just because other people have tolerated it, does not mean that it is tolerable, or that your feelings can be ignored. Everything is systemic, and when one person speaks out, almost always others follow.

---

Getting angry can make us feel simultaneously powerful (because we feel armoured up) and out of control as we 'let go' of our usual mechanisms for agreeableness. What it does not do is make us feel better because we have not addressed the underlying powerlessness that fuelled it and we also often feel guilty once our anger has subsided. Blame, and its simmering cousin contempt, are destructive because they impact the whole relationship, not just the person on their receiving end. Anger can lead to conflict because, as Gottman calls them, these behaviours are like 'the four horsemen of the relationship apocalypse'; and they never ride alone. For Gottman these are criticism (or blame), contempt, defensiveness and stonewalling (the state of freeze in which we retreat into ourselves and block out the outside world). Gottman's research shows that their power only comes when one horseman is met with another, creating an ongoing cycle of escalating conflict as the horsemen feed off each other. In other words, misusing our anger and projecting it out instead of owning it, can trigger an equally unskilful response from those around us.

If we go back to our fictitious couple, Petra and Rahim, this could look like:

Petra:  'I can't believe you're abandoning veganism. You've barely given it a chance! This is typical of you [criticism], I should have known you'd quit because [whining voice to show contempt] "I miss meat". You *never* stick with anything! [criticism] Meanwhile, they're cutting down the Amazon for your precious burgers!' [blame]

Rahim:  'I gave it a go didn't I? It's been impossible to find vegan options near my office [defensive]. It's not like I had a choice anyway [contempt]. As usual you did all the deciding for us!' [blame]

Can you see how Petra's anger sparked similarly unhelpful behaviours in Rahim? They could go on like that all night, and they will until one person is able to articulate their feelings and own them, as opposed to projecting them onto the other. It could instead have gone like this:

Petra:  'I can't believe you're abandoning veganism. You've barely given it a chance! This is typical of you, I should have known you'd quit because "I miss meat". You never stick with anything. Meanwhile they're cutting down the Amazon for your precious burgers!'

Rahim:  [taking responsibility] 'I'm sorry [a repair bid]. I know that you really wanted me to join you in this one. I can see how upset you are and I'm sorry that I made you feel that way. I tried to give it a go, but I just didn't enjoy it. I should have been more honest with you [being vulnerable]. I still think eating more responsibly is important. Can we work out how to do that in a way that works for us both?' [a repair bid]

Petra:  [sighing] 'OK … it's OK. I know that you weren't doing it to wind me up. I just really thought this could be something that we could do together [a dream]. But if you really hate it, it's OK.'

Rahim:  'Maybe we could find another thing we could do together to eat more sustainably? Or maybe it's just that you'll be doing it as a vegan and I'll be doing it as, I don't know, someone who only eats meat that's local or organic.'

Petra:  'That sounds like the start of an idea. I like that.'

You may be reading this and thinking 'Why didn't Rahim stand his ground? He didn't have to become a vegan!' and you would be right, he did not. He could easily have kept fighting to make Petra accept that point. But by accepting that he has hurt Petra, Rahim is making what John Gottman calls a 'repair bid', and he is telling her that he values his relationship with her more than he cares about being right. Instead of getting defensive, Rahim accepted the blame Petra threw at him and calmly put it down rather than hurling it back. Without something to volley back, Petra could not then stay angry and her rage dissipated. Once calm they were able to find common ground again. Petra was able to express her dream of them doing something about their diets *together*, which allowed Rahim to find a way to reposition the way they did that.

OVER TO YOU: What is your favourite 'horseman of the relationship apocalypse'? What can you do the next time you feel the urge to employ it?

### When not responding makes it worse

When someone lets their anger loose, on us or even directed at systems, it can be deeply uncomfortable for those who witness it. For many that is because getting angry is not socially acceptable, we don't know how to respond to such huge emotions and our response is to go quiet hoping the moment will pass and we can go back to 'normal' social interactions. Often passionate people feel shut down because their exuberance is seen as stepping too far out of line with a socially acceptable behaviour that is more measured. While in the example Rahim joined in with Petra's anger, he could just as easily have gone silent and shut Petra out – and that would be Gottman's fourth Horseman, stonewalling. Rather than diffusing the situation it would likely have triggered Petra just as much as criticism, contempt or defensiveness because it is still a refusal to engage, and would leave Petra feeling isolated and escalate her rage. This is not a stonewaller's intention of course, but rather a learned response to dealing with conflict.

For some of us, the force of someone's anger can be terrifying, even when it is not aimed directly at us at all. If we have a negative relationship with anger more broadly – for example, we see anger as destructive and a form of harm – we may find it hard to engage in conversation with someone who is angry, and may shut down, hoping that they will calm down if we don't add any fuel to the fire. In fact, stonewalling is just as much a lighter fuel as blame, and the angry person is likely to continue because our silence makes them feel alone and exposed in their pain. They may then reach for criticism and accuse us of not caring enough about the issue, not understanding that it is because anger is a no-go area for us. Instead we can say, 'I can hear how angry you are.' We may also explain 'I find it hard to hear anger, and though you don't mean it to, it upsets me' to contract how we need to interact in order to feel safe, and not make someone wrong for expressing frustration.

OVER TO YOU: What is your existing relationship to anger and what would you like it to be?

## Step two: helping others understand their anger

Anger need not damage our relationships. It is an emotion that we can harness and use its underlying energy to fuel us. To do that, we need to be able to sit with anger, examine it and understand how we might be triggered into using it against the wrong targets. Here are some methods that we know help people, listed from basic to advanced. Test the ones that feel most natural to you, and over time experiment with newer, more challenging techniques.

### Basic tools

- Don't make the person wrong for feeling angry, avoid statements that might sound judgemental or defensive such as 'OK, OK, calm down!' or 'Wow, someone's angry today!' Instead name the anger neutrally 'I hear how angry that makes you.'
- Help them to understand and process the anger by allowing them to explain what makes them angry. Resist jumping in with your own thoughts, but intermittently ask 'What are you thinking/feeling now?'
- Help them to stand back and observe the emotion rather than being caught in it. Suggest that they use this sentence 'I *notice* I'm feeling angry' instead of 'I'm feeling angry'. This technique, which comes from Acceptance and Commitment Therapy, creates the cognitive space to think clearly by separating the thought from our response to the thought, and with it the opportunity to respond differently.
- Avoid meeting anger with another 'horseman'. Notice if you are triggered into reaching for blame, contempt, defensiveness or stonewalling. Stay calm and slightly separate from the other person. Observe them (and your own emotions) and share your observations. Remind yourself that you are perfectly safe (unless you are physically threatened in which case you should leave).
- Normalize the experience of anger but not any unskilful behaviour. 'It is normal to feel frustrated.' Rather than 'It's normal to shout at people for taking an aeroplane.'
- Avoid colluding with the topic: 'You're right all politicians *are* useless' but instead acknowledge the legitimacy of the emotion: 'I can understand how frustrating this is.'
- Vent the anger by truly expressing it, and don't allow them to dismiss it. 'Honestly, it's not that big a deal.' 'It sounds like it is a big deal and that's OK. If you could say exactly how you feel to me now, knowing there would be no consequences, what would you say?'

### Intermediate tools

- Help them to be honest about their anger and congruent with it. Many people explain anger away, while we see it on their faces, bodies or the way they gesticulate. Naming that incongruence can make it safe for people to connect to their anger. 'You said you are not frustrated but I see your fist is clenched. If your fist could talk, what would it say?'
- Name the feeling and help them to notice the energy that comes with it, so that they can connect more viscerally to the experience of anger. 'You sound really thwarted. What does that feel like physically?' 'Like a knot in my stomach.' 'Can you focus on that knot for a moment, to see what it tells you?'

– Seek resonance in the dissonance by asking: 'What's useful about having this much energy? What could you do with it?' 'What would you like to do to put this anger to work?'
– Work with the values that are stepped on to regain a sense of agency and return to empowerment (see the conversation that follows for more on this).

### Advanced tools

– Scream and rage wordlessly/in silence by thrashing around. This is very effective where the other person feels they may be overheard if they shout out loud. It helps for both of you to do this together so that the person doesn't feel exposed.
– Get out of the head and tap into the intelligence of the body. This can be done simply by asking the person to sense the location of the anger in their body and pay attention to it. Ask: 'Where do you notice this feeling inside you? What's it like there?' 'What does this feeling want to tell you right now?'

## *In practice*

### *Coaching anger*

*Megan Fraser, coach and founding member of the Climate Change Coaches*

*Rachel's shoulders were tense and her brow deeply furrowed in our session today.*

> *I live near a lake that for 50 years has been abused. It's been owned by successive people who – for want of a better word – have raped it, to develop property. I've been fighting this for years ... and now we're taking the owner to court.*

*Her eyes were anguished and fierce. 'A group of men has just stepped in to undermine our plans to fight it. I'm so angry. I think it's the arrogance.'*
*I nodded, pausing to let her words sink in. Then I said, gently: 'I'm struck by the language you used here. Abused, raped. I'm curious ... what do these words bring up for you in this context?' She sighed heavily. 'I've always fought for the underdog. It's just who I am. And these words ... well, maybe they bring up "undefended".' From our previous coaching conversation, I knew that Rachel resonated hugely with the Celtic warrior queen, Boudica. As our session continued, we explored what role Boudica might be playing in Rachel's life right now.*
*Rachel's eyes lit up. 'She's definitely alive and kicking.' 'Where does Boudica live in your body?' I asked. Rachel closed her eyes and sat back. After a moment, she clasped both hands together on her breastbone. 'In my heart.' She paused, her hands remaining there. I placed my hands on my own heart, mirroring her body language. 'What do you notice in your heart right now as we talk about this raped, undefended lake?' I asked. 'That I will continue to give it my best efforts to try and protect it,' she said. 'And how does that feel?' I asked. 'It feels really sad for me. And also ... even though I'm working alongside men to protect this lake,*

*it's also men who are doing what they're doing. It's the arrogance – I'm sorry, but that's what gets me so angry. It's the male energy that just ... arrghh!' she exclaimed, as if she were about to explode. 'These men ... there's no connection from here down' she gestured, violently cutting a hand across her throat. 'It just cuts off beneath the head.' Then she moved her whole body violently, from side to side, as if to shake something off.*

*I mirrored the shaking. 'What's this?' I asked. 'It makes me so angry,' she cried. 'So angry! But I know it's not really that helpful to stay here. I know it's a force I can do something with.' I paused, aware of the rawness and depth of this emotion. 'I'm cautious not to step too soon into asking how we might channel this anger,' I said, gently. 'I sense that there's a lot here, wanting to make itself heard. How does that land with you?' Rachel nodded. 'I wonder ... if we were to really turn the volume up on this anger, what would it be saying?' Rachel ran her fingers through her hair. 'Why, after the lake was first purchased privately all those decades ago, has nobody stepped in before?' she burst out. 'And what have these men been thinking? It's like: come on, guys! Wake up and smell the coffee here!' She laughed again, sounding incredulous.*

*I noticed something incongruent so I asked, 'Can I share a bit about what I'm noticing?' Rachel nodded. 'I'm observing that there's a lot of tension and anger in your body – like you want to really shake these guys! And at the same time, I notice that you smile as you talk about it. I'm wondering what's going on there.' 'Hmm.' Rachel reflected. 'I've always worked in male-dominated environments, and have led with my masculine energy. I never realized how I saw the lake in quite feminine terms ... I suppose my language gave that away. I think this is about valuing the feminine in me, too.' As our session came to an end, Rachel's eyes were still fiercely determined, but she was sitting straighter. 'I feel more excited, determined to continue doing what I do, from a grounded place.' As we closed, I acknowledged the strength and warrior spirit that I saw in her and that so inspired her about Boudica.*

### Other coaching behaviours in the context of anger – noticing and challenging

In addition to the tools we described here, there are two key coaching behaviours that can help us to accompany people when they are angry: noticing and challenging. Noticing can help us to handle our own triggers (and make sure that we do not reach for one of our horsemen too), and we can consciously stand apart from the emotional field created between us and the other person. Stepping back we can observe, not just their words but the body language that accompanies them. This is the attuned listening from Chapter 5. Taking a few deep breaths to slow yourself down can remind you to step into observer mode, rather than getting caught in the emotion. Challenging is a skill that we can use anywhere, but it is particularly useful when someone is angry or wanting to enrol us in their disempowerment.

To challenge, first you need to dispel any myths you have about assertiveness being aggression and that challenge will result in conflict or breakdown. In challenging, you are offering insight without the clutter of a lengthy apology, holding true to the belief that the other person is 'good and right' (Dorothy, not the tornado). To soften the challenge, you can begin with: 'I don't think this blaming and ranting is true of who you really are, and so I'd like to challenge you on it. Is that OK?' Second, challenging requires us to develop strong relationships because we are more likely to accept challenges when we know that the other person respects and cares about us. So while you may want to challenge someone's powerlessness without any preamble, you will likely get pushback if you do. For example, perhaps a colleague who always has our back has taken us aside and told us that we were in the wrong and need to apologize to someone. We will accept a challenge from them precisely because we have a good relationship and because they speak to us with respect, as opposed to criticism or blame.

The skill of challenging can then help us (on the periphery of a conflict) to deescalate it, by helping the protagonist to understand their own behaviour. Typically, a coach asks permission to challenge and then offers an insight that may be confronting, but that can illuminate a blindspot or point to something incongruent of which the person themselves may be unaware. You could explain by saying:

> I think that there's a mismatch here between what you are saying and how you are feeling, and I wonder if bringing those two things together might help you understand it better. To do that, I need to tell you what I'm observing in you as you are speaking. Is that OK?

The important thing is not to proceed if someone says 'no'. It is crucial that both sides agree to it. A coach may share an observation or an intuition that they feel reveal a truer set of feelings. For example: 'You were saying how you have given this person the benefit of the doubt, but as you did so I thought I heard contempt in your voice. Was that there?'

### How to avoid colluding with unskilful anger

It can be tempting to join in when someone is ranting, in a show of solidarity or sympathy, and easy to get carried away. When we collude with someone else's fears it only deepens and perpetuates everyone's sense of powerlessness. 'Setting the world to rights' feels empowering in the moment, but the world is not listening. Instead we have spent time reminding ourselves that the world is a mess and convincing each other that there's nothing we can do to change it. When we have a partner-in-blame, our emotions become amplified and we end up in a holding pattern of behaviour where neither side explores their powerful anger underneath. We can part company deflated, our disempowerment deeper and our hearts heavier because voicing your anger in this way has led to no new awareness. We can still bond with the other person, but we need to help them find a sense of power from self-knowledge: 'I know, I read that [bad environmental news] too and felt so disappointed. How did it feel to you?' or even 'I want the people who govern us to stand by their word. What's most important to you about it?' In everyday conversations you may feel more comfortable sharing your own view first, while a coach might say: 'I can resonate with that too. What's it like to feel that way?'

Often what we say does not truly reflect how we feel, but our anger hijacks our speech and says: 'This government is so hopeless! How can they allow a new airport at a time like this!' When really we feel:

I feel so powerless when I see headlines that show things moving in the wrong direction. I'm so scared. I feel guilty for not doing more, and impotent because the people who are meant to protect us seem to be doing the opposite.

Notice the shift from one to the other. There is still frustration in the second example, but far lower amounts of blame and judgement. Being able to voice what is underneath our criticism can bring us into congruence with how we really feel, and then direct us towards action that feels meaningful. Instead of refusing to be with these feelings, we can allow ourselves to notice them and accept them and help others to do the same. This doesn't require us to become therapists but just to be caring, patient humans.

Let us put this idea to work in an imaginary conversation with our fictitious characters Tim (in charge of sustainability at his organization) and Nisha (our sustainability consultant). Tim has hired Nisha and considers her a kindred spirit. At the end of a long day, Tim is feeling pretty alone in his endeavours; he finds himself letting off steam about the news with Nisha. You can see in the following dialogues how Nisha's different response affects Tim's sense of agency:

Tim:     'Did you see the news? This government is so totally hopeless! How can they allow a new airport at a time like this!' [Nisha now has a choice of how she can respond. She can collude with Tim or she can open a door for his feelings to surface. She chooses the latter.]

Nisha:   [acknowledging and sharing]: 'I know, it feels like such a kick in the teeth doesn't it? I honestly feel so sad about it.'

Tim:     [beginning to calm] 'Yes. I'm really worried about what this means for our carbon targets, and also for the example it sets to other countries. How are we meant to tell a country like India what to do if we don't do it ourselves?'

Nisha:   'I'm worried too. [gently opening up] What worries you the most, the targets or the example we set for others?'

Tim:     'Well they're the same thing I guess. If we don't set an example then none of us meet our targets. It's about reducing carbon at the end of the day. We do what we can do at home but what about everyone else?' [Note here Tim's lack of belief in the system.]

Nisha:   [encouraging opening up] 'Yeah, I really get it. I often feel like that. How do you feel? Is it OK to ask that?'

Tim:     [sitting with his emotions more] 'Yes, it's fine. I don't know that I feel lost, more angry and frustrated. And I suppose also impotent. Like there's nothing I can do about it.'

Nisha:   'It's not easy to feel like that, huh?' [empathizing]

Tim:     [settling] 'No, it's not. I'm like a bear with a sore head a lot of the time. I hate not being able to have answers, or solve things myself.'

Nisha:   [opening up further] 'It's hard feeling out of control.'

Tim:     [accepting more] 'Yes it is.'

OVER TO YOU: What are you learning about how you collude with someone's else's anger?

## In practice

### How can anger help in the fight against climate breakdown?

*Scott Johnson, founder of Kung Fu Accounting (the UK's first B-Corporation account-ing company), Kung Fu teacher and founder of the Climate-Conscious Accountants Network*

*I'm terrified for the future we are leaving our children. I'm scared of the global impact of overconsumption, in terms of resource scarcity, and worried about the risk of wars over things like safe, clean water. I am furious that our elected officials act like nothing's wrong, when our planet is melting, burning and flooding. And I absolutely feel powerless to make any impact on the decision-making of our so-called leaders. But too much fear is overwhelming and debilitating, so I get angry.*

*When I teach our self-defence courses, we tell students that it is OK to be afraid, but it is not OK to be paralysed with fear because that will not help you defend yourself or protect your loved ones. Anger does something strange to our physiol-ogy – we tense up; jaw clamps shut, chest puffs out, forearms tighten and fists clench. It also changes our breathing. And that is millions of years of evolution helping the body prepare for what is about to happen. So we teach our students to harness an inner rage, using anger as a tool to face up to the opponent because just as anger helps control our fear, there is another state beyond anger, that allows you to control and then apply that anger at the appropriate time: calmness. To know when to use anger as a tool, you need to be able to evaluate the situation and choose the right moment to respond. You cannot do that from a starting position of anger. In the martial arts there's a progression; from fear, to anger, to calmness, to action. Using each new attribute to control and enhance the preceding one. When it comes to climate breakdown, my experience has been of a similar progression.*

*I raged daily against the destructive, extractive, divisive capitalist system that has been imposed upon us, which we all, to a greater or lesser extent, have to abide by.*

*But that, right there, was my little glint of hope … and I didn't see it until I stopped being angry at the system. With anger replaced by calmness I began to think logically, look at the things I could control, and then start organizing and sharing my thoughts, feelings, and findings. I realized that we all have a huge amount of control and influence, but we have been marketed to our whole lives to become compliant, unthinking consumers, so it is easy to miss it. But here it is: No one can tell me where to spend my money. No one forces me to buy from particular companies. No one tells me I must own a car, or eat meat, or have a brand new laptop every two years. They are all choices I can make, every day. And simply knowing that gives us all power.*

*As a result, at Kung Fu Accounting we changed our purchasing policies to specifically exclude buying anything new and prioritize local suppliers. As a result we are saving money and supporting our neighbours and the local economy. We found these changes quite easy to implement, but initially it took time and effort to do the research and uncover the alternatives. Once we'd started it occurred to me that we could help others to change their businesses too. And then, the penny dropped that every company in the world has a finance professional involved somewhere in their business. Which means that there is a unique opportunity to change the world, through their accountants. So I set up the Climate Conscious Accountants' Network, to do exactly that. Building a network of like-minded people in a very traditional industry has not been easy, and when my passion and anger spills over I have seen people back away a little from our aims. But when I calmly and logically highlight things we can all do to make an impact, I have had more success in getting people on board.*

*Applying anger can be like a surgical tool when used in exactly the right place at the right time and can enable us to demand change. But anger can often trigger a fear response, whereas calmness is not threatening. So by being calm, analytical, and logical I can help people see where their money goes, from what companies we use, to who we bank with, to where our pension is invested. And then make impactful changes. In a world where 'money talks', I want to help people to use theirs to send a message that extractivism and environmental destruction is not acceptable. We all have that power, in our wallets and bank accounts, when we harness our anger to act calmly.*

## Harnessing anger to create awareness and momentum

Although anger in response to a threat is common in many societies, we have not developed a culture of processing it in a healthy way. If we can understand and accept our emotional triggers and their impact on us we can dramatically improve the way we respond. Once we have helped someone to unhook from their anger, we can help them to reconnect to their sense of capability and meaning by digging into the values or sense of purpose beneath those feelings. Either we can ask them a values-based question like 'what's important to you about that?' or we can flip over the picture that they have painted to create the positive from the negative and show us the yearning beneath the anger. If Tim and Nisha were to continue their conversation, she might help Tim to regain his sense of capability, by tapping into his values. It could look like this:

Tim:     'I'm like a bear with a sore head a lot of the time. I hate not being able to have answers, or solve things myself.'

Nisha:  'It's hard feeling out of control.'

Tim:     'Yes it is.'

Nisha:  'What is important to you about solving things?'

Tim:     'Feeling like we're getting things done.'

Nisha:  'Like you're having traction or progress?'

Tim:     'Yes! Progress! Like all of these things we're doing are making a difference. Are moving us forward, not back.'

Nisha:  'Have you seen any examples recently that show progress is happening?'

Tim: 'Yes, when I read Positive News I see lots of things. It's the mainstream media that makes me feel hopeless. Maybe I should stop looking at the news, I just can't help myself. I feel like I'm ducking my responsibility if I stop following it.'

Nisha: [challenging unsupportive behaviour and tapping into the value of responsibility] 'OK, but what else could you do that feels responsible?'

Tim: 'I guess I could share the stories of what is being done, maybe also write to my MP, I haven't done that in a while.'

Nisha: 'Do you know who your local MP is?'

Tim: 'Yes, and the stupid thing is, they're even a cabinet minister. I should write a letter.'

Nisha: [contracting a challenge] 'I've got a bolder suggestion. Are you up for it?'

Tim: 'Go on, why not!'

Nisha: 'What about booking a slot in their constituency surgery and seeing them face to face? You have a lot of passion for this, you've got kids, and you're also doing the hard work in this company. You're a really credible voice on this one.'

Tim: 'I'd not thought of that, and I'd not seen it that way. Yeah, OK, I will!'

Nisha: [contracting accountability] 'Great, do you think you can do it in the next month?'

Tim: 'I'll try and do it next week! OK, maybe in the next fortnight.'

Nisha: 'Great! And you don't need it, but if you want me to help you prepare for the conversation I'll happily pretend to be the cabinet minister for you.'

Tim: 'Thanks! I might actually take you up on that.'

Notice how Nisha was able to help Tim connect to his values by asking him what mattered to him about solving things? Once Tim was connected to 'progress' he was able to translate his anger into the energy of 'getting things done', and not be triggered by the idea of things going backwards. Nisha then used the skill of challenging to get Tim to question the inner critic that told him it was his duty to read news articles that made him angry. She used one of Tim's values to do that and asked him to find another way to bring the value to life. Nisha didn't tell Tim what to do, but built on his idea. If she had started the conversation with the MP idea, it would have come too soon in the process for Tim. Instead Nisha focused on addressing Tim's disempowerment before suggesting the MP. In doing so, she rebuilt Tim's sense of agency and systemic belief. Tim would not approach an MP if he thought he did not have anything useful to say (individual agency), or if he thought the MP would dismiss the concerns of a citizen (systemic belief). You may think this conversation is too perfect to be real, but we have had conversations with total strangers on stage in front of hundreds of people that are just as seamless. When you help someone to feel normal for being angry, to understand why they are angry, to step into their values and then to design steps forward that feel workable to them, their anger can transform into intention and action.

If you arrive at this book seething with rage, don't make yourself feel bad for that. This is a situation that merits strong emotions. We can, however, cultivate a healthier relationship with anger, rather than pushing it away into judgement and blame, which rarely helps us to transform mindsets or enrol others. Rather, judgement and blame repel others from being in a relationship with us. Imagine instead a world in which everyone was able to skilfully handle their emotions. A world in which anger existed and was used as a force for change, not harm. It would be a place in which we were able to accept how we feel when we experience discomfort, own our frustrations and not take them out on others. We could be at peace with the part that we have played in the climate crisis and have compassion for the fact that others have not joined us yet. That would be a world where true collaboration and climate action is possible.

# 14 Making space for climate grief

Most of us would say that we are good at supporting others when they are sad, but when it comes to the deep emotion of grief, the stakes seem much higher. It can be a familiar yet unsettling landscape – difficult to talk about and uncomfortable listening to others describe theirs. Because of this, we subconsciously shy away from the subject, sticking to 'safe' conversations in which no one shows vulnerability or incapability. We are 'fine, really' and on top of things. If you find yourself landing into this chapter wanting to skip past it or wondering whether this subject is really relevant to you and the conversations you need to have, then stay awhile. While every human heart has the capacity for grief, many of us feel too ashamed, reticent or unsafe to share it. You may discover that along with the rest of us, you and the people you know have also been pushing grief out of sight.

When we find it hard to be with someone else's grief, it is often because we care and want to take the pain away from them, while knowing that we cannot. We often slip into giving advice, not because solutions are actually needed but because we want to rescue the other person. Our desire to be helpful makes us look for the upside to balance the dark picture they are sharing, as if we cannot trust raw emotions to hang in the air. In the process we unwittingly deny them the opportunity to truly express their pain; perhaps the one thing that will alleviate it. Alternatively, unhelpful learned behaviours that mean we are not able to be with our own difficult emotions could mean that listening to others' grief triggers our own experience of loss, that may be painful and present even after many years. If we believe that grief is damaging to us, or to our relationships (perhaps we were taught to 'put on a brave face' or 'stop wallowing') then we will back away from it, wherever and with whoever it shows up.

We cannot paper over the cracks that grief causes, however. It finds a way no matter what. As we engage with the climate crisis we may experience a whole range of emotions from self-blame, to scarcity, overwhelm, anger and aggression. In our experience it is grief that sits below almost all of them, which is why, having travelled through all of those feelings, we come to grief last. Emily Dickinson's poem on this subject begins 'I measure every Grief I meet / With narrow, probing, eyes – / I wonder if It weighs like Mine – / Or has an Easier size.'[69] To help others with their feelings of loss, we need to have a well-developed relationship with it ourselves. We cover this in more detail in Chapter 21. Here we point you simply to reflect on grief, and to our changing planet in particular.

OVER TO YOU: What is your reaction to other people sharing their feelings of grief with you?

## A different form of grief

As a relatively recently established concept, climate grief does not have the same social acceptance as personal grief – whether human loss, or even that of a job or a home. In fact, it is something many do not even recognize … yet. We have coached people whose friends have flatly refused to accept that it exists and have belittled them for expressing such pain. This may be in part because (as we said in Part A) we have become so separate from the natural world that we do not grieve the burning down of a rainforest as we would our own house. While we might grieve the loss of identity that comes with being made redundant or retiring, we do not see an equivalence with grieving the loss of identity as 'someone responsible' when we realize the extent of the damage that has happened on our watch. This can leave people who are experiencing climate grief feeling extremely isolated and unable to talk about it.

People who have done a lot of sitting with these feelings, however, can tell you that they feel these losses just as, if not more, keenly than the loss of a family member. This is personal to them, in ways that others may struggle to understand. Someone with climate grief feels a depth of connection that goes beyond their lived experience. It is possible that someone in a tower block can deeply grieve the melting of the glaciers and someone else can shed tears for the Amazon while standing in their local forest. To the outside observer this is histrionics or a bit 'over the top', but for the person experiencing it, it is potent and real. Climate grief is slow and cumulative grief without an end in sight. Unlike grief for the death of a loved one, which will never leave us, but which may become muted overtime so we can live alongside it, grief for the climate can be resharpened continuously by new events and catastrophes, as we lose more of the things that we value and love.

Climate grief can take different forms:

- Present moment grief: for the losses happening right now (perhaps on an abstract scale or distant to us but nevertheless current), e.g. wildfires in Europe.
- Personalized grief, as if we have lost a part of ourselves through the loss or destruction of the natural world or a landscape with which we so closely identify, e.g. the cutting down of a tree under which we used to hang out as kids.
- Anticipatory grief for the lost or blighted future more broadly or for the things within it specifically that will be lost, such as distinct animals or habitats. This is a similar grief to the type experienced when a loved one is given a terminal diagnosis. Even while they are still with us we feel an intense pain for the shortness of time remaining.

## In practice

### Creating ways to voice feelings of climate anxiety and grief

*Antonia Godber, climate activist and doula*

*When I first woke up to how bad things are, I felt like I had been plunged into freezing water. It was like I was sitting on a bus with everyone I love, having just been told that the bus was headed for the edge of a cliff. I sat on that bus, and watched all my loved ones carrying on as normal in their seats. Surrounded by people yet desperately alone, I felt an immense and debilitating fear, and an acute rage. It was like I had fallen down inside a deep, dark pit. My need to talk about it was utterly overwhelming. I was lucky enough to be able to access a coach,[70] and she helped me to see that if I faced, and allowed the waves of grief to pass through me, they would do just that. She didn't think my concerns were silly. She didn't try to quieten me. She acknowledged my pain completely, and let me just cry. She was the only person who clambered down into my deep pit, and sat down next to me in the darkness. She helped me to see that my grief was a very normal – and healthy – reaction to the situation. And so not only did she make me feel less alone, she made me feel human again.*

*Coaching helped me to understand that I was reacting in exactly the way a person should react when faced with a global emergency. I was able to sit in, feel, process and then eventually be grateful for the pain I was feeling. I had been feeling like the odd one out – my friends and family were worried that I was having some kind of mental breakdown. I felt pathetic for caring as much as I did. Indulgent, even. But my coach took all of that self-doubt away, and helped me see that my feelings were valid – and potentially very useful. Coaching helped me realize that far from being something I had to simply endure, like an injury, my climate grief could become my power. I began to see that I could help other people.*

*People began contacting me out of the blue, asking for advice with their own feelings of grief for the earth. Because of the pain I had been able to process through coaching, I was able to reassure them that I had felt those feelings too, that they were valid and that they would become more manageable in time. I have worked with new parents for years as an antenatal teacher and a birth and postnatal doula, so it was natural to support new parents in their climate grief as well. I now run support sessions both in person and online. All sorts of people from all sorts of backgrounds turn up. I share photos of climate destruction and ask people to write their greatest fears into the chat box. This little tool enables the darkest feelings to tumble out of people almost effortlessly; it is incredible to witness. And as they type, they read all the fears of others. They all say that it is an immense relief to see that others share their grave concerns. 'I thought I was the only one' they often say as we discuss the complex feelings of watching the world burn while trying to live meaningful lives and make a difference.*

*I use lots of metaphors – many I learned from my coaching sessions – to help people wrap their heads around this most complex of problems. Many express the blind panic that they feel for their children as they anticipate the food and water shortages – and violence – that may be a consequence of climate and societal breakdown. We talk about how guilt and shame are such poor and short-term motivators for action, and I invite each participant to let these feelings go by the end of the session. We discuss each individual's potential to make a difference in ways that work for them.*

*These things are impossibly hard to talk about, so the safe spaces these sessions create are proving to be hugely valuable. Parents tell me that they feel much less alone after talking to others who feel the same way. People leave saying that they feel uplifted, empowered, energized and much less alone for having been able to share their darkest feelings. Running these sessions feels really fulfilling. I am so grateful that coaching enabled me to adapt my skills in this way. I can now hold the hands of people who felt and feel just as I did and still do. I can be for someone else the person that I needed when I was struck with climate grief the first time, and it feels wonderful.*

## Our differing familiarity with climate grief

Some people find themselves hurled into climate grief, as a result of suddenly 'waking up' to climate change. It can feel like crossing from one existence to another. Yesterday I thought the world was largely OK, and today it feels like we are doomed. It is like walking into the doctor's office for a minor ailment and leaving with a terminal diagnosis. We did not expect it, did not plan for it and feel helpless in its path. Just as with any grief, we may cycle through many other emotions from anger and fear, to bargaining and denial. And just as all grief is not about death, climate grief may not be about our own survival but about the loss of the familiarity of the world as we know it. We can feel so unbalanced that we may scramble to reinforce our old reality and soothe ourselves that the crisis is not real. Equally, we may begin to see the whole world through new eyes. For the first time, bumble bees or garden birds appear in technicolour, as either heroic survivors or victims of humankind's carelessness. It can even be hard to hear our children share their hopes and dreams, because our vision of the future has become so bleak. We can lose our temper with others and with the trivialities of present-day life. We can find ourselves wanting to be silent, separate and deeply sad.

For others, climate grief may be an old familiar friend humming away in the background of their work and lives for years. Activists, scientists, ecologists and even concerned citizens have been watching the slow car crash of the climate crisis unfolding with a deep sense of pain and loss. Many of them, however, have been conditioned by their work culture not to express emotion, forced to leave climate grief unacknowledged or pushed shamefully away. Kimberly Nicholas wrote what many have told us in coaching, that those working on the facts of the climate crisis were never trained in how to deal with the emotions that came with their work.

> Bearing witness to the demise or death of what we love has started to look ... like the job description for an environmental scientist ... My dispassionate training has not prepared me for the increasingly frequent emotional crises ... sadness is valid; it need not dictate my actions single-handedly, but it deserves acknowledgment.[71]

## Ways to support someone who is experiencing climate grief

First, notice that we say 'ways to support' and not 'tools'. Helping someone with climate grief is not a 'doing'; instead you need an open ear. It requires us to lean into how we are being, and consciously not do much at all. 'Being' supportively with others as they experience a (possibly new) kind of pain and loss is powerful, as Antonia's story shows. This is because staying connected to ourselves and others helps us to be with big emotions and can neutralize their debilitating power over us. We are not looking to banish grief, but to find an easier way to accommodate it. We need someone to stand with us in our grief, not try to make us feel better. This may not feel like doing very much at all. When someone is feeling lonely in their grief, human connection is profound.

When we resist facing grief we can cause unease within ourselves and become separate from others. Many of us don't even recognize we are in grief and thus the healing process cannot begin. We cannot selectively switch off our emotions, like a fuse board. When we numb our painful emotions, we also numb the positive ones. Coaching creates a safe enough space for the coachee to create and own their own internal process, in which they can develop a deep understanding of their emotions, beliefs, behaviours and bodily reactions. The coach simply listens and acknowledges their courage, enabling them to draw their own conclusions. The impact of this is profound. Almost always, we are not looking for conversations about grief to include ideas for 'what I can do about it', as that is yet more avoidance; racing into 'sorting it out' instead of sitting with it. If action does come from these discussions, it is more likely to be further reflection or, where the grief is prolonged and debilitating, perhaps seeking a grief counsellor (and we will look at who else can help and when to refer someone shortly).

You do not need to be formally coaching someone to help them to sit with their climate grief. Many of us have listened when a friend or loved one talks about a loss and in doing so we are creating that container in which they feel safe to keep speaking. Here are some methods that we know support people, listed from basic to advanced. Test the ones that feel most natural to you, and over time experiment with newer, more challenging techniques. Before we show you successful ways to support someone, we need to first take some behaviours *off* the table.

### What *not* to do when someone is experiencing climate grief

The common things that coachees tell us do not work for them are as follows:

- Do not assume that grief will be debilitating or chronic. We are all living with grief on some level. This is not a sign that someone is falling apart. Remind yourself (though not them, as this will seem like 'fixing') of their resourcefulness and what they have overcome in life.
- Do not try to jolly the person along or 'look on the bright side' with platitudes that begin with 'at least' or 'I'm sure it's not that bad really'. This is not the time to change their mind.
- Avoid doing things that may seem to be 'fixing'. Don't offer advice, make suggestions, or turn the conversation to yourself by talking about your own grief as a way of empathizing. At each step, silently ask yourself 'where is the spotlight?' It needs to be on them.
- Avoid the overuse of models, e.g. the Kubler Ross 5 stages of grief. People grieve in different ways and there is no correct way. Models distract us back to logical thinking, and away from our hearts and bodies, where grief can be fully experienced and processed.

### Basic tools

- Simply listen and be there. Don't change the subject or dismiss their sadness. Be quiet and they will speak.
- Create intimacy by listening without judgement and asking (a few) open questions. Be gentle and caring in your manner. This can be done with nods and warmth.
- Contract how they would like you to help, and be honest: 'I feel like I want to reassure you or give you advice to make it better, but that's probably not what you want. How can I be really supportive right now?'
- Acknowledge them for who they are and their courage to explore this. 'You really care', 'You are deeply committed', 'It is courageous to talk about these things'. This helps you reframe grief as a form of resilience not weakness: 'I see how much you love our planet.'
- Acknowledge their climate grief. Many people have been told that they are strange for this type of grief, so a powerful way to help someone to fully experience their feelings is to reassure them that it is rational to feel a sense of loss when so much *is* being lost.

### Intermediate tools

- Who and what we grieve tells us a lot about our values. We can help people gain a sense of why they are grieving by asking 'what is important to you about the things you feel you've lost/you are grieving?' and the reply will be a value or a belief about the world.
- Explore the heart of the matter. 'What has been lost that you feel most keenly?'
- Help them to release the voice of grief as a shameful 'shadow self', and invite them to speak from their grief. 'What if your feelings/grief could speak, what would they say?'
- Explore the dream that is being grieved. 'What did you dream of/imagine in the future that now feels unlikely or impossible?'
- Ask them to focus on something they can see in nature and what they notice about it, and allow them simply to be with it, as a way of deeply connecting. If you wish to, you can move from here to the advanced skill, a conversation with nature.

### Advanced tools

- Have a conversation with nature. Invite them to focus on something that they can see and then ask that natural element what it would tell them to think/feel/be in relation to their grief. Enquiring of nature is effective at drawing out a person's own emotional thoughts.
- Ask them if they have an image that accompanies their grief. 'If you had to draw this feeling, what would you draw?' 'If you gave your grief a form, what form would it take?'
- Visualization – invite them to walk through a different world of their own design, in which they may feel more comfortable to express their grief. Choosing places that radiate love or calmness can be helpful here, though don't make them wrong if they choose another mood. As they move forward through the landscape, allow them to express their feelings.

– Help them to connect to the place in their body where the grief most power-fully resides and notice it. 'If you had to point to a place in your body where you feel this, where would that be?' 'What do you feel now?' Don't rush this, let the person guide you with how long they need to stay.

OVER TO YOU: Look at the below common expressions of climate grief. Notice your instinctive reactions and write down what you usually do. What might be a more helpful and compassionate response?

– 'I feel personally responsible and find myself apologizing in tears to plants and animals.'
– 'I feel so keenly what we have lost.'
– 'I feel so much despair and so overwhelmed in public spaces like shopping centres and airports. All these people are buying stuff and the world is burning.'
– 'My friends are worried about me. They tell me I'm obsessed and should calm down.'
– 'My husband doesn't want to talk about it any more. He says he hates seeing me like this, but I don't know how else to feel.'
– 'I feel despair, just despair.'
– 'I keep thinking about our children and the world they're inheriting and I'm so scared.'

## When to refer someone to specialist help

Many of us, coaches included, can feel that helping someone with grief is outside of our roles. As coaches, there are certainly circumstances in which our ethical code would instruct us to refer a client to a different form of support, such as therapy, and *none of us*, coaches or otherwise, can engage in therapy without being trained to do so. For some people, climate grief has become coupled to another past loss, which may be traumatic or painful. If working with their cli-mate emotions requires revisiting a past traumatic event, then this is outside of our scope of practice and we must refer them to an appropriate professional. Professional coaches are trained to help people in the now and the future, and so coaches and concerned others can help people to notice how their past grief affects them in the present moment, but we must be careful not to ask questions that steer the other person to explore the past. Instead we might ask: 'You said that the last time you grieved was when your mum died. I wonder what feelings are here now?' We may also invite them to connect to their past feelings, rather than to examine their past experience, as a means of supporting their current experience: 'You said that this feeling reminds you of when your mum died. How can knowing that help you now?' Again, we want to be careful here that our ques-tions do not steer the other person to share more about the past than they feel comfortable with or we are equipped to handle.

Whether you are a coach or not, you will know if you are overstepping your capability if you feel out of your depth because the subject has grown to encom-pass more than climate grief or if you find yourself steering the conversation towards your own interest and curiosity, not following the other person's train of thought. While we have said that listening to this subject requires us to move past our discomfort with big feelings, it is not a sign of weakness to say that you don't feel able to help, provided you do not make the other person feel ashamed for sharing their feelings. Notice that there is a difference between someone express-ing painful *feelings* and someone sharing details of a traumatic *event*. If you are in any doubt at all, gently ask the person if they feel that exploring their feelings with a trained therapist might help, and consider helping them to find someone if they seem overwhelmed by the thought of looking.

The Climate Psychology Alliance has lots of resources on their website that can help. Do remember, however, that no one was ever hurt by simply being qui-etly listened to, accepted for their pain and respected for their courage in sharing it. More often it is the opposite that causes hurt: when we brush people off or give thoughtless, unwanted advice, or change the subject so abruptly that we make a person sharing their vulnerability feel awkward or ashamed. If someone is expressing pain and loss as a result of the climate crisis, we can all make time to listen in silence as they articulate it. Listening is enough.

## In Practice

### Coaching and referring someone with climate grief

*Megan Kennedy-Woodard, co-director of Climate Psychologists and co-author of Turn the Tide on Climate Anxiety: Sustainable Action for Your Mental Health and the Planet*

*While my clients rarely come to me looking for help with climate grief, it is fre-quently in the background, often labelled as anxiety, anger or burnout. When they begin to investigate more deeply their emotions that the climate crisis pro-vokes, that is when grief is named. Before we speak, I am often prompted by 'the email'. This is a message prior to our session to warn me that they are feeling quite emotional, really angry or a little upset. It is usually apologetic. I always take a moment to breathe this in. The fact that they feel they need to warn me that this is coming sheds an important light on how difficult it can be to find a place to talk about climate grief.*

*I've noticed there is often a spike of client contact and presentation of climate grief with the release of a big negative news: the latest IPCC report, wildfire season, or some other catalyst that reminds my clients of the challenges and the fragility of this improbable planet. How lucky we are. How precious it is. What we are losing. I'm also aware of how these events may impact me personally. When working with grief, this can be more of a challenge. I increase my attention towards self-care during these times so that my emotions do not spill into sessions. For example, I'm mindful not to engage with social media or news that could be triggering for me prior to our session. I remind myself of my own expe-rience of climate grief and what helped. The need for validation, holding space and containment is essential, reminding myself of what is important to me and steeling my commitment to my actions.*

*As a coach, giving permission to allow my clients' feelings to process can be hugely helpful in supporting people to remain resilient in the face of The Doom.*

*When we think about grief, we usually think of direct loss; of someone who has lost a person close to them. The indirect or sometimes anticipated grief that can occur around climate can feel more confusing. Often clients feel guilty or that their experience of grief is not valid or justified. They are mourning the loss of nature, animals or of their hopes of safe and happy futures. In an early session, a client expressed real anger at themselves for feeling grief. They said they felt guilty, and that they did not have a right to these emotions. I found this to be a recurring feeling among clients. As coaches, we can honour the fact that we are privy to a moment that can potentially reveal core values and distinct objects of care. Often, my clients have become quite tearful as they describe their pain and sadness about the destruction happening. Within those sessions, I often hold long moments for pause and reflection. We talk about it and discuss how they think they can move forward. We endeavour to make plans to move away from the feelings of paralysis that can arise. I'll ask clients what they fear losing, what matters to them, essentially what they personally fear is most under threat or aligned with their values (meaning), what they feel their strengths are (ease), where they think they can have the biggest effect (impact). With this Ease, Impact and Meaning model, we can help clients to formulate plans and narratives that help them to support themselves and the planet. Leaning into self-care, identifying local, tangible tasks like volunteering in nature and finding communities to talk about these emotions have all helped my clients. Asking my clients to schedule and commit to these tools often helps transition the grief to more motivating emotions.*

*We, as coaches, can support this; however, part of climate coaching is working out where the line is between coaching and therapy. This can feel tricky with climate grief. Though we do not need to shy away from climate grief, it is important to really look closely at 'what is what?' There have been times when it felt appropriate to refer on to a psychologist; for example, with some clients, climate grief was bringing up past bereavements, and with others, unresolved traumas that benefitted from exploring these experiences more deeply with a psychologist. It is important that we feel honest with ourselves, that we can trust our training to know when this time comes but also to be mindful not to pathologize this pain. As coaches we are in a wonderful position to help clients move from that feeling of 'stuck in grief' to sustainable climate action; meaning action that is both sustainable for them and for the planet. I often think about the words of Jamie Anderson, 'Grief is just love with nowhere to go', but as climate coaches, we remember that for climate grief, there are millions of places for that love to go. This pain reveals love and this love can inspire the climate action we all need. I am inspired by my clients that have moved from climate-action paralysis to forming grass roots organizations, non-profits and feeling proud and motivated by their work.*

## Grief in the current paradigm

In the industrialized world of consumption and the advertising that drives it, decay is a terrifying concept. Whether we deny explicit decay in the form of ageing or the more subtle decay of symbols of status (your car is too old, your clothes not up to the minute), we privilege the modern and the shiny over the old and well loved. Some argue that we have been conditioned to channel our sadness or dissatisfaction into consumption. Rather than sitting with difficult feelings, we are encouraged to have a bit of 'retail therapy'. For many people, modern life can feel breathless and lived slightly in the future, lurching from one to-do list to the next. In this context then it is little wonder that we struggle with grief. Grief requires patience, silence and 'being with', and we are programmed into rushing, broadcasting and 'doing about'. By taking a stand for simply being with grief, we are ushering a new paradigm into being, in which there is time to simply be with each other, to accept decay as part of the natural cycle of things, and to experience emotions instead of numbing them.

## The gifts from grief

We are often guilty of looking at grief as a negative, yet grief is an expression of love. It is because we care and love deeply that we grieve our planet's past or future loss. It is possible to gently hold grief as a resource, or a wellspring from which to draw love and care. Rather than avoiding it, we can use grief generatively, and support others to find the action that comes from acknowledging their love. For those hurled into climate grief, its systemic, globalized nature can create a powerful new perspective on the way they frame old troubles. Whatever felt hard before we knew about climate change, can appear more manageable in comparison. While that might seem bleak, it is possible to frame this in the guise of self-compassion. 'Notice that before you knew this, you would beat yourself up for not working hard enough, being clever enough, etc., and now that feels much less potent. I wonder what compassion you might offer yourself now?' This new self-compassion can be a tool when they find themselves anxious, stressed or experiencing scarcity.

As a result of increased awareness on the subject of climate grief, those working in climate and environmental science can work in a more sympathetic emotional landscape. Leehi Yona, a PhD candidate at Stanford, framed the challenge for the new generation of climate scientists thus: 'I think our role, above all, is to be stewards of grief, to hold the hand of society as we enter the unknown space of the climate crisis.'[72] Facing our grief does not destroy us, however painful it may feel at the time. What tears us and our relationships apart is refusing to face it. When we are able to be with grief we realize the depths of our resilience. By helping ourselves and others to simply notice and feel our sense of loss, we can remind ourselves of our power to overcome sadness, heartbreak and deep grief for the beautiful world that we have ourselves changed.

Part C
# A coaching approach in systems change

# 15 How to help organizations to commit to change

In this part of the book, we look at systems change, first from the perspective of organizations and then of wider social movements. To begin with we focus on traditional organizations because these require the greatest degree of change and present the most challenges in terms of transformation. While we are thinking mostly of for-profit companies, much of what we discuss is relevant to the way that we can make and cascade decisions and plan change in volunteer organizations, local governments, institutions and even communities. Systems change is a subject on which many hundreds, if not thousands, of books have been written, so here we limit ourselves to the emotional wrestling that has to happen to achieve this change and how a climate change coaching approach can enable shifts in formal and informal groups and ultimately in the systems of which they are a part. We will show you how to avoid creating roadblocks and instead create successful, sustainable systems change. In the next chapter, we will look at how organizations can bring people on board (as well as alienate them) and create psychological safety so that teams can come to terms with the complex challenge of climate change as well as organizational transformation. We share stories of the common pitfalls that can occur and offer coaching 'remedies' to overcome them, through case studies, examples and key coaching questions.

It would be convenient to assume that organizational change happens in a linear fashion. In reality, things rarely occur in such an orderly way; change is an ongoing process with many different entry points. As we cycle through change we may return to or repeat many stages as the process evolves. Change does not always start at the top and cascade down, and systems rarely move as one homogenous entity; different parts of a system change at different rates. A climate agenda can begin at many different points in an organization, from the board, to the grassroots, to satellite teams. When a certain team within the business has connected the dots between its function and mitigating the climate crisis, they will agitate for change. Change may also be sparked by shifting consumer demands or by industry or regulatory changes in which bigger forces start to apply pressure.[73] Here we will share some ingredients that we believe support organizational change, and offer a way of diagnosing where an organization is on their change journey, and how to help it move forward.

## Deciding to change

A first step in any change is making the decision to act. Often (as Nick Ceasar from NatWest writes in the following chapter), to succeed in making wholesale changes, it is useful to make clear, public commitments to change, to create powerful external accountability.[74] Coming to a point at which leaders are comfortable with making such a public commitment can be a lengthy and often barely visible process. Organizations are composed of people who make decisions, and just as individuals change when they are ready, so too do organizations. The transtheoretical model[75] (TTM) of individual change can also be a useful framework to explain systems change here. The model includes six key stages: pre-contemplation, contemplation, preparation, action, maintenance and at times, relapse. Here we will focus on the first three of these – moving from not considering our role in climate change (pre-contemplation) to overcoming ambivalence (contemplation) and then devising what our potential role can be (preparation) – because these are such important stages for genuine change. Often these stages are not visible to the outside world, which only sees the public declaration of change as a starting point, but they are key to laying solid foundations for what comes next.

### Stage one: pre-contemplation

During pre-contemplation an organization does not see the environment as a key priority, is not thinking seriously about the need to change to address the climate crisis and is likely to deny that they have a role to play. Organizations at this stage may make changes if forced to by law and regulation, but the focus is likely to be upon finding ways to side-step changes and continue with 'business as usual'. If you are a coach or consultant working with or within this stage of organization, it may be difficult to get traction for more proactive approaches, and you may be told that the company is doing what it can, or doing enough, or that acting could damage a fragile competitive advantage. Does this mean that we abandon these organizations? Not necessarily, but it is worth considering where we place our time and energy, given that these are a limited resource. One thing we can do here is diagnose the reason that the company is not engaging, which could be at leadership level or a feature of the wider system in which they operate. For example, perhaps leaders do not feel connected to the imperative for change because their competitors are not or maybe they fear a backlash from shareholders or investors, who are not supportive. Rather than characterizing these businesses as bad, a more compassionate and useful approach would instead be to get a better idea of what is holding them back from change and then finding ways to address those factors, either through mindset change inside the organization or systems change outside of it. It can be easy to fall into the trap of assuming that, when working with organizations at this precontemplation stage, we have to take them all the way along the change journey in one go. Instead, you may be just one companion on a long road to change. Your job here may simply be to help them to connect to the issue and understand that they can have a role to play. Think of it as planting a seed. This may pave the way for someone else, or even you at a later stage, to engage again when they enter the next stage: contemplation.

### Stage two: contemplation

At this point key people within an organization are becoming aware of the consequences and cost of current behaviours and are contemplating change. They may have seen their competitors or clients changing and realize that the financial survival of the organization depends on making a shift. While there may be customers and stakeholders making a strong case for change, this stage is crucially characterized by ambivalence – the internal conflict of whether to change or not. Leaders are becoming aware of the reasons to do things differently and are open to understanding the cost of maintaining the status quo but are also cognizant of the costs and risks of change and may be vacillating between the two. This is true of any change, but as psychologist and strategist Renee Lertzman points out this is not a sign of lack of care, more it is coming to terms with the gravity of the situation while hoping they may not have to.

> Ambivalence is not acknowledged and recognized. It tends to run the show. That's where inaction and paralysis comes in … a lot of what we are seeing out there is a combination of anxiety and ambivalence, not a lack of care or concern.[76]

For many of us, the scale of change required is a huge mental leap that we may experience in cycles of acceptance and avoidance.

As organizations are all too often aware, there are costs involved in creating organizational change. However, as the world slowly wakes up to the need to address climate change, and to the challenges that it will create, taking action now is increasingly being seen as an act of 'future-proofing' organizations so that they are fit for purpose. One individual who we spoke to anonymously described trying to help his board grasp the need to pivot their waste management company's business model because their biggest client had announced that it would become zero waste within five years.[77] He could see that this huge risk posed a fantastic opportunity to move with the times, but he also understood that it was a giant imaginative leap to conceive of the company differently.

Organizations at this stage may also engage in actions to defend or distract from their current position and behaviours, using greenwashing to buy time or put off the decision to change. While their interest may be piqued, this is a stage in which the environment jostles for position among a number of more traditional priorities and can easily fall into the 'doing good' list, as opposed to being a feature of what will keep the organization relevant or functional. This may be the stage in which the advertising agency is tasked with positioning what the organization already does as ethical or environmentally aware. It may be when internal self-organizing green groups are given a mandate to make only piecemeal changes. We spoke, for example, to one green team who had been told that they could test out environmentally conscious initiatives in all but the organization's most critical locations. They had also been given the mandate of the board but no funds, so remained limited in what they could do.[78] While it may seem tempting to simply legislate and regulate organizations to force change at this stage, this alone will not create the level of change we need to see. If an organization bypasses the contemplation stage, there is a risk that it will only act because it must; 'going through the motions' because it 'should' and lacking sufficient energy or drive to push through the challenges necessarily presented by the change. Without genuine contemplation, change may be superficial and incremental rather than transformational.

Coaches, advisors and change consultants may well be invited into organizations at this stage to help give the appearance of taking action, when the organization's leadership is not yet ready to. This may take the form of a sustainability audit, a staff survey or a senior leadership strategy day. Just as with individuals, until an organization has decided that change is either imperative or inevitable, coaches and consultants may have a hard time making true change at this stage, led as they are by the agenda of the client. One way in which outside agents can potentially influence is to bring its senior leaders together with those from organizations that are already on the journey to change. This leverages our human desire to remain current organizationally as well as individually – today's CFO in one organization may want to be the CFO in another company tomorrow – and addresses concerns about not wanting to be an outlier in the system. These soft influencing opportunities, such as conferences, networking and mastermind groups can more safely (or less confrontingly) bring leaders 'in from the cold' among their peers who have undertaken similar change. The Business Declares initiative is one such example of this, and Fiona Ellis, its founder, has written here.

You might also be wondering how there are so many examples of a new chief executive officer joining a company and almost immediately making a clear commitment to address the climate crisis, as if circumventing this process of contemplation. A new broom clearly can sweep a company towards commitment and action; however, consider how this person was hired and how they have been able to be so bold, so fast.[79] It is almost always the case that for these change-makers to be recruited the key decision-makers who appointed them were already at the contemplation or preparation stages and were looking for someone with the confidence to enact their nascent ideas and put them into action. In other words, the new broom was pushing on an already open door.

### Stage three: preparation

A significant number of the organizations that are leading the way in tackling climate change have spent much time in contemplation and preparation, really understanding the pros and cons before announcing a commitment, to make sure that their vision stands up. Linda Freiner, from Zurich Insurance, explains this lengthy process below. For people outside of the organization who feel the urgency of the crisis, this preparation stage can be extremely frustrating and may look a lot like resistance to change. The outside observer will want the organization to move quicker. However, for the commitment to change to be genuine, robust and internalized, this part of the change process is important, because it is the moment when an organization moves beyond generic *extrinsic* reasons or motivators for change, to making a connection to its own *intrinsic* motivation and becoming truly committed.

Whether we are working as coaches in organizations at the contemplation and preparation stage, or are internal coaches or changemakers within them, the best way to help an organization to move through this to commitment is to expect, normalize and engage with the ambivalence and doubts in the system, rather than suppressing or bypassing them. Using the coaching skill of acknowledging (not judging) misgivings and feelings of scarcity and overwhelm, can give stakeholders an opportunity to express their concerns (and we explore this more in the next chapter). A coaching approach can open up a space to ask questions, road test ideas and take a more constructive, problem-solving approach, ultimately clearing the way to commitment and action. This is not the same as just listening to naysayers and stopping there, but rather is normalizing concerns and showing positive regard when we listen. Demonstrating curiosity rather than judgement makes it possible to identify the needs and beliefs that exist beneath their concerns so that those values and conditions can be accommodated in the change.

Some key coaching questions can help to move teams from contemplation to preparation:

– Zoom out and consider this period of time in the context of humankind's legacy in the world. What is the impact we want the organization to have at this time?
– What can we currently feel proud of? What would we like to feel proud of in the future?
– What is the legacy that we wish to leave for future generations?
– What is the world that we want to be a part of, and what role do we want to play?
– What will the impact be on us of changing and of not changing?
– What will the short-, medium- and long-term impact be upon each part of our organization, and what challenges and concerns might each function have?
– If we were able to look future people, plants and animals 'in the eye', what would we want to be able to say about our contribution as an organization?

## In practice

### The journey to committing to climate at Zurich Insurance Group

*Linda Freiner, group head of sustainability, Zurich Insurance Group*

*I joined Zurich in 2013. Climate change is part of Zurich's balance sheet because we take on a lot of the physical risk through our business. However, addressing climate change has come to mean a lot more to us than that. We have been carbon neutral since 2014. First, we started looking at our own operations, for example changing the way we travel for business, and later implemented many other measures, including creating an internal carbon fund that employees could apply for, both for internal and external solutions. But as an insurance company, our operations are only a piece of the puzzle on our journey to decarbonization, and these things were not going to deliver the big impact we need. In 2019, we committed to the UN Business Ambition for 1.5°C Pledge, a Paris Agreement for the private sector. We were the first insurer to have made this commitment.*

*At the beginning of our sustainability journey, it was about convincing all members of the executive committee that setting meaningful targets was the way forward. Transforming the business is not simply a matter of will. It is also related to technical and legal challenges. For example, to decarbonize our underwriting portfolio, we want to set science-based targets. However, there are currently no industry-wide standards for measuring emissions from insurance underwriting. To change that, we are working together with peers under the umbrella of the UN Principles for Sustainable Insurance and created the Net Zero Insurance Alliance. Once we have a widely accepted methodology, analysis needs to be done on the transition pathways of the underwriting book and how we should expect the carbon emissions to develop in line with our path to a 1.5°C future. The final step is to translate that into realistic, actionable steps and work with our customers to facilitate the transition.*

*When we started looking at climate action 8 years ago, we focused on investment management as this is an integrated function with fewer decision-makers and clearly defined scope. We then looked at our underwriting business, both retail and commercial, and there we are benefiting from changes in society and consumer patterns more broadly. Five years ago, when I started talking to people about what we need to do on the product and services side, I realized that there was more work ahead of us. But since consumer behaviours had already changed in other industries already, why shouldn't it happen in insurance too? Today, we are also seeing it in insurance, and we have data that show that sustainability is a growing topic among insurance customers. As a company, we need to focus on the next generation of customers and what products and services they expect from us. It may be that our core insurance products do not fundamentally change, but we can develop add-ons to existing products that incentivize customers to make more sustainable choices. For example, could we incentivize customers with a better premium if they buy an electric car over a petrol one?*

*What I did not expect when we started looking at our value proposition was that it would lead us to a new organizational purpose: to create a brighter future together. And part of that purpose is a planet promise. Integrating our climate action into the purpose of the company and into our customer value proposition is powerful. We will undoubtedly still have to have to make tough choices, but this will give us a compass to navigate them.*

*Key to my team's ability to make progress has been support from inside the business. In the middle of this process, I found a key supporter in the executive committee who was also passionate about this topic and knew how to inspire fellow leaders. There is now a younger generation of leaders, born in the 70s and 80s, who see this topic very differently from those born in the 50s and 60s. However, those older leaders also feel pressure from their own children and younger relatives to use their sphere of influence. As a result, we have found that many of our leaders are becoming advocates. These changes in corporate leadership are very important for the overall success of commitments, such as the ones we have made.*

*I am not a scientist and I had little understanding of the insurance industry before I came here. I tried to form a team with a good mix of people from within the company, as well as hiring subject matter experts. We need to ensure we have both scientific and business knowledge. I manage a small team who act as catalysts and support the rest of the company. I like to influence without authority. My strength in this journey has been to really get to know people, understand how they see the topic and to help them want to be part of this journey we are on. I do not waste a minute. I have tried to get more people involved, including customers, to learn where they are coming from. I hope I have helped people to really identify with the transformation of the company, with what is important to employees and with what customers will expect from us in the future.*

*Before I had children, I ran ultra-marathons in the mountains. That is my personality, and it is probably why I work on this topic! I am quite resilient and I keep fighting for the long game of it. When we agreed on our climate ambition, it was simply the right thing to do, and that is why it happened quite easily. The awareness and the understanding of the topic was such that we were ready to take a big step. I have found the ability to be patient with these processes to be a quality that is important. My resilience drops when people divert the conversation, take things out of context or stall because they have a different agenda. That is when I feel like I am wasting my time. But when I see that I have won somebody over with whom I have previously had really challenging conversations, I feel a real sense of pride.*

## Using the preparation period to connect to the problem and to purpose

To motivate organizations to move beyond thinking to doing, it is useful not only to consider the risks of not changing but the benefits of doing so too. Rather than being hostage to the future, organizations can meet and shape the future that they want, by creating a compelling purpose and vision that can drive action and deliver true transformation. Educator and computer pioneer Alan Kay coined the phrase 'The key way to predict the future is to invent it'.[80] As humankind embarks upon this unprecedented level of organizational reimagining, purpose is one of the most powerful tools we possess. As we explained in Part B, purpose unites the essence of who we are with what we do and is our reason for existing. This is just as true for systems as it is for individuals. A clear, compelling organizational purpose builds motivation into a system, creating meaning and energy for the people working within it. Linda Freiner described Zurich's purpose as a compass to guide decision-making, as it can help to clarify the organization's direction when dilemmas arise. It also builds staff resilience for the inevitable challenges that change creates, because it provides a strong sense of meaning.

Increasingly, there is a movement away from organizations existing for the sole purpose of making money for owners and shareholders, towards holding a wider purpose at their core.[81] Some organizations, by the nature of what they do, are purposeful from their inception, but others are only just beginning to realize the importance of this, not only for moral reasons but because a strong social purpose is increasingly required to recruit the most talented staff.[82] Leaders and organizations that move beyond a transactional, short-term reason for existing towards a purpose and vision that leaves a social and environmental legacy, gain benefits that go far beyond the financial gains of the organization.[83] These include:

– increases in employee loyalty and retention[84]
– an increase in intrinsic motivation for action by staff leading to increased commitment, performance and effectiveness of those within the organization[85]

- creation of a positive brand image and increases in customer loyalty because people connect to and care about a similar purpose to that of the business[86]
- an increase in the ability to 'pivot' at times of challenge and to find a way forwards even when the world around them changes[87]

Just as 'values' has become a dirty word in some organizations, so too purpose can fall victim to being either bland and meaningless or ambitious but empty. Whether intentional or not, it seems all too easy to confuse purpose with products or to declare a progressive purpose in our marketing but do the opposite behind the scenes. Yet meaning and purpose have never been so important; they not only motivate us to action but deepen our resilience in times of hopelessness.[88] We saw in 2020 during the Covid-19 pandemic that individuals and organizations began to ask deeper questions about what their work was for and to seek a more concrete, relevant form of social imperative in the face of a social catastrophe.[89] So we believe the climate crises will begin to have the same effect in encouraging us all to ask what environmental imperative we serve with our work and our lives. An organization can, of course, be purposeful without addressing climate change; many are. However, as we shall go on to show you, if you have undertaken a stakeholder exercise to create a true picture of the relationships between your organization and the wider world, it becomes hard not to include a desire to do no harm beyond the organizations' borders.[90] The more the well-being of the wider world is aligned with an organization's purpose for existing, the more successful the organization will be in addressing climate change. This creates a *revolutionary* organizational purpose, rather than an *evolutionary* one, that has real power to motivate people inside and outside of the organization, and we know that high motivation is important during the discomfort of change.

Coaches and consultants can use the following prompts to help an organization to define a powerful purpose that put the needs of the wider world at the heart of what it does:

- What need(s) in the world are we drawn to meet as an organization? Why is that important to us, and why is it important to us *now*?
- What would meeting that need mean to us, culturally, as an organization? This is not a question about how it would work, but rather how it would *feel* to the organization's stakeholders.
- What do we stand for/want to stand for as an organization in the world at this time in history? What part do we want to play?
- What values are important to us? How would a new purpose connect to the organization's existing core values? Digging into values as part of purpose and vision setting helps people to connect deeply to why change is relevant to them.

## Crafting a compelling vision for change

The changes needed to address the climate crisis are complicated and complex and as such require a long-term commitment, not a series of short-term tactics. Designing a compelling vision to work towards is a key element in making this possible, as it creates a meaningful narrative to which people can connect. This forms a counterbalance to the inevitable sense of disempowerment, scarcity and overwhelm that this subject can create in teams. Connecting to a vision enables us to move towards a place of purpose and fulfilment, where we can access more creative solutions to our problems and refuel our drive. There are, however, some pitfalls to overcome in creating this kind of north star. Individuals and teams, gripped by the urgency of the crisis, can feel that they need to immediately get into designing actions, rather than stepping back to consider how to change their modes of thinking. As these are the same modes that have got us to this point, as we start to imagine an alternative organizational future it can be beneficial in the long term to consider the new paradigm in which our organizations will operate. For example, if we leave behind the idea of the never-ending growth of our economies and become more regenerative and circular, how does that new mindset change our organizational strategies? Many organizations have not worked through these bigger questions and are instead assuming that they will operate within a similar paradigm to our current one and can simply reduce their emissions to zero, otherwise operating as they currently do.[91] Others, such as NatWest Bank and Zurich Insurance, written about here, have considered very alternative futures and are attempting to be ready for them. In all cases, however, visioning processes will need to be iterative and continuous, as even the best forecasters do not have crystal balls. Nevertheless, those teams that have at least broached these questions will find it easier to shift and easier to articulate a positive vision within the new paradigm because it will not be such a shock to the system.

The more vivid and alive our vision can be, the easier it is for the people within our organization to picture it and make it a reality. While the initial reason for change may be motivated by what we wish to avoid (for example, increased insurance claims caused by climate change), to develop a vision based on this would result in a narrow focus and potentially many missed opportunities for staff engagement. Instead, a proactive framing will have radically different outcomes. For example, a team could consider: what is a positive alternative to a world of increased insurance claims, and what part would our organization play in creating that? As with purpose, a truly powerful vision must consider the wider context or system that the organization is part of, beyond simply considering the company's bottom line, shareholder profits and customer experience. The wider we go, the greater the resonance and the more powerful the vision will also be for the organization's staff and customers. A vision of 'ensuring the organization is fit for the future' is far less compelling than one in which 'the organization works for the benefit of the next generation as much as this one'.

Visioning requires coaches and consultants to help teams to temporarily disengage their rational, thinking brains, so that they can access a different kind of emotional intelligence, one that will create a deep kind of meaning and connection.[92] If a vision is developed in a purely cognitive way, not only will the outcomes be more limited but it will ultimately hold less motivation for the organization as a whole. It is also hard to be wildly creative if we are constrained or distracted by the transactional day-to-day needs of the organization and pressures upon us personally. Changing the immediate environment, particularly spending time in nature, is a powerful way to set the mental scene for climate-positive visioning, not least for staff who, as we said in Chapter 1, may suffer from a kind of what Richard Louv defined as 'Nature-Deficit Disorder',[93] in which we are dislocated from the natural world. Coaches and consultants may consider hosting these exercises away from the office, or using more subtle approaches such as sending plant kits to team members, or playing the sounds of nature during a session.[94] Coaching questions, with their focus on 'who we are as an organization', not just 'what we do', and on a team's emotional, not only rational knowledge, can help teams create a vision that connects with their 'being', which in organizations is where culture resides.

Some useful questions include:

– 'What is our dream for the world in which we wish to live? What does that world look and sound like? How does it feel to live in that world, and what are the rules or norms there?' Help people to connect with this by giving examples of the sensory nature of different environments, for example 'concrete', 'competitive', 'noisy', 'tranquil', 'generous' and 'verdant', and encourage them to choose adjectives that describe the sensory qualities of their envisioned new world.
– Now that we can imagine a world that we would be happy and proud to inhabit, how would we like our organization to be a part of creating that world?
– What is the *dream* for how our organization would behave, operationally and culturally? This is less about granular operations and more about behaviours. For example, in a calmer world, would our staff behave less transactionally because they have time to think and listen to each other?
– What values are important to us, and what values would we be able to live more fully as an organization if we fulfilled this vision?
– What is the organization's stake in this future vision? Why is it important for us to achieve this vision? Why is doing this important *now*? In this you may find teams drawing on a darker vision of a world in which the organization does not change and suffers for it, as much as connecting to the positive opportunity if it does change, and that is fine. As with individuals, the shock of the darker vision of the climate crisis can be motivating if we can move from shock into resonant action.

## In practice

### Helping organizations to declare a climate emergency

*Fiona Ellis, professional coach and director, Business Declares*

*I became aware about climate change in the early 90s, yet in my career as a coach and leadership development consultant, I was frustrated by how little time I was spending addressing the climate crisis in my work. After the 2019 IPCC report, when a wave of climate protest led by youth and Extinction Rebellion erupted in the UK, I had a conversation with Gail Bradbrook (one of Extinction Rebellion's Founders), where I said, 'It's important to pressure governments but what about business?' I found myself becoming an activist, facilitating a group of business leaders who wanted to act on the climate emergency. Together with John Elkington, we wrote a letter in the* London Financial Times *supporting the Easter Extinction Rebellion protests in the UK. In September 2019, we set up Business Declares as a not-for-profit network for businesses wanting to declare an emergency and accelerate their action on the climate and ecological emergency.*

*In the last, hectic, 2 years, as a team of volunteers, we have worked with a range of businesses to declare, make climate transition plans and accelerate action to net zero in 2030 not 2050. This can be a huge step for an organization. Some organizations, such as Ecology Building Society and Triodos Bank, initially told us 'This is our whole raison d'être, why do we need to declare?' Others were anxious that in raising their heads above the parapet and declaring, they would be attacked for not being perfect. We had a really helpful conversation with COOK Food about being honest about the journey ahead. They went on to write the following lines in their declaration that were then adopted by many other businesses. 'We are certainly not claiming to be perfect, neither are we judging others, instead we are a coalition of the willing looking to bring solutions.'*

*I remember the power of our launch event at the Institute of Chartered Accountants of England & Wales, held on the day of the Global Climate Strike in 2019, where I nervously spoke from the heart about the need to act 'if not now when, if not me who?' Afterwards, walking through the crowds of youth protesters in London, with the CEO of Ecology Building Society, both of us were visibly moved. I think that day influenced him to become a member, recognizing the time to declare had come. More recently, he wrote a Letter to the Earth as part of a campaign that we supported.*

*The mood and awareness have changed hugely over the last 2 years and during COP26 in 2021, the media, TV and social channels were full of recognition of the emergency, following the UN's announcement of a Code Red for humanity. Many were talking about how to take action. Leading businesses are being seen to step up in campaigns such as Race to Zero and to address the biodiversity emergency. There are now challenges over net zero pledges, the need to verify commitments and whether greenwashing is taking place. Taking action does not result from just absorbing the facts and the science, although this is critical to understand the depth and scale of the emergency. To act and speak out, leaders need to be moved at an emotional level and feel a sense of solidarity with others, and to take a step that can feel risky. Business Declares is just one way to achieve this, and I am very proud to be a part of it.*

## Deciding who to involve in determining the scale and type of change, pre-announcement

This process of defining the purpose and setting the vision may be done within a small group of senior leaders or taken more widely to staff to test the waters for an announcement. How organizations navigate this shift from preparation to announcement varies depending on their scale and culture. Organizations need to decide how much preparation to undertake to feel confident in the announcement they then make, and that necessitates determining who to involve in the planning. Organizations that are used to making big strategic decisions in small, senior huddles, may act differently when it comes to setting a strategy in favour of the environment because of the charge attached to environmental action and concerns about external judgement and criticism. Many feel that they need to do much more preparation and involve far more people than they might for a more anodyne announcement.[95] Even though, as we have said, for many organizations this announcement is about decarbonizing their operations (which could be said to be more pedestrian than transforming their business model), the heightened external atmosphere in which any environmental declaration is made, can make senior teams nervous.

For some organizations, who to involve is also a cultural question. Those with a strong culture of collective decision-making – for example, volunteer organizations – may canvas the opinions of everyone, and then synthesize ideas down, or may float a specific idea and seek support or dissent. More traditional and smaller organizations – typically medium-sized SMEs[96] – will often seek the support from a large number (and sometimes all) of their employees before taking a decision.[97] With fewer than a few hundred employees this is not too onerous a task and sets the scene for change early on, while bringing lots of useful information into the decision-making process. For much larger organizations, where there are thousands or even tens of thousands of staff, this canvassing is unfeasible, as surveying at that scale may only yield bland results. At this size of organization it is more likely for contemplation and preparation to take place in smaller, more senior groups, who announce the change to staff once a decision has been taken.[98] Nevertheless, in both cases, the granularity of planning how the change is implemented and scaled usually lies not in the C-suite but with the management level. We look at this in detail in Chapter 17.

As part of preparation, organizations usually consider the impact of a change on their traditional stakeholders, such as customers, shareholders, business partners and suppliers, through stakeholder mapping. If they only do that, they are likely to get only a limited, short-term evaluation of the potential impact of a big change. An example might be that a company simply reduces its waste and energy consumption because that saves money, does not inconvenience customers and does not require a huge shift in the way it works with suppliers. The knock-on effect of this more limited version of stakeholder mapping is that addressing the climate crisis is likely to be seen more as an isolated change initiative, rather than an organizational transformation. When change is seen as something for just a few people or functions, it is not embedded at the heart of the organization and is viewed as separate from its overall purpose and strategic direction. This has led, for example, to the formation of separate units within organizations that are single-handedly tasked with implementing the change initiative while everyone else continues as before. This can create fundamental conflicts of interest between business targets and sustainability targets, which teams are then forced to navigate every day, creating uncertainty and draining precious time and energy.

To conduct a truer picture of an organization's stakeholders, a wider climate change coaching lens is useful, because this brings in stakeholders beyond the immediate business to consider the ripple effects of its actions on families, communities and the wider environment. This wider lens enables teams to take into account stakeholders who are traditionally underrepresented and yet often the most affected by the climate crisis itself. It is especially important for organizations to recognize that they are making not just organizational decisions but decisions that affect communities (and even countries) in which they have operations, customers and suppliers, and to involve people from those affected geographies, both from within their workforce and also local leaders from those countries and communities. It is very easy for preparation to become an activity dominated by those in the majority, who will be the most insulated from the worst effects of the crisis. This allows an organization to connect to the longer-term impact and legacy of its actions. By zooming out and taking a view of the whole interconnected system of which we and our organization are a part, it becomes much harder to act without integrity in a way that harms the whole.

Here are some activities to help organizations with this wider stakeholder analysis:

## Activity 1

Draw a series of concentric circles and ask the team to place different stakeholders in each layer, depending on how important they think they are. Staff will most likely be placed at the centre, along with customers, and often the environment and community are in the outer circle. Now ask them why they chose to put the various groups where they put them. Invite them to change the order around and see what that does to the way that the company perceives the stakeholders and what decisions would flow from the new perception. The activity here is less the plotting and more the discussion that happens around it.

## Activity 2

Offer a simple visualization exercise, either as a guided visualization or as a drawing exercise in which teams draw individually. The theme could be either seeing the organization as an indigenous settlement or seeing it as an ancient (i.e. biodiverse) forest. Once the team has connected with that place, ask them:

- What are the norms and laws here?
- What is the impact that this place wishes to have in the present?
- How does it want to leave the world for future generations?
- If they spoke to their ancestors, what would they advise them is their role in the world?

## Activity 3

This is an activity in which people speak from the perspective of different stakeholders. This works just as well in person as virtually and can help to ground teams into the human aspect of a change, while understanding precise needs and challenges. First ask the team to list all of the stakeholders that they can think of, asking them to expand beyond the immediate to the wider system of those impacted by the company. Make sure that teams know that stakeholders do not have to be human, so that the environment can be represented. Then ask individuals to choose a stakeholder to play, and then raise their hands, tell the group which stakeholder they are playing and explain how they (the stakeholder they are playing) feels and what they think about the organization's plans to change.

People can speak from multiple stakeholders' perspectives, and several people can embody the same type of stakeholder (for example, 'new parents' or 'senior managers'). Importantly, the person speaking the perspective does not need to be in the role they are playing, and, in fact, ideally they would not be (e.g. someone playing a parent may not be in real life). Indeed, part of the utility of this exercise is that a senior manager may better understand the perspective of a community elder by speaking from the elder's perspective. This also elicits ideas from the team that speak to what they value as much as it expands their limited definition of stakeholders. Go around the group until everyone has spoken at least once, ideally twice. Ask each person, what, in the role they are playing:

- '... is most important to you?'
- '... could the organization do that would be most helpful to you?'
- '... do you know about the organization that the organization itself does not know?'

## Power and influence: internal changemakers and their barriers to success

Until this point, we have described change that has been green-lighted from the top and looked at how change professionals such as coaches and consultants can help those key decision-makers. Now, however, we want to look at how decisions can be influenced from around the system because when we want to influence change it is helpful to understand how our position in the power structure of the organization alters the approach we may take to creating change. In many organizations, particularly those with traditional top-down hierarchical structures, the ability or opportunity to influence the purpose and strategic direction of an organization is directly proportionate to the person's level of seniority – the closer they are to the top of the organization, the greater their level of influence and ability to embed sustainability into the heart of the organization. However, this does not mean that only senior management have the ability to catalyse or implement change.

Some organizations have been influenced to consider their environmental impact by organically formed 'green teams' who are groups of disparate staff that coalesce around this issue and lobby for it to be put on the table in the C-suite.[99] Also, as Linda Freiner described, senior leaders are increasingly influenced by their teenage or early-career children from outside the organization, who are alarmed by the ecological crisis and challenge their parents to use their power more overtly. In less hierarchical organizations, like matrix organizations (where power is more evenly distributed and everyone contributes to the direction of the organization), seniority plays a much smaller influencing role, and it is the power in numbers that counts, when influencing direction. When individuals group together like shoals of fish the whole becomes greater than the sum of its parts, and they have the ability to shift the system. Also not to be overlooked are those individuals who hold an informal seniority and rank because of their knowledge, long service or how respected they are by others. These 'natural leaders' can come from any part of the organization and are likely to influence at the 'water cooler' more than the board table. If you are trying to drive an agenda inside an organization, these informal leaders can become powerful allies and advocates for change, who can also make you aware of potential relational pitfalls to avoid that may not be obvious in the formal structure and operating rules of the organization. An example of this might be an informal leader whose long service means that they know of animosity between certain leaders who you need to enrol.

There are other less formal aspects of power and influence that relate to the culture of an organization, such as the level to which diverse voices are listened to and valued. For example, is the person or group calling for change in the minority or in the majority in the organization and also the society in which the organization operates? This may be in relation to gender, ethnicity, age, disability, neurological preference, or anything that may mean a person's perspective, experiences, priorities or way of working is different from the majority in any given system. While being in a minority can sometimes increase social capital because that particular quality is rare, it is more often the case that being in the minority can make us feel less secure and more like an outlier, especially when up against unacknowledged, unnoticed and unaddressed behavioural norms that play out as rank and privilege. All of this can make it feel like influencing change is much harder to do. In these cases, seeking out others who share our desire for change can help us to influence more confidently and counterbalance feelings of isolation that being in the minority might create.[100]

Sometimes referred to as 'social and symbolic capital',[101] factors like levels of education, being articulate and able to argue in a discussion, popularity with a group and even confidence and imperviousness to judgement, can easily lead to power imbalances, in favour of those with the most education and erudition, who may not also necessarily have the best ideas or be most aware of the needs and wishes of staff. This can be effectively managed if acknowledged openly and intentionally addressed, and here a coaching approach of assuming positive intent can help. Some key ways to level the playing field and minimize imbalance in out-dated structures are:

- Creating a high level of psychological safety within the organization so that people are more likely to speak out about what matters to them and feel heard (which we cover in the next chapter).
- Bringing staff together in confidential groups to safely gather opinions that can feed into the planning process, to overcome biased cultural norms and power differences.
- Using coaching tools to boost confidence and build the resilience of individuals and teams as they create change and suffer inevitable setbacks.
- Applying a broader coaching mindset to communication that emphasizes mutual respect, rigorous listening, openness to influence and contracting around what is and is not acceptable, to enable people to voice dissent without personal attack.

## Preparing to launch – moving towards the announcement stage

As you move towards announcement, it helps to take stock of where the organization is and whether those who need to be, are clear about what they are embarking on. If you miss this step, you risk someone derailing the process later. An example of this would be that the Director of Finance does not realize that this means a shift in how they allocate cost-centres, not just a change for the marketing department. A relational approach of ensuring that everyone who will be leading the change feels prepared (both emotionally and practically) makes this much easier to do. In team coaching, this would be a contracting discussion in which the team would clarify and agree first to the change and then to their individual roles within that. This session benefits from the presence of an external party (who could be a coach, consultant or even internal coach), as this part of the discussion is when people may feel threatened by the change and turn to unskilful behaviours such as defensiveness, which a coach or an outsider can deescalate. Some useful questions can be:

1. What level of change is being committed to? For example, does change look like small, transactional actions, like carbon offsetting or is the organization ready to make more transformational change in either operations or of its entire business model that results in rethinking its purpose and way of doing business in the world?
2. Is the organization ready for what you defined in number one? If the organization were ready to begin preparations to move forwards, what would that look like, operationally and culturally?
3. What will the organization look like on the other side of the change and adjustment period? How will it navigate and behave during the change period? What permission will it seek from stakeholders to allow it to change well? These could be practical steps like loan deferment or more cultural agreements like 'shifting from a culture of low risk to one that learns from mistakes'.
4. What role will each of their changemakers play, and is everyone clear on how those pieces fit together into a whole? Even though at this stage of the journey, detailed implementation plans may not have been worked out, this is a fundamental part of the discussion as it helps people to land the 'what' into the 'how', and to get to grips with the practical implications on their roles.

In this chapter we have explored the practical and emotional difficulties that organizations may face in making climate-positive change. We hope we have begun to show you how, just as with individual change, feeling conflicted, fearing judgement and worries about making mistakes can also stall change at an organizational level. In the next chapter we assume that an organization has made a decision to change and consider how organizations can announce this and take big groups of others with them on that change journey, such that it stands a higher chance of success and a lower likelihood of being derailed.

# 16 | How to bring people with you in change

Climate change requires organizational change on a global scale, and yet time after time ambitious change initiatives fall short of expectations because they fail to bring the people within organizations along with them. This can be because the motivation has not been successfully created or because those inside the organization lack the empowerment or ability to change their behaviours or to challenge systems that are not working. If your organization has moved through the rocky terrain of ambivalence, through contemplation and preparation and is ready to commit, it is worth pausing to decide how to announce the new vision and plan to everyone, internally and externally, not least because of the context in which you are making this change. Here we will also look at how change initiatives happen more broadly, as well as the impact on staff of becoming more aware about climate change itself and how organizations can support staff during this double change process of internal change in service of external survival. No one person will respond in the same way to the news that the organization and the climate are dramatically changing. Some will be motivated, some threatened, some cynical and some will experience a burst of existential anxiety as they come to terms with the climate crisis (sometimes for the first time) because the organization (which for many is a constant or an anchor in their lives) is taking it seriously and sounding the alarm. In this chapter we will look at how a coaching approach can help teams to feel secure and how to connect staff and stakeholders emotionally to this big goal. We cover how to make it feel safe for staff to voice dissent and what internal capacity you may want to build to support people as the change journey unfolds.

## Announcing climate-related organizational change

So you are now ready to launch your change process with the organization. This communications piece is important because it marks a clear stake in the ground and creates permission to start doing things differently. So far, so easy; however, this opens up both opportunities and challenges, some of which accompany traditional change processes, and some of which are particular to globalized threats like the climate crisis. First, before it even leaves the C-suite, the gravity of the announcement can be overwhelming for senior leaders who have until now been working behind closed doors. However committed people are – and we find that by this stage they are – there can still be hesitation about how the announcement will be received. Often organizations can be reticent about clearly stating a big, bold direction out of fear of negative reactions from internal or external stakeholders. As Ellie Austin touches on, leaders can be concerned about media accusations of green-washing or finger-pointing at the organization's previous environmental record or that their plans do not go far enough. The confidence of those announcing the change can be intuited by their audiences, so it is worth ensuring senior leaders feel confident and connected to the reasons for the change, so that they are authentic and sound committed. It is also useful to have a clear plan for how information will be cascaded and how those with a role in stakeholder relations will field questions. Waiting until the announcement to have these answers will result in key leaders being overwhelmed with questions from the management level and inevitably becoming bottle-necks for answers flowing back down.

At this stage, there are two clear ways to allay fears:

- Climate change coaches, consultants and trusted friends of the organization (such as Business Declares) can help senior leaders to better understand the nature of their reticence and develop new perspectives around it; for example by creating the Cook Food framing of 'we are just at the start of our journey' that Fiona Ellis wrote about.
- Leaders can work through informal networks to canvas opinions from key external stakeholders (such as institutional investors) behind the scenes, so that their announcement is framed in a way in which it can be publicly supported by those stakeholders, without, of course, losing its punch.
- Strategists can help organizations to plan how they will practically share the information accurately as well as widely.

When we share this kind of externally oriented change with staff and stakeholders, we can engage a greater degree of employee loyalty and connect better to customers, who can attach satisfaction to buying from or working for an organization that is more ethical and sustainable.[102] Briefing staff before external stakeholders also means that they can become ambassadors for the change. While it is possible to bring in change under the radar (and organizations may be able to soft launch a big initiative), failure to internally announce something on this scale robs staff of the opportunity to internalize the change and connect to the climate crisis itself. These missing pieces can create confusion, misunderstanding and resistance that waste time, and without a formal process of engaging with the climate crisis, staff may feel isolated with difficult emotions. A useful coaching tool to help with cascading information is to place different internal functions or audiences around a table, on separate pieces of paper, and to then 'visit' each group in turn, asking: 'what is great about this commitment for them?' and 'what problems does this create for them?' This is less to elicit ideas than it is to identify gaps in knowledge. If the answer to the first question is consistently the same (for example 'because it will make them feel proud') then there may be a lack of true understanding about what the change will entail, and a better sense of the impact of the change is required before an announcement is made. Coaches and consultants helping at this stage can help leaders to assess the consequences of each decision they make in relation to disseminating information and ask questions about the organization's existing resources, such as whether they already have systems in place for cascading other information, such as Unipart, who are featured in the next chapter.

## The increased need for internal support when making climate-related change

While all change inside an organization can be unsettling for staff, most changes are driven by the organization's needs rather than those of the wider world or by existential threat to our very survival as a species. Restructures and new strategies are a familiar part of organizational life, and while they can be upsetting, they do not fundamentally shake our perspective on the wider world. Change that is sparked by the climate crisis has a very different quality and has the potential to surface and even create anxiety. This is because the organization will need to situate the change they are making in the context of climate change itself, and in explaining the urgency with which they feel they need to act, they will also be expressing the seriousness of the crisis. Staff that have not yet fully realized the extent of the threat we face, may need significant support. Many organizations are now employing outside consultancies to provide climate education to teams. Make Tomorrow (founded by Gillian Benjamin, whose story we feature in Chapter 20) is an example of a specialist climate education company in this field, and there are now units within bigger brands providing similar services, such Axa Climate Education. The UK media company, ITV, is a well-known example of a company that provided online climate crisis training to all of its 6,000+ staff.[103] This is undoubtedly a useful way to bridge the gap between the company 'doing good' and 'doing something imperative', and can build motivation and a burning platform. However, companies would be advised not to overlook the anxiety that this can raise and make sure that staff have a place to process it, or else it could become a means of sparking disempowering 'water cooler' chat that can undermine the wider change effort (this is a team version of colluding with each other's fears that we described in Chapter 13). Staff with an existing concern around the climate may find that they are both proud to be part of taking action and at the same time feel haunted by climate anxiety. Other staff who were previously not engaged may feel shaken by what they learn. This is where climate change coaches and other helping professionals can play a strong role.

When organizations make big internal change in service to the climate crisis they essentially connect to two change processes at once: a familiar change management programme nested within a far broader change that threatens our survival. Feelings of existential threat coupled with a much more tangible worry about what will happen to our jobs and teams as a result of a new direction can create two powerful downward pressures on employees. Two forces of change start acting on staff simultaneously, of 'what will my job involve now?' and 'will my family be safe?' If that sounds farfetched, we have already witnessed it. During the Covid-19 pandemic in the UK, we spoke with organizations whose staff were coming to terms with radical, overnight changes in their roles and departments, while also working under the constant worry of potential infection, food shortages and a feeling that the world was falling apart. When it comes to the climate crisis, for many of us in Europe the threat to our survival is located in a dark future, but for those living in wildfire-prone areas or living under threat of dramatic flooding, the threat is very present, as Elizabeth Bechard, who contributed here, wrote in her own book:

one of my Bay Area friends [in California described] the sense of panic at feeling unsafe in her own home amidst the wildfires. She posed a question that was simultaneously rhetorical and literal: where can we go to be safe?[104]

Aside from the well-publicized examples of North America and Australia, any organization that has a team or supply chain in Asia or the Middle East will employ staff experiencing both the immediate and the long-term pressure of climate impacts, alongside the internal changes that the company is making. This can significantly affect stress levels and with that concentration and the ability to keep things in perspective.[105]

This is not to argue against making change, rather it is to argue strongly for internal capacity to meet these emotional needs. This capacity can come from within the middle management cohort who may already be trained in good listening skills or who can be additionally trained in climate change coaching skills. Bringing curiosity to the conversation, rather than trying to force the outcome, helps to make people feel heard and like their stake in the project counts. However, managers also need support because they can often feel like the 'jam in a sandwich' caught between the vision and strategy from above, on the one hand, and delivering on the day-to-day challenges of their team on the other. Unsurprisingly, this drains vital energy for implementing new initiatives and leaves little space for attending to their own and their team's emotional health. This can stagnate progress and the team can return to a business-as-usual mentality. While in any change process, feeling that you have given up or failed can damage team confidence, we can only imagine the feelings that might surface for the team that fails to meet the change agenda of an organization that has taken time to explain the grave consequences of the climate crisis.

Managers can play a vitally important role, but they often feel uncomfortable in emotional conversations with their teams[106] (not least because they are also likely to be going through their own). They may also feel (as we have explained in Part B) that they need to have all the answers or that if they open up a conversation about their team's needs they will then have to 'give' people everything they request. Employing climate change coaching skills of listening deeply, asking powerful, open-ended, customized questions and working with people's overwhelm and scarcity, can help managers and team leaders to navigate these conversations more skilfully, without recourse to solving or fixing. Additionally, managers can contract with staff exactly what the conversation is for and what they both expect, which can help managers to feel more confident. For example, a manager may wish to make it clear that they are not there to provide answers, and a staff member may say that they simply want somewhere safe to share their feelings. In this way, managers can agree (or contract) a space in which employees can work through their emotional reactions to change and design solutions together. This in itself can start to shift the organization's behaviours toward collaboration if the previous mode was 'command and control'.

Managers are a useful support for staff; however, they are necessarily not impartial, and staff may not always feel comfortable sharing concerns with them. This is where external or internal coaches can come in. Smaller organizations, or those with no internal capacity, may hire in climate change coaches to work with specific teams or instead to provide 'clinics' or 'climate cafes' in which staff can drop in (virtually or in person). These can be a very flexible way for staff to access coaching in the same way that they might visit the in-house counsellor. Larger organizations may alternatively have pools of internal coaches and can signpost employees to them for climate change conversations. This can be beneficial because internal coaches will understand the context of the organizational change more deeply. NatWest's coaches, who have been trained in climate change coaching,[107] went on to create digital flyers advertising their willingness to speak confidentially with staff about anything that was concerning them either about the climate crisis itself or in relation to NatWest's strategy of being a more purposeful business. Formal coaching can also be useful because some staff may not be fully conscious of their feelings and a trained climate change coach can help to deepen their awareness. Where perhaps a staff member may enter a session frustrated about a new way of working, they and the coach may discover that underneath it is a sense of despair about the environment.[108] It is important to note that in many countries in which climate impacts are not yet deeply felt, people may still have quite a cerebral attitude to the crisis, which is often compounded by the bias of organizations towards 'head' and away from 'heart'. Coaching can reconnect these two places so that staff can be more congruent with their feelings, and many internal coaches have told us that when invited to, staff will readily discuss their anxieties about the future of life on earth.

Some might worry that opening the lid on these conversations could encourage staff to deepen anxiety, feel more powerless and ultimately stall the change process. However, the opposite is true. Creating safe spaces in which staff do not feel judged and are allowed to express their bigger feelings without interruption (both about the internal change and the crisis itself) can help people to process those feelings and gain the clarity to move forward more easily. Sharing concerns also diminishes feelings of isolation and allows those feelings out safely, as opposed to suppressing them into unskilful behaviours like anger and irritability, which will damage teams at the very time that they need to come together. As with so many things, relationships are at the heart of the success of organizational change,[109] both in terms of our relationship with ourselves, our colleagues, our customers and stakeholders, and ultimately our planet and the crisis it faces. In this context, staff can sometimes find it hard to maintain those relationships (for example, retreating from colleagues because they feel unsettled). Providing adequate support can help colleagues to stay connected, even when it is difficult.

Developing your climate change coaching capacity can help with:

- supporting staff to grapple with their feelings of systemic disempowerment not only individual lack of agency, so they feel capable of making organisational change
- helping teams to be more able to work with uncertainty and complexity
- underpinning efforts to create psychological safety and normalizing fear of failure
- supporting managers to smooth the journey of change by being able to read the emotional field of the team and attend to staff who are finding it hard
- creating safe, confidential reflective spaces for the key changemakers to re-energize themselves and make sense of setbacks.

## In practice

### Broadening and deepening change processes using coaching within NatWest Bank

*Nick Ceasar, lead coach for NatWest Leadership and Coaching Faculty*

*In February 2020, NatWest's CEO Alison Rose unveiled the bank's new purpose to champion the potential of people, families and businesses with the three focus areas of supporting enterprise, building confidence through education and combating climate change. By the time of COP26 in Glasgow the bank had made a significant start by joining three high-profile initiatives, setting targets to provide £100bn in climate and sustainable finance funding, achieving NetZero carbon emissions in our own operations and publishing the* Springboard to Sustainable Recovery Report *into how to support UK SMEs in the low carbon transition. From an organizational change perspective such commitments are encouraging because they place a level of external scrutiny on leaders not only to deliver but to continually ratchet up their commitments. Yet, from a systems perspective, public commitment is just a fraction of the story. Huge levels of collaboration and intelligent risk taking are required to build transition plans and bring the hearts, hands and mindsets of colleagues, customers, stakeholders and shareholders on the journey.*

*We knew that we needed to reduce the experience of profound dissonance, ensure our commitments led to sustained and thoughtful collaborative action, and that any difficult decisions landed on the right side of history. Nowhere was this internal change more important than in the internal workings and personal transformations of our leaders and colleagues for whom an upgrade of their operating principles and values would be needed alongside a set of new extended capabilities that would more commonly be found in the multilateral, think tank and civil society sectors. Such upgrades are seldom linear, smooth or predictable and require significant midwifery on the part of the learning professionals that support them. These learning professionals need to know how to create trusted and safe environments, support and challenge with non-judgemental compassion and create momentum for concrete, lasting and systemically located action. This means paying a lot of attention to coach capability and CPD, benchmarking with emerging best practices and ethics as well as the boundaries and limits of their approach. To that end, in mid-2020 we invested in Climate Change Coaching training[110] for 23 of our coaches so that they were equipped to support the emotional responses of our leaders as they confront the profound implications of their role in perpetuating and responding to the climate crisis and then use systemic coaching techniques to enable impactful action. Here are five examples of how we embedded coaching into all aspects of our leadership transition:*

#### 1. Leadership development through coaching

*Cultivating the right kind of leadership calls for less traditional, knowledge-based training and more identity and systems-based interventions that encourage inquiry into self, purpose and systemic change. Our leadership coaches were taught to use tools and techniques to develop our senior leaders in this way. The coaches also had to role model desired leadership behaviours such as: reflective and mindful attentiveness; creating safety through careful contracting, signposting and showing vulnerability; exploring broader ranges of knowing; and using metacognition, systemic questioning and action planning. Many of these practices have been transposed into other areas of organization life to aid purposeful decision-making and action that is more in line with the bank's new ambitions.*

### 2. Technical training for leaders, supported by peer coaching

*Several hundred leaders undertook knowledge-based, university-level training to understand the science, policy and sector-based responses necessary to halt and adapt to climate change. For these colleagues peer-coaching groups were facilitated to take the 'headiness' of this training and find resonance with the heart, the gut and the hands in order to take action with confidence, support and congruence.*

### 3. 121 climate coaching

*More confidentially, we offered one-to-one coaching support where leaders could rely on complete confidentiality in order to work through the emotional, behavioural and identity level landscape. Here they could locate themselves in the story of our times, galvanize their agency and motivation and identify the differences they were most able to collaborate around for the greatest good.*

### 4. Inclusive decisions

*A long-standing partnership with Blueprint For A Better Business led to us using their framework for more inclusive and thoughtful decision-making. While many colleagues had exposure to the framework, practical application was at its best when facilitated by the coaches, helping leaders and teams to take time to tune into each stakeholder's needs and test decision-making and decision transparency through the stakeholder lens.*

### 5. Using coaching with other players – coaching at COP

*During COP26, seven of our leaders hosted round tables and used coaching tools to generate insight and big business commitment in aiding the small to medium enterprise sector transition to a low carbon future. The outputs of this session are being used to maintain momentum on this topic.*

*It can be tempting to bring our old ways of thinking, being and acting to this problem, but this is a crisis that is locked into our contemporary values and belief systems that sees the world as a series of disconnected parts and nature as independent from humans. As former Governor of the Bank of England Mark Carney has said, unless we change what we value, we cannot be a basis for the change we need to see. Focusing on action without attending to the structures of mind that sit behind that action risks us going nowhere and for NatWest to be able to support needs us to align and transition at every conceivable level. Real momentum needs whole systems to shift and that requires us showing up fundamentally differently to the challenge with new values, beliefs, and capabilities. Coaching has the power to facilitate this showing up with congruence and conviction and to help us to align with the true task at hand.*

## Creating an opportunity for staff to connect more powerfully to your organization

Beyond mitigating pushback, the announcement of a bold environmental commitment presents a fantastic opportunity for an organization to connect to its staff. We spend more time working than doing any other single activity, yet we often assume that people are not as affected by what happens at work as they are by their home lives. In an old, much-told fable, a man stumbles upon two stonemasons building the first few rows of an enormous wall. It is pouring with rain and the stones that they are using are heavy; one of them is grumbling and the other whistling. The man asks the grumbling stonemason what he is doing. 'Building this wretched wall day after day for terrible money in this awful weather, until my hands are raw from the cold and my back aching from the weight,' he replies. Curious, the man asks the whistling stonemason why he is so cheerful when the weather is so abysmal and the work so obviously thankless. 'Because I'm building a cathedral,' he answers proudly. Too many of us think we are just building a wall: only 53 per cent of over 13,000 employees surveyed in a 2020 poll said that they felt engaged in their work. Some in the climate movement now use this metaphor of cathedral building[111] to represent the many small parts that each of us play in such an enormous endeavour, to overcome the overwhelm that naturally accompanies it. Connecting staff to meaning not only of the new direction for the organization, but also to the climate challenge itself, helps staff to commit to action beyond the things that are fun or easy to do, but that require compromise and discomfort.

This is where it helps to think of the vision and values of an organization as the roots of a large tree, anchoring it into the earth, enabling it to stay strong and weather storms, and draw up the necessary nutrients from the soil so the whole tree can grow and develop. In order for leaves, flowers and fruit to grow they must be connected to the nutrient source, not cut off from it. In this same way, each department, team and individual in the organization needs to be connected to the overall vision, purpose, values and strategy so each can bring their own unique contribution of skills, values and perspectives to the challenges ahead. It is beneficial for each team to have the space and opportunity to understand how the vision holds meaning for them and to see themselves in the future organization. All of the aspects of understanding personal meaning described in Part B help to do this, from digging into hopes, dreams, purpose and passions to also recognizing talents, priorities and needs. The announcement of the change is a great opportunity for managers to do this with their teams.

Effective coaching questions here are:

– What are your/our hopes and aspirations for your/our future in this organization?
– How does this new vision challenge those aspirations and support those hopes?

- When do you/we feel the most fulfilled and satisfied in your/our work and how can this new direction offer more of that?
- What attracted you to this organization/this work, and how does this new direction affect that?
- What is our stake in this new vision?
- What are you/we really good at? How can we connect what we do day to day with the bigger goal of the organization?
- Which aspects of your/our work do you/we enjoy most, and how can this new direction offer opportunities to do more of these?
- What will you/we find challenging, and what do you/we need to do to support you/us with that?
- Do you/we have any concerns about your/our role in relation to the change? How can we work with that?
- What do we need to agree/do next to take this from an aspiration to a reality? What permission or mandate might we need to bring this to life?

As these aspirations are discussed, a coaching approach of listening with curiosity, asking open, powerful questions and being mutually respectful can build the resonance of the speaker. Allow the energy to build as people spend time reminiscing about why they joined the organization or what they currently enjoy about their work. This is not for nothing. It is the equivalent of asking a couple in marriage counselling why they fell in love in the first place, and it has the same effect of transporting people out of their current stresses and pressures and returning them to their original intentions and dreams. From this more hopeful starting point we can better build on the new vision and develop new ways of working, than we can from current feelings of being overstretched, which many staff may be. The health warning here is not to lapse into a transactional alternative to this more meaningful process, in which rather than allowing people to fully express their 'dreaminess' we opt for a box-ticking exercise. This is similar to that of a badly executed performance management process in which we do not feel seen or heard, and the conversation has little energy and is rushed to conclusion. In addition to building emotional commitment, activities at this stage that draw on the intuitive parts of our brains can be like having a valuable 'early warning navigational system' alerting us when something is off kilter. Doing a 'pulse check' of everyone's emotional responses to the change adds another layer of information as the change settles in. This is in contrast to the other activities that the organization will be doing – such as translating a vision into a concrete strategy and action plan – which tend to rely on the executive function of the frontal lobe, where we plan, organize and problem solve.

Coaching questions to gauge the emotional temperature around change include:

- What is your initial reaction to this?
- What do you notice about your energy levels as we talk about this?
- Do you still feel motivated about doing this?
- Does this still hold meaning for you?
- How have you described this to your spouse/family/friends?
- What would make you fall in love with this change?

## Making it safe to share unpopular opinions and hearing from diverse perspectives for a climate action culture

Not all staff will be overjoyed at the news of the internal change, even if they broadly support climate action. This could be because staff are still at the stage of 'someone else should' on our flow chart in Chapter 2, or it could be because we have overlooked a negative consequence for a particular group of staff that may in fact turn this good news into very bad news for them. It could also be because, like many in the current paradigm of fast-paced growth, the organization operates too leanly with little 'wintering' or celebration and has failed to recognize the impact of stress on already overstretched staff, for whom any change can be the last straw. Rather than squashing this dissent, leaders can listen for the useful information within it, though that can be easier said than done. Organizations often do not plan extensively for these more emotional responses to change, which can make them seem like a nuisance when they do occur. Dissent can be disheartening, but an emotional reaction, even a negative one, signals that the different parts of the system are assimilating and adjusting. By accepting this and engaging in it with curiosity not judgement, we can better understand the genuine causes behind it and from there the solutions to move us all forward.

Leaders and managers in the system often feel ill-equipped to work with these reactions, seeing them as blocks on the roadmap. A compassionate way to reframe these reactions is that people are trying to meet their needs and find their place in the new strategy. A coaching approach here would be to build the emotional health of the team into the change process, to support the organization to create behaviours fit for a new paradigm in which care and rest are features. More immediately, coaches can help managers to work more positively with these reactions and seek out systemic information from them. If everyone in the team reports wanting to act on climate change but disagrees with a change to a specific way of working, it could be that we need to go back to the drawing board in this particular case. Staff who present a more traditional and institutional style can also often be marginalized during change[112] because they care for and champion aspects of the old world order that is being swept away. They may complain that new ways are 'not how we do it here' or that 'we tried that in X year and it didn't work', and they are often dismissed as simply resistant to change. We ignore these voices at our cost because they can give us useful institutional knowledge and help us to see what needs to be retained, as well as how to help them to get on board.

Powerful relationships act as 'containers' that allow people to successfully navigate change, and to create them we need to build 'psychological safety'.[113] We can do this using the key coaching principles that we shared in Part B. Our fundamental starting position is one of seeing the person as resourceful and whole. This creates 'psychological safety' because the person feels accepted for who they are and enables them to engage with change without feeling threatened. Timothy Clark[114] defines psychological safety as 'the lubricating oil of human collaboration', and he argues that when interacting in social settings such as the workplace, we continuously engage in 'threat detection' to monitor if the group dynamics make it safe to be ourselves. For Clark, psychological safety means that it is not 'expensive to be yourself'[115] either socially, emotionally, politically or economically.

Clark identifies four levels of psychological safety in which we:

1. Feel included: This is the starting point for succeeding in adapting to change. If people do not feel included then the system will not be able to take everyone on the journey.
2. Feel safe to learn: Being open to new ideas requires us to be vulnerable. Teams need to learn and adapt during change, rather than reverting to known ways of working.
3. Feel safe to contribute: Staff need to feel safe enough to contribute and to do this the ground rules need to be clear, to allow all voices to be heard and expectations to be clear in terms of final decision-making powers, so that people understand how their ideas will or will not be used and do not feel dismissed.
4. Feel safe to challenge the status quo: This is crucial for systems succeeding with complex challenges, where we must challenge not just the technical aspects of what we do but also the cultural and behavioural ones.

Timothy Clark argues that, depending on the psychological safety of a social setting, we have two modes of response:

- Performance response – in which we play to win; make discretionary effort; and contribute skills, knowledge and experience without reservation.
- Survival response – in which we play *not* to lose; use a mindset of self-preservation and loss avoidance to manage personal risk; use our creative and productive capacities for self-protection; and reduce our interactions, by saying, doing and trying less and making fewer attempts to avoid mistakes.

You can see how a sense of safety leads us to the 'performance response' while lack of it takes us into those threat-response behaviours that we discussed earlier. The climate crisis plays a role in organizational psychological safety both in terms of its existential threat, as well as in its ability to create rank for those with climate or sustainability expertise. Others may censor themselves and feel 'not clever enough' to contribute to strategic discussions. Subject matter experts within organizations need to recognize that this dissonance can affect whether their colleagues fully contribute to climate-related change processes. Climate change coaches can facilitate activities such as constellations in which all voices can speak and in which judgement is absent.[116]

A healthy system is a diverse one, whether we are talking about organizations or any other living system. One of the greatest threats to an organization's ability to address this challenge is either not having, or not embracing all of the diverse perspectives in finding a way forward. Building psychological safety is key to an organization's ability to embrace diverse perspectives, experiences and skills, and to enable them to contribute fully and creatively to find solutions to these complex problems. No system is perfect and each has its blind spots, and often leaders and managers intend to be inclusive but are unsure how to translate that into the day to day. It is not easy for all voices to be heard when others may get more airtime due to their gender, ethnicity, rank, role or even geographical location.

Consciously creating processes for diverse representation makes it much more likely that these will be achieved.

   Coaching questions to help teams reflect on the system of which they are a part:

–  Which voices does this system routinely hear from? Which does it not?
–  How are emotions handled in this system?
–  Can anger or sadness be safely expressed here?
–  How does the system respond to differences within it?
–  How does this system work with disempowerment?
–  Does this system allow for periods of time when people need to rest and restore?
–  About what do we generally hold fixed mindsets in this system? About what do we generally hold growth mindsets?
–  Are there certain ways of being in this system that are accepted and others that are not? In other words, is there psychological safety within the whole system, or only in sub-parts of the system?
–  How has the organization handled big change in the past? How might this affect people's reactions to change and be a source of potential resistance? If change initiatives have previously stalled, some staff may be slow to buy in, preferring to 'test it and see' before fully committing to disrupting the status quo.

Taking a coaching approach, particularly where concerns and doubts are expressed, allows each group to be able to assimilate the change and move into action. This also includes accountability, and so at the end of these conversations it is useful to then set mutual agreements for next steps (similar to those described for individuals in Chapter 12) so that everyone is clear about what happens next to bring these aspirational conversations to life. Setting agreements is a useful way of creating 'Rules of Engagement' for people to know how best to work with each other across functions. You can do this in a meeting by:

–  Attending to behaviours: Asking 'What behaviours or attitudes will make it easier to do this?'; 'What kind of an atmosphere will help us to thrive and succeed?'; 'How will we know if we have these things in place?'; 'Which values do we want to bring to life in the way we work?' – for example 'honesty' or 'we need to be able to make mistakes without blaming or criticizing each other'.
–  Creating a process: It is helpful to translate these ideas into processes, for example: 'For each new idea we test, we will decide how long we will run the test for before we review it, so we have time to learn from what happens rather than ditching something that is not working too soon.'
–  Agreeing accountability: Asking 'What will we do if one of us cannot meet a deadline?'
–  Designing an agreement for conflict: Asking 'How do we want to behave when things get difficult?'; 'How will we successfully challenge each other without damaging our relationships?' For example, 'If things get difficult let's not make it personal and remember that we are all trying our best to make this work.'

- Reviewing the agreement itself: It is helpful to be specific and ask, 'What is working well with our agreements, and what do we need to adjust?', rather than a general 'How is it going?' This allows not only for agreements to be upgraded but also for any festering issues to be acknowledged. For example, 'We said that if we missed deadlines then we'd have a meeting about it, but it keeps happening and we've not come together to discuss the impact of it.' On the flipside, reviewing how you are working together can also be a way to celebrate what is going well, and this increases the emotional positivity of the relationships, which as John Gottman[117] says increases the resilience of the relationship.

As an organization signals that change is coming and that familiar ways of operating are ending, there will be an inevitable unsettled period as people search for solid ground and create new ways of doing things. Performance inevitably dips as people assimilate the change and there is likely to be a corollary decline in confidence which can also affect motivation. This is all a normal part of change as people let go of the old and begin to internalize and make sense of the new. It is natural for people to feel uncertainty about their role in the change and the future, about their identity, their contribution, their ability and their place in a changing system. However, when we are making a change in service of our bigger environment and planet, we are likely to feel a greater sense of motivation, pressure and responsibility, and staff may also need help coming to terms with the crisis itself. Each part of the system will go through their own stage of reacting emotionally, testing, questioning, doubting, grieving as the new gradually emerges over time. If coaches, leaders, managers and staff throughout the organization are able to employ a coaching approach in their interactions with each other, they are more likely to come through this process with strong (if not stronger relationships) that can support good collaboration. They are also likely to create more positive memories of the change and of their ability to be part of impactful climate action.

# 17 Creating a climate action culture navigate the challenges of landing big climate actions

The new terrain of climate-related change poses different challenges for organizations that are used to driving change initiatives with a limited organizationally oriented goal (such as growth or cost reduction). Familiar change techniques may or may not work in a shifting external environment in which competitors, regulators and even the natural world itself are shifting beneath the organizations' feet. Not only is the organization attempting to turn its aspirations into reality – the organizational equivalent of the new year's resolution – but it is doing so in the context of (and driven by) an uncertain external environment. As visions get turned into plans, the management level in the organization will naturally have a lot of questions about how to meet the requirements of an economic reality that does not yet exist. Landing big ambitions in this context requires some familiar elements of translating values and ideas into concrete strategies, setting goals to meet those, and milestones to keep it on track, but it also requires some potentially new behaviours including a culture that embraces its mistakes as a form of learning together with the emotional resilience to operate in uncertainty. To activate the whole system to succeed in complex change, it is even more useful to hold in mind the two dimensions of the Flynn model of sustainable change with its being (motivation) and doing (action). While, as we have said, leaders and teams may wish to leap into action, it is useful to pause and assess the nature of both the context and your organization's norms first.

Here we look less at the mechanics of traditional organizational change, and more at the newer context of climate-related change, offering some coaching tools to create a 'climate action culture' and ensure that motivation and action stay aligned. While in the last chapter we looked at how to bring people with you emotionally in change, here we consider how teams begin to transform what the organization does. Leaders who assume that command and control will solve their problems, and that motivation no longer matters once plans are created, do so at their cost, both because staff respond badly to imposed change at all stages and because those at the top rarely understand the operational realities as well as those on the ground. The essence of engaging teams in the complex work of landing a climate ambition is to help them to better understand the context in which they are implementing, connect them to what the broader issue means for them and free them to take risks and design experimental practices that can be scaled.

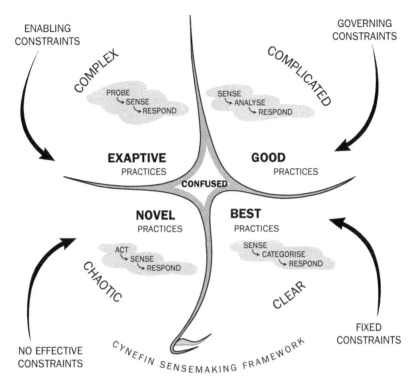

**Figure 17.1** The Cynefin framework[118]

## A new, complex terrain in which to solve problems

This is important, purposeful work, but it is not easy. We have spoken to many technical experts leading organizational change who admit they are very much still learning. Dramatically retooling existing businesses to be fit for the future is often pioneering work, as Linda Freiner noted in relation to decarbonizing Zurich's underwriting portfolio (just one aspect of their vision). This can involve the creation of structures and processes with few precedents to follow. If we enter this realm of change using our old approaches to planning and managing, we can make it harder for ourselves.

The Cynefin framework (see Fig. 17.1) is a response framework that explains why it can be hard for teams and organizations to grapple with this context, by helping us understand the nature of the terrain we are in and the best way to respond.

The framework describes five contexts, of which we will explore complicated and complex:

1. Clear contexts – in which cause and effect are easy to see. Here we see an issue (sense), put it into an existing category and take action (respond).
2. Complicated contexts – in which we need to analyse what we sense because problems are not immediately easy to understand. We may require experts to help us respond, by using known solutions that have worked elsewhere. Here

we do not apply best practice (one size fits all) but 'good practice', allowing that there will be several equally useful solutions, and recognizing that insisting on one creates resistance from teams.

3. Complex contexts – in which cause and effect are hard to see until they have happened, which are unpredictable and where the act of probing itself affects the system. Here it is safest to try small experiments rather than set sweeping policies.

4. Chaotic contexts – in which the best starting point is often acting to establish order, from which other sense-making can follow (i.e. during an emergency like a flood).

5. Confusion – which is the confusing state of not being able to see which of the four other contexts we are operating in. This is where we are much of the time, and unknowingly we often choose a course of action based on our personal preferences.

At first glance, many organizations may assume that climate-related change falls into the complicated context. Organizations decide on change and then look for experts to advise on how to create it. However, it is more often the case that climate-related change presents a *complex* context because the territory is relatively uncharted and the future unknown, and because this type of change involves not influencing the way that the organization behaves but also influencing the wider economy (and being influenced by it in turn). This is a context in which the mindset of acceptance and non-attachment are supremely helpful, to ensure that change leaders do not become fixated on a particular solution, and that specialist teams do not become identified and siloed as 'the ones that know' in the organization. Creating a culture of what Nick Ceasar from NatWest has coined as 'authentic dissonance' (the ability to be publicly honest about the difficulty that such big change presents, rather than attempting to pretend all is smooth-sailing)[119] can help teams to feel more comfortable with uncertainty and make those 'small bets' that probe for solutions. It is also important to heed the warning baked into the Cynefin framework, that our personal preferences can cloud our ability to even recognize which context we are operating in. If you are an engineer, it is likely that you will see problems through that lens, assuming that the solution is an engineering one.[120] Being able to step out of our identities and learned behaviours can help us to grapple with these complex contexts, and arguably this is where team coaching, with its ability to create a safe space to explore, can come into its own.

For those supporting organizations through this type of change, it is not enough to bring technical expertise alone, because it can become irrelevant or out of date as the operating environment shifts. Rather, it is that technical experts can use their knowledge to determine the right questions to ask, to navigate the complexity. For those advising companies, powerful, open and customized coaching questions can help to tease out information from teams that they may otherwise not think to share, but which illuminate the context and the way in which the company can respond to it. Professor Snowden and his team have found that coaches and consultants work best in this context when they facilitate the process of discussion without trying to solve the organizations problems themselves.[121] As a consultant or coach it can be very easy (and tempting) to become enrolled in the company's quest and jump in with them, but Professor Snowden cautions against this. He and his colleagues have seen that refusing to suggest ideas and instead holding the process so that the team creates the solutions, leads to a stronger sense of organizational ownership.

Complex contexts are often unpredictable, and according to Snowden and Boone, the best approach here is to 'Probe – Sense – Respond'. Rather than trying to control the situation or insist on a widespread plan of action, it is often best to be patient, look for patterns, and encourage a solution to emerge. It might be impossible to identify one 'correct' solution, or spot cause-and-effect relationships, in 'complex' situations. It can be helpful to conduct short or micro business experiments, which are low risk because of their scale, duration or intensity, and into which is built an acceptance of failure as part of the learning process. Equally, because failure in and of itself is edgy (especially for those who are ambitious), creating team 'contracts' around permission to fail and how they will use and attribute learning from failures, may make it feel safer for individuals to engage with projects that may not succeed. You may also put processes in place to guide the team's thinking; even a simple set of rules can lead to better solutions than no guidance at all. Breadth of perspective is also helpful here, and the more diverse the group of people you bring together (both in terms of length of service as well as of gender, ethnicity etc.) the better.

Climate change coaches, consultants and others working within organizations committed to addressing climate change can support teams to embrace qualities that encourage innovative problem-solving such as:

- growing a team's ability to work in the context of uncertainty
- helping teams to work creatively to find new solutions
- encouraging staff to step outside of comfort zones and familiar ways of working
- accepting that no one has all of the answers and that failure helps find solutions
- fostering collaboration across departments, specialisms, perspectives, experiences and ways of working, and dismantling a culture of siloed working
- incorporating information from a range of sources:
  - emotions as well as thoughts and facts
  - hearing from diverse stakeholders within the system (not just the ones that traditionally hold power or who agree with or confirm what we wish to hear)
  - bringing in the needs of wider stakeholders of the system including customers, communities affected by the organization and the planet.

Changemakers in organizations have told us that they have had to develop a lot of self-confidence to be able to publicly acknowledge and positively accept that they cannot 'fix' the problem for the rest of the organization, but that they need to work collaboratively and creatively with the organization to find a new way forward. Coaches can provide the kind of confidential space in which leaders can test out these behaviours and build their confidence, and the behaviours we covered in Chapter 5 of compassion (and self-compassion) and a growth mindset that is open to new ideas can support teams through challenges. However, the most important quality to develop when making any change is trust, and this is particularly important when the change is taking place in a complex context, in which competitors are also transforming, and which has the potential to lapse into chaos. Here again we see the importance of robust relationships, with ourselves and our own sense of identity, and with each other and the systems in which we live and work.

## In practice

### The challenge of organizational transformation

*Sheila O'Hara, IBM France*

*Companies today are faced with three major challenges, first, managing and estimating the impact of climate change on their business, second, estimating the impact that their business has on the environment and the climate, and third, implementing strategies to counteract the first two challenges. As most large, complex organizations are entwined with others in their sectors, as well as with their suppliers, this means transforming not just their own organization's activities but working across their sectoral ecosystem to also reform the way the system behaves.*

*A recent study by the IBM Institute for Business Value (IBV)[122] shows that the pandemic has fundamentally reshaped the relationship people have with sustainability, from shopping and investing to employer preferences. This shared sense of purpose between businesses, employees and consumers should make it easier to implement new business models that support more sustainable and responsible practices. These business models need to be collaborative, transparent and involve cross-enterprise networks. While technology has a major and positive role to play in the solutions that will need to be implemented, technology is the easy part. The difficult part is breaking down the silos of information and processes, distributing responsibility and determining incentives. This is where technology meets human behaviour.*

*One could argue that everyone's incentive is already clear... saving the planet. But, of course, it is not as simple as that. Having a shared sense of purpose is the first step. When you start to get into the details it is not as straightforward. Companies want to make decisions and choices that best suit their business objectives and protect their competitive advantage while also fulfilling the objectives of the wider ecosystem. Yet the distribution (and therefore public ownership) of responsibility is often a difficult pill to swallow. The new business models require companies to be more transparent and therefore be openly responsible for their activities, and this can take leadership courage.*

*An example of this can be found in a project IBM is working on with other technology and fashion industry partners for the UK fashion and textile industry. The goal is to design, prototype and pilot a new technology platform to help drive sustainability through increased transparency within the supply chain. Several well-known retailers in the fashion industry will be part of the initial pilot. The new technology platform will combine a number of emerging technologies like blockchain, AI and sensors to digitize the key processes in the supply chain creating a shared system of data that the different parties can trust and easily act upon. This will enable a better understanding of where and how fabric was processed, by whom and in what conditions. It will also make it possible to monitor production processes with the objective of reducing waste and optimizing stock.*

*Projects like this address one of the major obstacles preventing organizations from implementing more sustainable, responsible practices: a fundamental lack of transparency and visibility across the supply chain that lowers trust. Data is siloed within organizations, systems tend to operate in isolation and people in organizations have had little to no incentive to share data with the rest of their business ecosystems due to the significant manual effort and a lack of trust. Building trust and determining incentives encourages communities of participants to cooperate, which makes it worthwhile, and leads to the success of the network. Although trust is a qualitative attribute the characteristics of blockchain technology (like the fact that it is immutable and secure) reassures and builds trust. The alignment of purpose, collaboration and digital networks is a powerful combination. It is also important to devote time and energy to understanding what motivates and drives each of the companies in a network. Trust, transparency, visibility and collaboration are key to transforming the way that companies work, helping them to transition away from a linear and siloed way of working, to taking holistic, network-based responsibility for their activities to create a more sustainable world.*

## Keeping eyes on the prize – connecting practically to purpose, vision, values and each other

As each part of the organization turns towards action and creates its own strategies, goals and milestones, it is important to keep questioning how they link back to the new purpose, vision and values, using these as a compass to navigate decisions and dilemmas. Clarifying and checking in on vision, purpose values can ensure that our strategy and goals remain meaningful, so that instead of doing something because we have been instructed to, we are doing it because it aligns with our beliefs. This can also help to pull the thinking up from the micro detail to take a look at the bigger picture as it relates to the climate crisis itself, to imbue day-to-day changes with a sense of motivation and enable teams to get less 'bogged down' with the current paradigm and take steps towards creating a new one. Values give us organizational cultural guidelines as to how 'we do it around here', and can be revisited to ensure they fit the emerging context. Connecting to shared values can bond teams together during change. If values are diluted at the organizational scale, to create enough resonance to tackle change, teams can instead identify how organizational values are alive in their team and add in other values that they feel are particular to them. This is also a great opportunity for individuals to recognize (as Tom Crompton explained in Chapter 10) any misapprehensions they might have about colleagues being extrinsically more than intrinsically motivated, so that they can align around values that serve the world as opposed to just themselves. This is especially useful to building trust – because we perceive others to have our backs – and also to overcome perverse incentives against change (which we shall cover later in this chapter) such as financial incentives for continuing with old ways of working. Understanding that others can be as motivated as we are by 'doing the right thing', can help teams to travel together.

Some useful coaching questions at this stage can be:

– How does doing this fulfil our bigger vision for the planet?
– Is this micro-action in line with our macro purpose or our reason for existing?
– Does doing this as a team move us as an organization nearer to, or further away, from our vision of a liveable world?
– Does doing this enable us to operate in the way that fits the culture of our vision (i.e. the behaviours we employ in our future vision and paradigm)?
– Is this strategy in line with our vision? Will it deliver our vision?
– What is our/this team's part in achieving this vision for the organization and the planet?
– Does this action support or thwart us operating in line with our values?
– Which of these options is most in line with our values?
– If we were to consider this from a perspective of 'value x' (e.g. efficiency, integrity, creativity), what might that do to the ideas we have/decisions we take?
– If we were to do this in a way that modelled 'value x', what behaviours would be displaying/not displaying?

This is obviously most important to enable teams to succeed when the change starts to bite and be implemented, to make sure that individuals do not become lost, detached or disengaged. Doing so can also help to create a sense of belonging and loyalty that unlocks more solutions and makes it safe to test out new behaviours.

## Encouraging new behaviours, celebrating success

Once the organization has entered the action phase of implementing change a coaching approach can help it to think about the new behaviours that teams may need. This is an area that is often overlooked yet we often witness a chance for innovation being undermined because the old systems, processes and rewards baked into people's day-to-day jobs make it hard work to do things differently. As part of the cycle of change, we also have to consider what are the consequences of these new behaviours, and do they increase or decrease the motivation to embed change in the long term. In organizations, individual behaviours come together to create norms, which groups then maintain. Managers therefore do not need to be responsible for changing norms on their own. Asking teams for their ideas about how to change and celebrate new behaviours makes it a shared responsibility and the act of asking for ideas in itself is an opportunity to empower staff.

Because change takes effort we can naturally resist it – looking for excuses to do things the old way – so teams can agree to changing together as a means of holding each other to account. When we learn new ways of working, our old, familiar behaviours require less effort and are likely to be more rewarding in the moment than newer ones that require conscious thought and work. Whenever we embark on a new project, if the actions required are easily repeatable and the results are positive, we are likely to keep repeating them until they become embedded as a habit. However, if there are unintended negative or even neutral consequences our motivation will nose-dive, we will find excuses and our progress will falter. This is where the team comes in. If we have made an effort to do something new, but it goes unmeasured, unnoticed and unrewarded, we may be less motivated to do it again. Equally, if we cannot see how a new behaviour feeds up into the overall new vision, that too can feel demotivating. Celebration, positive feedback and staff development are all ways to change behaviours and, over time, norms. It takes time and effort for a new behaviour to become a habit, and a range of tools may be used including training, coaching, mentoring and peer support, acknowledging effort and celebrating success, until new neural pathways have developed and a habit formed. With climate-related change, organizations can also offer rewards and puncture points along the behaviour change journey that reinforce the new vision. These could involve funding environmental initiatives like tree planting or even sending teams on bespoke environmental awareness courses.[123]

We often hear from teams and organizations that 'we do not really celebrate things. We just move on to the next thing', or 'there is not much recognition around here'. This 'constant fruiting' (impossible in nature's systems) misses an opportunity to strengthen relationships and underline what is meaningful and worth celebrating. Much more than a 'nice to have', celebration should be seen as a crucial element to successful change, particularly when dealing with climate change, which can create such intense opposite emotions. Celebrating successes creates a sense of connection and fulfilment in a team that can fuel their motivation to get through difficult times. Psychologist John Gottman, calls this 'positive sentiment override'[124] and suggests that in order to build resilient, healthy relationships, particularly during times of change or conflict, we need to front load our positive to negative interactions by a ratio of at least 5:1 to help us navigate through change and challenge.

## Overcoming perverse incentives

For as much as staff may want to positively impact the planet, systems can work against them and leaders need to be aware of any negative consequences that might unwittingly make it harder to create new norms and for new processes to work. This requires teams to iron out conflicts or perverse incentives in the shift from the 'old' to the new way of operating. Where organizational change centres around a simple, single vision such as increasing profitability, this can be far easier than when it is a more complex change such as addressing often conflicting sustainability targets. A simple example is where a new directive is issued to move away from working with companies that have a negative environmental impact; however, the sales team are rewarded based on the number of short-term sales they make, and finding companies with transparent environmental records is complicated and time-consuming. The outcome is that the employees will now have to work much harder for their sales bonus, which will feel like a punishment rather than a reward. This kind of contradiction creates friction for people to navigate, draining motivation and diverting energy from the team that could be used more creatively to move the business forward. This is before the organization reckons with the compromises that are part and parcel of becoming more sustainable.

To address this, change agents can work directly with teams to identify the conflicts and use a creative problem-solving approach to overcome them. While some may be resolved at a local level, others may point to how the vision, strategy or structure of the organization need to shift. It is important that the people directly impacted are involved, as they are best placed to explain the conflicts and have the greatest stake in solving them. This process reduces the dissonance staff may feel when they are told, on the one hand, to do one thing but incentivized by the system to do another. Below are some coaching questions to identify outcomes and impacts of new behaviours:

- When you are doing your job day to day, is it easier to do it the old or the new way? Why?
- Do our systems and processes help or hinder you to work towards our vision?
- Can we resolve this here ourselves, or do we need to involve other teams?
- Can we resolve this at this level, or is this a conflict inherent in the vision/strategy/structure/goals of the wider organization?
- What are the next steps for taking action on this? What would an ideal outcome look like if we were to really be able to go full steam ahead with this?

## In practice

### How global logistics provider, Unipart, rolled out its net zero plans to over 8,000 employees

*Frank Nigriello, director of corporate affairs, Unipart*

*When the company set the strategic intent of becoming net zero before 2050, major programmes around green energy and converting the fleet to EV were introduced. These were in many ways the easy things to agree, requiring relatively few decision-makers. But we knew that to really reach the target would require a culture change, and that meant engaging the creative thinking and commitment of the majority of our 8,000 employees.*

*We have a strong track record of employee engagement, so we had some existing processes in place for this. Over the past 30 years, we have been training people in a creative problem-solving approach known as 'The Unipart Way'. Unipart's 'Our Contribution Counts' programme uses a six-stage methodology and peer coaching to enable teams to identify and solve problems in their respective areas. The six stages in creative problem-solving are:*

*1. Define the problem and the desired state.*
*2. Determine the root cause.*
*3. Explore options to address the root cause.*
*4. Design a solution.*

*5. Implement and test the solution.*

*6. Measure results.*

*In practice, teams meet daily for 15 to 20 minutes at a digital communication cell that provides all the essential information for the team's operation. Colleagues are encouraged to cite problems that are logged and addressed through problem-solving circles. Because employees are trained in the creative problem-solving approach, they can work through the process together in groups that vary from four to eight people, capturing the results of each stage on an electronic system. When the results are quantified and seen to be successful, the information is logged on Unipart's searchable knowledge management system and shared with the entire company. In this way, all employees know what Unipart knows when they need to know it.*

*We had invested in creating a culture where people solved problems at their own level and were encouraged to do so on a daily basis, and so when we set 'going green' as one of our strategic directions for 2021 we could leverage these processes straight away. To kick things off we launched 'Green Friday', a one-day event where teams were asked to study their work areas and identify ways of reducing carbon as well as eliminating plastic, reducing use of resources (like water and energy) and designing improvements that would make a difference to our environmental impact. We hoped to see some good ideas emerge, but we were surprised by the enthusiasm and creativity. More than 600 ideas were recorded across the group internationally, many of which were implemented immediately. In some teams, for instance, energy usage was reduced, in others single-use plastic was eliminated. Others proposed more complex solutions such as localized solar power generation, while other teams simply eliminated the use of printing. We then used knowledge management tools, like our internal version of Facebook, to share those ideas across the group to spread good practice quickly. Many of the changes were relatively small but highly effective. Because the improvements were designed and implemented by the teams who would use them, they had strong ownership and so tended to become 'the way we do things now' and were sustained over time.*

*Having signed up to the UN's Race to Zero campaign, Unipart also initiated a strong communication programme to reinforce its commitment and to report the company's progress at a corporate level as well as on each site. People are motivated by seeing progress but also respond to the challenge to correct situations when results are moving in the wrong direction as well as when changes are successful. This is an area where coaching can be very helpful as a coach can help and encourage a team to identify ways to improve through the Socratic method rather than imposing instruction. We have been using this process when it comes to eliminating waste for many years, and now we are applying lean thinking to green thinking.*

*We have also adapted one of our digital products, Paradigm Insight, to provide condition-based monitoring on the core 'green' metrics at each site. This enables us to see at a glance exactly what is happening in terms of carbon reduction and resource usage at each site, which in turn enables us to act on a problem before it becomes a crisis. Through team-based problem-solving, colleagues can also make incremental improvements and measure the impact. That will be an important tool in building momentum and making the company's commitment to net zero carbon a reality.*

## Organizational accountability – measuring impact and sharing feedback

Regular measurement, such as Unipart's system, which offers clear, immediate, objective feedback about progress towards a goal, is a powerful tool for keeping change on track[125] and increasing motivation in a group to achieve a goal.[126] Measurement matters both inside and outside organizations, where the data is publicly available, as Sheila O'Hara from IBM noted, because it strengthens accountability and creates the value that ensures initiatives continue. Jostling for position with peers over universally acknowledged measures can also stretch teams and even organizations to improve. At a workshop in 2016[127] Charlotte Sewell (then head of impact and learning at Cook Food) joked that B-Corporations good-naturedly vie for who can top the leader board in the B-Impact Assessment, and we can imagine teams competing internally along similar lines. In this case, this undoubtedly serves society and the environment as much as the competitiveness of the company's involved.

In Part A we talked about the importance of agency in people feeling empowered in the face of complex systemic issues like climate change. In this context, concrete feedback is an important building block of people's sense of agency and ownership during change. Measurement should provide teams with specific, objective and ideally fast feedback as a consequence of their actions. This should not, however, rely on opinion, judgement or criticism. Our motivation and sense of empowerment are not objective and constant, and they can be affected not just by our own mood but also by bad outer-world news. Therefore having feedback that is even, objective and consistent is an important counterbalance. How managers discuss the results of measurement or offer other forms of feedback, is critical to the organization keeping behaviour on track in a way that maintains motivation. A coaching approach known as 'appreciative enquiry'[128] is a powerful way of maintaining that sense of ownership while understanding the factors that contributed to the results. This method asks: 'what went well and what needs to be adjusted in the future?' alongside 'what are you pleased about, and would you change next time?' This method is without blame and judgement, which promotes that culture of learning from mistakes that we covered earlier and helps people to move away from 'playing it safe'. Teams that use this approach may also want to take it into the rest of their lives, in order to have more successful conversations about climate change with friends and family.

When quantifiable measurement is not available, there are other ways in which subjective observation and feedback can happen and still maintain a sense of agency. We spoke about the importance of psychological safety for our ability to change and grow, and feedback is crucial for that. This does not mean 'letting people get away with poor behaviour' or 'rescuing' or protecting team members from bad news' – that is ultimately demotivating and creates a 'parent–child' dynamic. Rather it is about giving feedback that empowers someone to perform better and have agency over their own growth. In advance, consciously remind yourself of work someone has done that you were pleased with or even proud of or inspired by, so that you enter the conversation seeing them in your mind's eye at their best. If you can hold this perspective into the conversation that follows, you are both more likely to believe that this is a question of some behaviours needing to change, not that the person is fundamentally dysfunctional, and the other person will not leave feeling criticized for who they are but rather supported to correct something that is holding them back. Of course, before giving feedback, you should also 'contract' with them how they like to receive it. This puts the power for the conversation in their hands. Follow up by asking if there is anything the other person needs in order to put the decisions you have taken together into action.

Useful coaching approaches for giving feedback:

– Make sure that the feedback is specific and objective – naming the concrete behaviour, rather than a character trait, your impressions or judgement.
– Observe what is happening and describe the situation specifically and objectively, without judgement: I saw … /I heard …/I noticed …/the situation is …
– Formulate a clear, positive, do-able request: Next time/Please will you …/are you willing to do this …?
– If you are the person giving the feedback, own how something affected you: Express your feelings (rather than guessing at theirs) using statements that begin with 'I feel …'
– Acknowledge your needs. Humans help each other more than we help organizations, so do not hide behind departmental targets, instead express how this person can better help you: My need is … /because I would like …/ I desire …/I need to see/feel/hear …

## A story of shared humanity and humility

To end this chapter, we wanted to share a story of collaboration and in many ways the ultimate coaching approach. In 1984 Toyota and General Motors partnered to see if it was possible to rescue a GM car plant in which GM had lost faith. They called the experiment NUMMI (New United Motor Manufacturing Inc).[129] The GM leaders claimed that the staff were lazy, incompetent and uncooperative. Toyota, meanwhile, was producing higher-quality cars for less cost than GM and offered to share the secret with them. With little to lose, General Motors agreed. Toyota took key managers from the GM plant over to Japan and taught them the 'Toyota way'. That method consisted of one simple word – 'teamwork' – but contained in that word were a huge set of behaviours. Whereas on the General Motors assembly line, staff were not trusted to halt the line if there was a problem with a car, on the Toyota line they were encouraged to do so and more. Whereas at GM staff were at loggerheads with management, at Toyota, a manager's first thought on approaching an assembly line worker who had halted production was 'what is it about this line that is making it hard for you to do your job?'

Fundamentally, Toyota saw its staff as good people experiencing a problem, as opposed to bad people who were thwarting the company's efforts. In other words, at Toyota, managers put aside ego and vengeance and were not activated by threats. Toyota's approach to the shared goal of producing a quality car was to help everyone feel empowered around it and to feel proud that they had played an equal role in the finished product. Former GM staff spoke of feeling pride for the first time at seeing their cars in the same local dealerships that had stocked them for years. Where at GM cars would come off the end of the line with wing mirrors missing (at which point they were expensive to fix), at Toyota, errors were addressed while they could be corrected at low cost. Toyota staff knew that they would not be blamed or bullied, and that if the managers had to make tough decisions, it was not because they were being malicious. The compassion flowed both ways and it dramatically affected the physical finished product. As you sit down to create shared work plans, or give feedback on progress, consider how you can reinforce this message about compassion not blame and openness to failure not fear of it, and that doing so is not just a way of making the change process feel more comfortable but also of landing concrete ambitions. It is likely that you have people in your team who have experienced the GM rather than the Toyota way, and require reassurance that you care more about them and reaching the goal together than you do about keeping score along the way.

# 18 | How the cumulative acts of individuals build to create social change and paradigm shifts

In this final chapter on systems change we examine social systems, how the cumulative acts of individuals create tipping points and how this can be supported by coaching skills and behaviours. We look at collective action and consider unhelpful systems behaviours and the impact of polarization, looking at the dynamics of relationships and our ability to engage across differences. Finally, we consider paradigm shifts beyond an organizational or sectoral level to more global shifts. Earlier we used the metaphor of a tree, deeply rooted into the earth, to help us think about how purpose can connect different elements of an organization together. Trees, however, do not stand alone, but are just one part of a much broader thriving ecosystem – a forest. Individuals, networks, groups, organizations and political parties are just some of the elements that make up our social forest, each element interconnected and interdependent on each other, the whole greater than the sum of its parts.

## Emergence: how individuals connect to create networks and new systems

People often stall on climate action because they fear their individual efforts are insignificant. In many ways there is truth to this, and there is evidence that a focus on individual change has been employed by big vested interests (some corporations, but also politicians) to distract from the change they themselves need to make.[130] However, discounting individual actions altogether misses their function. Individual action when combined together amplifies those joined up actions and engages and enrols yet more people, as individuals connect and work together to multiply their effects at a broader scale. This phenomenon is somewhat described through emergence theory where an entity acquires properties that its constituent parts do not individually embody. New behaviours or systems are only observed when the sum of the parts interact as a whole. Margaret Wheatley and Deborah Frieze's[131] use of emergence theory to describe networks and scaling social innovation beautifully mirrors how forests behave: 'networks result from self-organization, where individuals or species recognize their interdependence and organize in ways that support the diversity and viability of all. Networks create the conditions for emergence, which is how life changes.' They describe how change begins with local actions, which if unconnected to others, has little impact beyond that single action, but if connected to other actions can be transformational. They observe that if networks form and those networks then connect, then their actions become converted into intentional practices from which new, powerful and influential systems can emerge. These connected networks become hotbeds of learning and sharing so that knowledge and information can easily and quickly be shared between the group where there is a common interest to advance the field.[132]

A striking feature of these emergent connected networks is that they have qualities and capacities that did not belong to the individuals within them; the system organically develops characteristics of its own that allow it to influence better than it would if centrally planned by a few leaders, and 'possesses greater power and influence than is possible through planned, incremental change'.[133] We see this in the development of our older cities. No amount of planning or imagining could have created the energy and vibrancy of cities like New York, London, Delhi or Nairobi. Intellectually planned new urban environments often lack the buzz and creativity of neighbourhoods that emerged when communities connected together out of necessity and created something that combined all of their personalities and desires. For those leading climate networks, while there will necessarily need to be some rules and structure, it can be helpful to loosen the reins of control over the activities of members, and how they link with other networks, so that the ecosystem can form more organically and capitalize on opportunities for growth of which the centre is unaware. For leaders this means leaning into the coaching behaviours of non-attachment and acceptance of failure, so that collaboration and innovation can proliferate, free of centralized command-and-control.

## In practice

### Creating and sustaining a social movement

*Rowan Ryrie, mother of two young girls, environmental and human rights lawyer and founder member of Parents For Future UK and Parents For Future Global*

*One of the first lessons I learnt as a new mother was the extraordinary power of love; that love can be a source of incredible strength, and that you cannot do it all alone. Like many of us in Western societies – where we have been socialized to judge our self-worth by what we can do on our own – the experience of needing to rely on others was uncomfortable, but parenthood taught me that our resilience lies in the community of which we are a part. Parenting also taught me about vulnerability, about fear of what the future holds and importantly also about joy. All these emotions are given space within my climate activism with Parents for Future, where pretending them away makes no sense. Emotions – and most importantly of all, love – are at the heart of this work. While there is, of course, space for anger, while anger burns bright, it also burns out. Love persists. Once you have connected the climate crisis to the lives our kids will live – and the lives many children and adults are already living – it is very hard to put down the determination to act, or the love you feel.*

*I work with Parents For Future in both the UK and globally. We communicate with narratives and frames that are hard to disagree with: we focus on ideas of legacy, love, long-term thinking and wanting our children to be safe, and use those to push for specific policy demands. Research we have done has proven the potential of this emotive perspective in engaging audiences not easily reached by much climate communication. This is a movement led by women and that shows in how we communicate. We talk about togetherness – both intergenerationally and internationally – and we deeply value connection. At times connection has been given priority over strategy; leading to us even pausing strategy development processes to heal divides within a group and look at how we are working to ensure the 'doing' of the work is not undermining the connections we value. As movements that have been entirely driven by volunteers, without that committed, connected community, no work at all will happen.*

*Building connection and trust among communities of activists all over the country or all over the world who have never met – and may never meet – is hugely time-consuming. At its best it can be fantastically grounding to feel connected to like-minded, similarly committed parents all over the world, but it does not take many people to cause a lot of disruption in a group with a very open structure. While mobile platforms connect us, they are not without huge drawbacks. We are still waiting for the perfect technology for movement organizing, so making sure we have those relationships – and ways to heal them when tensions arise – is hugely important.*

*My personal engagement with climate activism was very much inspired by the youth school strike movement and Fridays For Future. I have learnt a lot and take enormous inspiration from the community the youth movement has built; from their ability to deeply listen to each other when uncomfortable conversations arise and to create transparent and safe spaces to question power both outside the movement and within it. Though learning about new tech options from teenagers makes me feel ancient! Both the youth and parent movements at international level have experienced similar challenges when it comes to retrofitting governance and decision-making to existing movements of hundreds or thousands of activists with different ideas about what the movement should be. Part of what makes a movement distinct is its energy and adaptability. That requires organizers to be OK with change and to be flexible. More traditional organizations like funders and NGOs do not always understand the sometimes chaotic nature of movements, and this can make working together harder. When structures are unclear, our relationships with all of our partners become more important than ever.*

*Connection gives many of us courage and security. Just as we see our children willing to climb higher on the climbing frame if they know we are watching them, so knowing people we respect are supporting us helps parent activists to take risks we would not have otherwise felt comfortable with. As a lawyer and an introvert the idea of giving a speech at a protest was laughable to me three years ago but I am connected to others who cheer me on. Connection also helps us stay safe; climate activism is not without personal risk and it can be exhausting. Being connected keeps everyone engaged because they feel that this work is not meant to be sung as a solo but as a choir. When one person needs to take a breath, others are there to keep the tune going.*

## A chain reaction that can tip the system – sharing and marking change

If we view individual acts as just a drop in a vast ocean we can quickly feel despondent; however, if we see them instead as a pebble being tossed into a pond, we appreciate the ripple that our actions can cause, and that we impact others by dint of being part of a dynamic social structure, (inter)connected to everything around us. As conservationist John Muir said, 'When we try to pick out anything by itself, we find it hitched to everything else in the universe.'[134] In order for this rippling out to happen and cascade into systemic change we must connect, share and mark our successes, not keep them quietly to ourselves for fear of causing conflict. Simply being satisfied with changing our electricity provider or cutting down the meat we eat will not change the system – it is talking about it and sharing the benefits of change within our sphere of influence that is the first step to normalizing new behaviour, making it easier for others to change and turning collective dissonance into resonance.[135]

The value of sharing is more pronounced when we do it outside of our groups who 'get it' the way we do. Those groups are already on the same page, and while sharing there can provide momentum, we are not enrolling new people. When we are able to reach beyond these groups, to our friends, family and colleagues, we have the chance to bring new people safely into a change process, because we already have some of those potent ingredients for doing so: mutual respect and connection. Everett Rogers' 1962 Law of Diffusion of Innovation[136] provides a well-known framing for how new ideas spread through a social system, and how niche interests 'tip' into the mainstream. Roberts describes diffusion as 'a social process through which people talking to people spread innovation'. He categorized individuals into five profiles, each of whom play a part in bringing a new innovation[137] to the market:

1. Innovators – (2.5 per cent of the population) who invent new ideas.
2. Early adopters – (13.5 per cent of the population) who take risks on the new ideas first.
3. Early majority – (34 per cent of the population) who wait to see the early adopters prove an innovation works and then move to adopt it themselves.
4. Late majority – (34 per cent of the population) who buy into an idea when it has become mainstream and de-risked.
5. Laggards – (16 per cent of the population) who resist the new until they are forced to do otherwise (for example, their current product ceases to be produced).

Roberts suggested that a tipping point is reached when an idea is adopted by 15 to 18 per cent of the population (i.e. when the early majority begin taking up the new idea). At this point the new behaviour or idea becomes normalized and widespread, until finally embraced by the late majority, when it becomes a self-sustaining practice. In the context of climate change, new technologies and ideas to shift behaviour are emerging all the time, some of which are comfortably enjoying the acceptance of the early majority. Plant-based milk is a good example here; once a niche item sold in speciality shops, it is now easily found on supermarket shelves. This is the joint impact of innovators creating a credible product, early adopters buying enough of it that it can become more available, and the early majority responding to its normalization alongside dairy on their supermarket shelves. This kind of change involves not just consumers, but individuals within advertising, the media, government and industry who see a change happening and build on it.

That is all very well when it comes to oat milk, but surely it must be different when we are talking about social change? Surprisingly, not really. Researchers from the University of Pennsylvania[138] found that to change social norms requires a dedicated group of 25 per cent of the population who are persistently pushing for a new norm. These are early users taking up a new behaviour or practice in a coordinated way that can help challenge and eventually replace the unsustainable norm. While it is reassuring that we do not need to reach 100 per cent, we should take heed that many innovations never manage to cross what organizational theorist and management consultant Geoffrey Moore calls 'the chasm'. This is the critical juncture that bridges the early adopters and the early majority and is fundamentally about confidence. Early adopters may revel in trying new things, but the early majority want the confidence of seeing lots of people try something before they jump in. The person who tries oat milk because their friend does, the café or supermarket that offers oat milk because their competitors do, and even the government that changes regulation on milk substitutes because more progressive countries have already done so, may all be examples of early majority actors. It is powerful to instil a sense of confidence (or empowerment) in individuals that new ideas are 'safe', whether that is your mum trying almond milk in her coffee, or the executives at McDonald's who internally supported public calls to remove plastic toys from happy meals. Both need to feel a sense of safety and inevitability – that it will happen sometime so why not now? If in the early stages of introducing an innovation or new behaviour, those early adopters lose faith that their actions are meaningful or worthwhile and stop adopting or talking about it, the chain reaction of influencing ends and the hope of diffusing a new idea throughout a population can quickly fade. You can see how connection to others and self-empowerment are both important here, as is scaffolding someone's belief in the system's ability to change. Changemakers can also usefully identify with which part of the population their change currently resides and who they should be enrolling next.

## Deciding which groups to prioritize in your influencing

While many worry about converting the laggards, it is the early majority that often deserve our greater attention; full of potential yet not fully confident or converted to the idea. Laggards eventually have to move with the times anyway, mainly because their needs are no longer served (because the system has changed), and as the herd moves some will opt not to be left behind. Laggards will only stop buying petrol cars when there are none left to buy, no matter how good electric cars become or how financially attractive. The early majority by comparison are waiting to see that electric vehicles are reliable and easy to charge. They watch early adopters leaping into Teslas and wait … to see how long electric cars last and how hard it is to take long journeys. Companies do the same thing, waiting to see if someone else's innovation catches on before launching their own version.[139] Once satisfied that an idea is safe (or if incentivized to try it thereby lowering the risk),[140] the early majority will pick up an idea. While early adopters are likely to be visionaries and flexible risk-takers, the early majority tend to be more pragmatic and risk-averse, and less flexible. It is important not to assume that the early majority do not care as much as the early adopters. Often they do care but they face barriers to converting their concern into action. Being busy with a job and young children, for example, can mean that many parents, although deeply worried about climate change, may not be the first to run out and buy electric cars, whether it is because they have less disposable income or because it requires a lot of energy to integrate a new idea into family life. Don't discount this group's interest based on their buying signals.

We have spoken to many people who think that to be worthwhile, climate work should be really hard and mean enrolling the laggards who are really tricky to change. But why would we want our work to be so intentionally difficult and what will that do to our chances of sustaining it into the long term? Do we really want to spend our days arguing with people whose minds we are unlikely to change, like trying to grow a seed in nutrient-devoid dust? While it would feel satisfying to watch a plant grow there, we will spend a lot of energy (and waste a lot of other seeds) in the effort. That is energy that we could have more strategically used growing tens of healthy plants in soil that was ready for us. Crossing the chasm from early adopters to early majority is *easier* than converting laggards, but it is still not easy. Here again, a climate change coaching approach of listening (rather than broadcasting facts) can help us to gain insight from the early majority who can tell us what would make them feel confident to change.

A coaching approach of curiosity can also help because we respond differently depending on the type of innovation, our attachment to the old or the perceived risk we attach to the new. You may be an early adopter of one idea and a laggard of another, easily embracing vegetarianism but only moving away from cooking with gas when new gas hobs are no longer sold. These categories are not fixed personality types into which people neatly fit but rather stances that we adopt in relation to ideas. We can recognize the circumstances that encourage us into each category, to build compassion and appreciation for others, which will help us to listen non-judgementally. Before we can take someone across the chasm, we first need to meet them where they are and be curious as to why it is hard for them to move, and what values and needs must be satisfied to allow them to embrace change. This is as true for politicians and CEOs as it is for your next door neighbour. A global movement consists of many individuals connecting well together and helping each other to change, until those voices become the majority. While there will always be 'laggards', they become outliers, making it more and more difficult for that position to be maintained.

OVER TO YOU: Write a list of the ideas or innovations in relation to which you are or have been:

– an innovator
– an early adopter

- in the early majority
- in the late majority
- a laggard.

Next to each one, write what it was about the innovation that put you in that part of the model.

## Out with the old and in with the new – how to help one system morph into another

The two loops model compiled by strategic systems designer Cassie Robinson, brings together the ideas of various systems thinkers[141] including Deborah Frieze and Margaret Wheatley, and defines innovators as 'pioneers', who are willing to shun the old order so that an entirely new system can be born (see Fig. 18.1). We see that when a dominant system peaks and then begins to decline (such as the fossil fuel industry), pioneers will pop up with alternatives to the old system, and if supported and connected with other like-minded first movers, can become resilient enough not to be curtailed or absorbed by the old system. With that robustness the pioneers can help a new alternative system to emerge. Like all living systems, this process requires that the old system be hospiced with compassion and care, as people transition to the new emergent system.

Here again we see that a coaching approach can be useful, not only in supporting the pioneers but in behaving with compassion towards the old systems actors, and rather than celebrating their painful demise, to help them accept rather than fight it. We can so easily become champions of the new system that we forget that the old system was once equally valued. If we do not support old systems actors

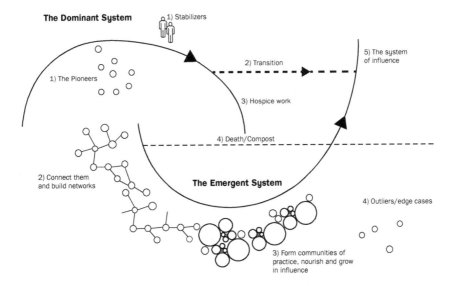

**Figure 18.1** A two loops model of dominant and emergent systems

through this process then we risk them undermining attempts to be hospiced. The model above shows how many different actions create systems change, all of which require deep relationships. We need to work well on our piece, connect to the other pieces, and trust that a new system will emerge as a result of what that connection makes possible.

## The ripple of belief and the contagion of giving up – helping networks to stay together

We can think about collective climate action through the metaphor of a sports team: there's no game with only one player and we need lots of unsung heroes, who do not score goals, but who hand out drinks at half time or massage injured players. Equally, we need supporters who do not disassociate from us because we have lost a few matches. Networks can feel like this; a group of tightly and loosely connected people all of whom are necessary to success. If one or two people stop believing that the team can win, they will have little effect, but if the majority stop, it is very hard to keep the collective faith. Moreover it is far easier to unravel a great opportunity with contagious disbelief than it is to build strong belief in the first place. With so much at stake, many find safety in becoming critical 'supporters', sitting in the stands with their arms folded, but our impact can be significant if we enrol others in disbelief. What helps a team succeed is a commitment to keep trying and a belief that they can win eventually.

This does not, of course, mean that everyone in the network should 'shut up and put up' or pretend to believe when it is cripplingly hard to do so. In those moments, however, we often inadvertently 'catch' or spread disempowerment, inadvertently colluding in their dark vision (as we explained in Chapter 13). When we trade disempowering feelings, we may find temporary relief, but we will ultimately feel worse. Few of us really want to feel this disempowered. It is a bit like someone attending an Alcoholics Anonymous meeting and explaining why they cannot stop drinking … yet they still came along. If someone truly doesn't care about the climate they will not be talking about it. When someone shares their worries we can acknowledge them for caring enough to feel that way, and normalize how hard it is. Instead of unpacking our own powerlessness, we can say, 'That sounds tough. What would help?' We can similarly help networks to stay bonded by naming and acknowledging their collective sense of disempowerment as a shared experience that connects them together. 'It sounds like you all care and are feeling frustrated … ?' With groups, we often need to do nothing more than name the feelings, for the group to then move themselves back into possibility.

This is not to say that groups should be forced to stay together if the relationships they have are costing their health or ability to make a sustainable impact. Rather it is that when groups split, they can do it consciously and skilfully, rather than with shame, blame and criticism, so that people are helped to leave well, with as many relationships as possible intact. This more productive splitting off will enable groups to collaborate as separate entities to take on the system together. If we can employ compassion and listen to each other when we are losing faith without colluding and eroding collective belief, then we are more likely to keep the existing players on the pitch, while we are out there recruiting new ones. Otherwise this erosion of belief will mean that reaching the magic 25 per cent tipping point is delayed, because we keep having to replace the original 25 per cent. This has been happening for years as movements have fragmented after setbacks, doubting their own and the system's ability to change, and as more bullish parts of the old system have spotted an opportunity to squash the emergence of the new. Remember that each conversation has the potential to inspire belief or invoke the opposite.

OVER TO YOU: Think about a successful network that you know about. What makes the network work so well?

## Bridging divides – polarization and coalition-building

Empowering others who ultimately agree with us is one thing, but in order to reform our systems we increasingly need to find ways to work in coalitions and change initiatives with those who hold fundamentally different views to us. The charity Larger Us frames this well:

> We are poised between two futures. One: a breakdown future of chaos, conflict and collapse. The other: a breakthrough future of safety, restoration, and flourishing. The choice between them starts inside our minds. Whether we are able to manage our mental states, at a moment of extraordinary turbulence. Whether we react to people who disagree with us with empathy or enmity.[142]

It is hard to fight the urge to push other groups away, when our instinct is to club together and defend our castle. Yet we are social animals; from the moment we are born our basic survival depends upon our ability to bond and connect so that we can work together as a group to find food, shelter and safety. Our brains have developed a fundamental need to belong along with primal, automatic responses to ensure this. Part of the way we do this is through the production of the powerful hormone oxytocin that enables us to form strong attachments to people in our close circle. The flip side of this is that, often without even realizing it, we can respond to people who are not in our close groups, or who feel 'different' from us, as a threat. To protect not just ourselves but the systems to which we feel a strong belonging, we often respond with defensive or offensive behaviours. In the context of climate change, these adaptive tendencies can play out in us shouting our differences, and undermining our collective response to the crisis itself. When we unskilfully shame or blame people we become complicit in widening the divide between 'us and them' and entrenching existing positions rather than enabling us to work together towards common ground. In climate change coaching we see different views as voices of the same system, and that everyone holds a different, equally rational part of the system. Our job is to understand what has led someone to feel the way they do and consider how they may feel justified for the views they hold. This is not to dismiss unskilful, unpalatable or even illegal views, but rather to apply compassion to understand why those views exist, because our best chance of changing minds is by first understanding.

What then is the role of disruptors, who purposefully shake things up? Disrupting the system is an important aspect of systems change, particularly when the status quo has resisted shifting for some time. Disruptive groups can wake up the system to the need for change, but may not be able (or see their role) to then cooperate on solutions. Rather, disruptors normalize the views of others who previously were considered too radical. The 'radical flank effect' suggests that actions by those who are more radical can expand the frame of what is considered possible,[143] so that older voices that were once seen as too revolutionary become reframed as moderate in comparison to the more radical new disruptors. An example given for this is that Naomi Klein's book *This Changes Everything* in fact did not shift public discourse but did allow Bill McKibben's work to become more palatable[144] and achieve traction that it may not have done without Klein's work. While divergent thinking moves systems forward, we can hold different views skilfully and without mudslinging; we still need cooperation across different groups and sub-populations. This does not mean that disruptors stop disrupting. Rather it means that everyone respects each other's motivations and roles in the change. As Martin Luther King so beautifully said, we are 'caught in an inescapable network of mutuality, tied in a single garment of destiny. Whatever affects one directly affects us all indirectly'.[145]

## In practice

### Do we really need political consensus to get things done?[146]

*Alex Evans, founder of Larger Us*

*I have often thought that it would be disastrous if the politics of climate change ended up even more polarized. In the US – where opinion on climate is split even more starkly than on abortion – the result has been decades of flip flopping. These kinds of wild swings are frustrating on any issue but especially bad on climate given that it gets cumulatively worse the more emissions we put into the air. But increasingly, many activists are unsure whether consensus is really what they want. Is there a risk that, rather than driving change, consensus instead creates a cosy complacency in politics, in which not enough change takes place and greenwashing is rife? Instead, some wonder whether what we really need is clear dividing lines between political parties to shift the window of the politically possible, and if that means polarizing the debate: so be it.*

*I have a lot of sympathy with this view: I can feel the clock ticking too. But I cannot believe that polarization solves this. We need to understand that we cannot just win this with our core base. We have to win people over from across the political spectrum. As climate expert George Marshall suggests, climate is simply too big to be tackled without a widespread commitment to act from across society, given the sheer scale of the changes it means for our lifestyles.*

*Instead, we should recognize there's a coalition ready to be built. As research from Climate Outreach, More In Common and European Climate Foundation shows,[147] there's agreement across the political spectrum that climate change is real and needs a global response. Our task is to activate potential members of this coalition from across the political spectrum. Winning them over will entail humility, curiosity and a ton of active listening: all skills that are central to the campaigning technique known as deep canvassing. We also need to recognize that it is this task – building a coalition that spans the political spectrum – that will really engage political leaders. Doing the work of building up a national movement of voters from left and right who will vote for whoever is most ambitious on climate, leads to real change.*

*Above all, we should resist the temptation to polarize. The populist right is actively trying to make climate the next culture war, because they know that, far from building ambition on climate action, polarization is the fastest and most effective way to destroy it and preserve the status quo. With the world's net zero plans in the balance, we all need to wise up to the trap being set for us. At the same time, climate activists cannot do this on our own. We need a real partner who can convert the political space that they create into power. We need politicians who show why and how bridge building across political divides leads not to wishy washy, lowest common denominator policies but instead to genuine, long-term transformation.*

*If we can rise to that challenge, then maybe we can build an alliance that works for a larger us, rather than a them-and-us, and for climate justice rather than climate delay. That would be a prize worth winning.*

## Climate change coaching tools to find common ground

There is much to be gained in actively connecting with people and groups who may feel differently about climate change than we do. A coaching approach can help us to do this:

- Find common ground and build on it – focus on what unites you, not only what divides you, even if what unites you is a negative – for example, 'this is exhausting' is still unity. The more points of agreement, the more you create mutual understanding.
- Recognize the importance of language and avoid language that you know will inflame. For example, those on the right of politics may respond better to ideas of conservation and stewardship than massive systems transformation.
- Put your differences 'out in front' so that you can acknowledge them as the problem that you need to work together to solve, rather than seeing the people themselves as a problem. Seek a commitment to work together to find a way forward.
- Notice if John Gottman's 'four horsemen' of criticism, defensiveness, contempt and stonewalling appear and seek an agreement to discuss differences without using these behaviours. Allow emotions and needs to be expressed instead. Use 'I feel …' rather than 'you are/did, etc'.
- Notice the 'emotional field' between the groups. Can you 'cut the air with a knife' or does it feel friendly? Allowing emotions to be expressed in a skilful way, without mudslinging but rather with 'I' statements that can calm a charged emotional field.
- Don't make each other wrong for having strong emotions about the topic. Notice that this indicates how much you care about it.

- Listen and repeat back what the other group is saying so that they feel heard. Approach the conversation openly and listen to differences with curiosity, seeking to understand rather than change the other person's view.
- Acknowledge the values that are important to the other person/group. From here try to find a way forward that honours the different values.
- If it is really difficult to find common ground, ask 'Can we find just 5 per cent in what the other group says that we agree with?' Once we have found that sliver of agreement, we have begun the process of mutual understanding.

When we approach our interactions with curiosity we can shift minds and bring people with us. Each interaction where we can connect and build a relationship rather than creating or furthering a divide has the potential to create a ripple effect.

## In practice

### Working across divides

*Elizabeth Lloyd (née Wainwright), district councillor, coach and writer*

*I am a coach, and I am also an elected Green Party district councillor in Devon in the UK – a place that has historically voted Conservative at local and general elections. Discussion and curiosity have been vital to my work as a councillor. In 2019, when I was voted in, I think many people were unsure of me, and chose to put me into the various boxes that, in their minds, went with being a member of the Green Party. Dismantling the assumptions and barriers that can get in the way of meaningful dialogue and collaboration has become part of my work. For example, this has involved meeting farmers and explaining how I want to support their work because, when done well, it can support climate, biodiversity and the local economy. National headlines and the framing of certain stories – e.g. 'farmers vs. environmentalists' – can get in the way of possibility. Part of the work of being a local councillor is about working with real people, to find locally appropriate ways forward. This involves putting the relationship I have with people first, through listening and non-judgement – something that my training as a coach really helps me with.*

*I still get frustrated – with the slow pace of change; with people who do not listen; with politicians that know how to claim credit without having done the work. I think that good work happens conversation-by-conversation – by mapping cares, strengths, assets, and then working with others to use these and make good things happen. I notice this frustration can make me defensive, and I can hold a grudge, which is to do with fear – fear that the status quo will persist; that nature will continue to decline, the climate will continue to worsen and inequality will rise. When I feel these responses rise up in me, I know I need a break, so I get outside – walking, reading and the natural world are cornerstones of the things that sustain me.*

*Sometimes disagreement on how to approach topics runs down political party lines and it feels that proposals from certain councillors are rejected without having been properly heard, based purely on the party the councillors belong to. It is easy to focus on the things that make us different, but it is a choice to be unwilling to see the things we have in common. I have found being curious really helps. I have asked fellow councillors and residents about the 'why' behind their proposals, and let curiosity guide me, rather than my assumptions. Having these conversations on neutral ground (in a coffee shop or on the phone) helps too because people seem less able to fall into the 'theatre' of the Council chamber. It is in these conversations that we find compromises, and mutual cares, and can share stories from residents that open up new possibilities.*

*Things are shifting. I have gained support for various Council 'motions' (a declaration that we want to change or do something) including to declare a climate emergency; support the Climate and Ecological Emergency Bill; and to improve inclusivity and community engagement. This has only been possible by listening, finding areas of overlap and common ground, and by building relationships around the difficult work, not just following due process. This is where approaches like coaching help.*

*Local politics feels better suited to 'play the long game' than national politics. People tend to stay connected to a place and care about the land and people where they are, and so patience is important. Setting goals and celebrating each small step towards them helps. Being elected has been an opportunity to use a coaching approach in the 'real world'. I would love to see more councillors using these tools, because local politics is a frontline of change. Coaching has felt like an 'oil' that has greased my conversations with possibility and opened up relationships that might previously have never existed. Nationally, if we are to achieve our net zero targets, over half of carbon savings will come from behavioural change and individual choices. Local authorities are a crucial partner in enabling these changes, and in my experience, coaching is a vital ingredient in doing the difficult work of dialogue and collaboration.*

## How social change creates paradigm shift – telling ourselves new stories of life on earth

If we succeed in creating change in our systems and our relationships, we will change the paradigms that we live by. In science, Thomas Kuhn[148] described paradigms as overarching theories, or a world view, but this has come to be used in everyday language to mean a perspective or mindset through which we see and shape the world. Of course, there is not only one paradigm for the whole globe (even in these globalized times) and a paradigm can also be the norms, values and worldview of a *particular* social group or society. Pierre Bourdieu[149] used the term 'habitus' to describe the 'schemes of perception' that shape cultural understanding of the world, reminding us how much a paradigm is a human construct. These are human stories that allow us to make decisions and as such, they are within our gift to rewrite.

In *Active Hope*, Joanna Macy wrote:

In choosing our story, we not only cast our vote of influence over the… world future generations inherit… we also affect our own lives… When we find a good story and fully give ourselves to it, that story can act through us, breathing new life into everything we do.[150]

The paradigm currently lived by much of the world is one of seeing ourselves as distinct from nature and able to use (or exploit) it for our human needs. This has become ingrained into the way we run our economies. But it is, nevertheless, just a story; palpable and real but not the truth. As a world view it is deeply flawed, and we can begin to craft new stories, whether through the individual conversations we have, by sharing the reasons for the action we take (or do not take, where that action fits with the old paradigm), by joining together to create new identities and by thinking systemically. Thinking systemically automatically helps us to think beyond ourselves to more sustainable actions that place us as part of the wider ecosystem of nature, rather than seeing ourselves as separate from it. The systemic perspective fundamentally changes the nature of our relationship with the natural and wider world. In this way we can create new stories for our societies.

While we may exist in the current paradigm, those working to regenerate our planet are undeniably building a new one. Socrates said that 'The secret of change is to focus all of your energy, not on fighting the old, but on building the new' and many are. We can help them by supporting their efforts and divesting from the elements of the old paradigm that are not serving us. We can also do it by unhooking the meaning that we have attached to our actions in the current paradigm, and rather than trying to change short-range behaviours, we can instead attempt to make a long-range shift in meaning. To bring this to life, consider the real-world issue of fast fashion that encourages people to continually buy clothes, resulting in a disposable fashion culture, significant environmental impacts of production and a huge amount of textile waste. Paradigms of meaning that could shift over time in relation to this might be: redefining what 'leisure time' means to move it away from centres of consumption (shopping malls) and towards centres of well-being (forests); or addressing the dissonance that drives us into consuming when we feel low, lacking in self-esteem or anxious, by redefining contentment and happiness as something we find inside ourselves and not something we purchase. Notice that in the new paradigm people can still have leisure time and can still find ways to feel good when they feel low, but the shift is in not just how they do those things but also in the creation of a new story of leisure and contentment.

What this also means is that as well as lobbying for changes in the fashion industry, we can also model the behaviours of the new paradigm in relation to contentment and leisure right now. That might mean telling people that we are choosing the forest over the shops on a Saturday or sitting with our sadness instead of reaching for the online shopping site when we feel low. Importantly, we don't have to wait for the new paradigm to come into existence to start modelling the behaviours that underpin it. For example, if we believe that in the new paradigm we can have time for community and family, then in this one we can model that by leaving work on time, volunteering as a school governor or using our weekends to be with our families not our phones. That will also help us achieve our more practical goals, when, for example, we lobby for a four-day work week and can draw on the sense of meaning that our own experience of prioritizing non-work time has given us. It is important also that in all of this talk of modelling we don't set a trap of perfection for ourselves. It is not that we must become the new paradigm in everything that we do, while living in the existing one. That is next to impossible and exhausting. As Rebecca Solnit wrote: 'As citizens massed together, we have the power to affect change ... It is not the things we refrain from doing, but those things we do passionately, and together, that will count the most ... Values and emotions are contagious ...'[151] We can choose to enjoy aspects of the new story now and demonstrate that enjoyment to others. This is not about privation but about standing in our sense of purpose and values, and feeling a sense of confidence that doing those things alone is a nudge of the collective worldview.

Part D

# Coach yourself

## How to sustain yourself while catalysing climate action

# 19 How to thrive not survive

Working on climate change can be like travelling between sunny uplands and dark, boggy lowlands. It can provide us with deep motivation and inspiration and fill us with a sense of usefulness and contribution to the world. But it can also be work that depletes, exhausts and burns us out. We can despair at humanity's folly or our own lack of significant impact, despite working all the hours available. We can feel unable to take a break when the work is so necessary – so never-ending. We can leave conversations feeling frustrated and defeated because our own emotional response to the crisis has clashed with someone else's. There is no doubt that when it comes to climate change we all have an emotional stake.

If we don't recognize that our own emotional needs play a role in our effective-ness, then we cannot only impede change in others – because we are not con-scious about how attached we are to them changing – but we also risk burning out because we devalue our need for rest and restoration. That can affect our resilience, for better and worse. In this part of the book, we will help you to become your own personal coach, to develop your own toolkit to help you thrive and coach yourself through the emotional difficulties of this work, so that you can find the joy, love and hope within it. Here we will prompt you to reflect on some of the big ideas we discussed in Part B and how they might show up in you, offering suggestions to return you to equilibrium. In the next chapter we look at what you can do when it is too late for that, and you are burning out. In the last chapter we turn to love, personal grief and hope. Before all of this, we look first to the foundation of who you are, your identity.

## Cultivating a rich inner ground

Margaret Wheatley observed that when you work to create a revolution in the outside world you

> have to be prepared with a rich inner ground, to face this time of groundless-ness … And that gives you the capacity to persevere in a way that goals, illu-sions and progress and all of these things that open us up to despair [do not].[152]

In other words, in order to stay the course, and work in the context of tumultuous change, we need to cultivate a deep sense of self, unconnected to the successes or failures of our work that can sustain us in our pursuit of change. While we will not go into an in-depth discussion about identity or indeed identity politics here, we do want to look at how we can all be mindful of cultivating a sense of self that is not dependent on what we do, but that is rather grounded in our 'being' and in how others experience us. It is all too easy to attach tightly to an identity such as expert, activist, even climate change coach, and then to feel threatened, scared or adrift when the world moves on and no longer needs us in the same way or when someone else comes on to the scene who rivals us. It can be hard to truly celebrate when we are triggered by a challenge to our place in the universe. Yet while we can easily be replaced in terms of what we do, no one can take our presence from us, and when we lean back into who we are as a definition of ourselves, we relax about competition and power, and step aside from the fray. We can instead turn to our solid inner ground, in which we know and trust ourselves.

If that sounds helpful, then the path to it is both straightforward and yet not always easy; it simply requires being ourselves and letting go of attachment. This means working on our sense of acceptance, including accepting maddening things like the state of the environment right now, of how we collectively got here, and of how we are individually culpable. It crucially also means acceptance of the fact that we may not be able to change the trajectory on which we are travelling and we may not succeed. These things are hard to accept, yet only when we let go of attachment to specific hopes can we also let go of the fears that attach to them. We can still go about our work and lives, but we feel less of the tension of *having* to change the world. Many of these ideas come from Buddhism, which encourages us to hold our existence lightly. Buddhism's withdrawal into self can be hard to square with the pressing, global crises that we face. Yet as David Loy explains below, many of the things that Buddhism originally sought to address in individuals have now become systemic, and there is a way to see the pursuit of personal liberation as part of our work to address crises like climate change. As David explains, it is not about retreating, but rather about responding moment to moment to what is happening around us, rather than attaching to fixed ideas of who we are and what we need to do.

## In practice

### Acceptance and responding appropriately: Buddhism in times of ecological crisis

*David Loy, Zen Buddhist teacher, co-founder of the Rocky Mountain EcoDharma Retreat Centre and author of EcoDharma*

*Some Buddhists think of social engagement as a distraction from the real practice. Buddhism has often been presented as an individual path, to be travelled alone and apart. That may have suited the times in which Buddhism began; however, alongside our own personal transformation, we can realize that the problems that Buddhism traditionally speaks about, the three 'poisons' of greed, ill will and delusion, are all problems that have become institutionalized and are at the roots of the wider problems we face today. For example, greed means you never have enough, and our economic system institutionalizes greed, not only from the consumer standpoint but from a corporation perspective: companies are never profitable enough or their market share is never big enough. As Joanna Macy says, the world has a role to play in our enlightenment, because engaging and getting out there is actually part of our own self transformation. Those who benefit most from this activity are ourselves, because in doing so we are able to leave behind self-centred habits and preoccupations. I see our engagement with the world as essential to our self-transformation, as well as to the benefit of humankind.*

*Last year, Noam Chomsky said in an interview that the world is at the most dangerous moment in human history.[153] If that's true, then everything we do has to be evaluated in terms of whether or not it is responding appropriately to that. Chan Master Yunmen, when asked by a student, 'What's the fruit from a lifetime of practice?' replied: 'Responding appropriately.' Buddhist training teaches us to respond moment by moment, situation by situation, and that is especially important in this time of ecological disaster. This is at the heart of acceptance; that we accept that the way things are is the way things are and do our best to respond appropriately. It is not about transcending the world, or transcending the problems of the world. Acceptance is accepting that there is a problem that has to be addressed, rather than hiding from it, or fighting it. Our problem with the idea of acceptance is that it comes up against barriers like, 'I don't want to think about that', or 'it makes me depressed' or 'it's not real'.*

*The Buddha said that the actions of an enlightened person are nirasa, 'without expectation', which means that our task today is to do the very best we can right here and now, without being able to know whether anything we do is going to make any difference whatsoever. We need to be as strategic as possible, but in the end, we do not know what the effect of our actions will be. It does not mean we are any less committed. It means accepting that there are other causal factors that are going to affect whether and how things happen. It is not about optimism or pessimism, or even hope versus despair. There is no doubt that people are freaking out now and I think there's a lot of despair, mostly unconscious. We must open up to our grief, but that's not the same as despair. We have to start from where we are, living in the present tense. Maybe the human species is going to disappear; that is a possibility. We should do the best we can to keep that from happening. But hope and despair can be a distraction from me doing the best that I can.*

*We do need to work to change 'what might happen', but not in a way that takes us away from 'what's going on right now'. For many, the present only has meaning as a way to get to somewhere else in the future. Ironically, that is a source of the problems we now face. For example, the idea of progress is often part of this way of thinking because we end up preoccupied with never-ending growth. We are sacrificing many living things in order to get to this wonderful future. This is connected to what I call 'lack projects', when we feel 'I'm not good enough' and then we look outside ourselves for answers. 'If only I had enough money ... If only I were more famous ...' The feeling of 'if only' means we aren't living here and now, because all of these lack projects are preoccupied with gaining something in the future. 'I'm not good enough right now, but in the future, everything will be OK.' That's a trap because that future never arrives.*

*This is where mindfulness or meditation can help: it grounds us in the present and helps us to notice all the ways we tend to run away from the present thoughts. This in turn can help create mental space for gratitude and compassion, including self-compassion. We all know activists who get frustrated, burned out and often drop out. But if we can ground our activism in some kind of contemplative practice, that can make a huge difference, not only to our general well-being but to our creativity.*

*There's a famous story of a Zen master and a government official who visited to learn about Zen. The official was full of ideas and opinions and as he spoke, the Zen master poured tea, but he kept pouring and pouring, until the cup was overflowing. The official said 'What are you doing?' and the Zen master replied 'This is your mind. You do not have room for anything new.' We need to be able to let go of our certainties in order to become more aware of what's actually happening, and then respond appropriately.*

## Proactive self-care – fit your own oxygen mask first

A bit like fixing your roof in the summer, not winter, it is a good idea to build healthy practices now, so that when you really need them, they are ready to fall back on. It takes anywhere from 30 days to 3 months to establish new habits and redraw neural pathways so that a behaviour no longer requires the same willpower. If we wait until we are experiencing a stress response, building these new pathways can seem overwhelming and can increase our stress levels. Here are some useful components for building your own resilience toolkit, that you can start to cultivate straight away.

### Adopt a mindset that allows for joy and ease

The mindset you adopt will sharpen or blunt the instruments in your toolbox. For example, it is no use setting strong boundaries around working hours that you do not enforce because you don't believe it is possible to switch work off. We create myths about what is possible and impossible all the time. You can also imagine them as an 'operating manual for how things are'. The striking thing is that no one has the same manual. We have all invented our own and think it is the master copy. You may know someone who has a hard rule about not working weekends. Their manual might say that 'the world accepts the need for family time'. Similarly, you might have a colleague for whom a busy day is just two meetings, when your idea of busy is not having a break for lunch. All of our behaviours are supported by our operating manuals or our mindsets. Carol Dweck has written about the difference between growth and fixed mindsets. Growth mindsets are those that accept that change is possible, and fixed mindsets assume that 'this is the way it is/I am'.[154] It would be tempting to assume that people apply either a fixed or growth mindset across everything they do, but it is much more nuanced than that. In reality, all of us face situations in which we employ a growth mindset – which, for example, help us develop in our careers – just as all of us have sneaky fixed mindsets that sit behind unhelpful behaviours.

We are going to assume that you often employ a growth mindset and are open to change. We are also going to acknowledge, however, that very many of us are raised in societies that prize hard work above much else, and that you may have inherited or subconsciously developed some fixed mindsets about your allowance of rest and enjoyment. Consider your mindset about ease: is it something that you are only allowed to have on holiday or something that other people with less important jobs can have? Is it something that only people who aren't committed can have, or something that happens 'when this big piece of work is done'? And now what about joy? Is joy something that you regularly make room for or is it something flippant, for which we don't have time? If that is so, what is your mindset there?

Often when we talk with people about the mindsets that prevent them from caring for themselves, we discover that the purpose that those mindsets play is to protect them from feeling pain. It might be the pain of feeling that you are not 'doing enough' or the anguish of missing out. To avoid that pain we then work all hours and attend every rally, in order to protect ourselves from suffering those feelings. A different way of approaching this comes from Buddhism. One of the Buddha's four noble truths is that no matter what we do, we can never achieve a state of satisfaction if we want things to be other than they are already. However much we might like to believe it, we cannot outrun dissatisfaction or ever work hard enough to satisfy ourselves. Instead we can change our relationship to dissatisfaction, such that we can accept it as a fact of life and coexist with it. This takes shifting from a mindset of 'overcoming pain by action' to a mindset of 'letting go of the need for things to be a certain way'. We cannot escape the difficulties of modern life or the anguish of the climate crisis. Instead we can imagine dissatisfaction as a dinner guest, who we will provide a place for at the table, but who doesn't get to dominate the conversation. The other guests, joy, ease, contribution and our other values, get to speak up too.

OVER TO YOU: What mindset will support you to maintain a balanced relationship with work?

### Set effective boundaries and be at peace with what is realistic and comfortable

In Chapter 7 we spoke about contracting. Another phrase for this is boundary setting, which can help you to maintain a healthy balance in relationships with others and with work and life. Commonly, people feel a healthy sense of equilibrium in some relationships, and out of balance in others. Many of us take on too much, sometimes through lack of trust in others' capability but more often out of a sense of responsibility or rescuing. Imbalance does not only happen when you take on the lion's share, however. We can experience an imbalance from the guilt of not doing enough, when we are in a relationship with someone who appears to be quicker, more capable or more engaged than us. It is important to remember that all of us have different needs and capacities at different times. Just because you have no time now, does not mean that you will not have time forever. For example, a community organizer who has young children at home now may have more time when they start school. We also all have different ways in which we like to engage with our work. One activist we knew took the step of buying a cheap second phone, on which she had several large-member WhatsApp groups, which she would physically put in a cupboard when she was at home with her family, as a piece of self-contracting to stop it distracting her. By being clear on your boundaries and articulating what you need, you can return yourself to equilibrium, from which place of calm you can help others to find theirs.

OVER TO YOU:

- What working environment helps you to feel a sense of calm and competence?
- What times of the day do you work at your best and how might you protect them so that they are available for your most important work?
- How do you like your life and work to be structured? Loose or strictly defined? What would you tell others about this?
- How do you like to collaborate and what is the most successful way to share tasks and ideas with you?
- How do you like to be held to account for tasks? What would drive you crazy?
- What boundaries do you want to set around communications and responding to them?

### Identify your resourcefulness and ability to manage complexity

We have discussed the way that scarcity and overwhelm reduce our imaginative capacity. Many people tell us that there 'is not enough time' for rest, which while having some objective truth, is often a manifestation of anxiety. Consider your own relationship to scarcity and overwhelm, and how you have managed those feelings to stay in action. What is your special recipe for handling them? Some people have told us that they purposefully stop and take deep breaths; others that they have a nap to refresh their brain. In a study of 18,000 people looking into what rest represented to them, the number one most calming pursuit was reading.[155] Perhaps you have a phrase that helps you, like 'one thing at a time'?

OVER TO YOU:

*Scarcity:* What is your experience of 'not enough'? Is it personal – a feeling of you not being good enough – or systemic – a feeling of there not being enough people, resources, political will or time? How have you successfully managed these feelings? Draw a line vertically down the middle of a piece of paper, on the left write a list of your feelings of scarcity. On the right, note all the examples you can think of when you have been good enough and when there have been enough resources. The examples do not have to relate to the environment. If you feel a persistent sense of not being good enough, consider seeking support, formally (in the shape of a coach or therapist) or informally from friends.

*Overwhelm:* How have you successfully handled complex or large amounts of information in the past? What is your way of navigating new and complex situations, and handling overwhelm? How would you tell others to help you when you feel that way?

### Understand your values

Clarifying your values can help you identify what brings you the most aliveness or ease. Here we want to look less at achievement-related values, such as contribution or responsibility, and more at the joy-related values that bring you happiness and contentment. These values are often sacrificed in service of our work (paid and unpaid), and yet they are a foundation of our resilience in just the same way as our more service-related values are. It is useful to consider the circumstances in which you feel a sense of unabashed joy and what values were present there. Perhaps you experienced true *connection* with someone, soaked in the *tranquillity* of a place or were able to be uncharacteristically *playful*. Not all of our values get as much airtime as we might like, and people are often surprised by how much they crave these lesser-activated values.

OVER TO YOU: What are your lesser-activated values, and what permission do you need to give yourself to enjoy them more regularly?

## Scaffold your sense of belief with positivity

There are some mighty yet wonderfully simple ways to build your own sense of belief and possibility. Here are just a few that we like:

*Gratitude practices:* these are a powerful tool for shifting individual or collective mood. You may simply start the day or even a conversation by saying what you are grateful for or create a detailed list at the end of each day. You might start each meal with gratitude (almost like a version of 'saying grace') with each person saying what they are grateful for. One headteacher we know, who took over a failing school, began each staff meeting with a round of gratitude to consciously boost the team's mood and found it improved the ideas they came up with to fix their crisis.

*A 'seeing the good' walk:* a development from simple gratitude practices, this is a stroll in which you consciously notice positive things. This could be as pedestrian as a working street lamp or as arresting as the changing colour of leaves in a forest. We are so good at seeing the negative that this method helps us rebalance with the positive and shift our outlook.

*Morning journals:* a huge number of people start their day with a few minutes of simple journaling. It does not have to take long, and it does not have to be legible. It is less about the output and more about the intentionality we set when we reflect before the day starts.

*Affirmations:* it seems like these should be very artificial, yet even the most reticent of our coachees has told us that taking a moment each day to repeat a phrase has a big impact on their subsequent behaviour. The trick is to make the phrase feel real to you, rather than something cliched. While it is perfectly OK to say 'I'm a strong and beautiful person,' you can just as powerfully say 'I am someone who finishes their day at 5 p.m.' Both work very well.

*Creating a nurturing habit:* we imagine you would not dream of not cleaning your teeth every morning. What if you also wouldn't start your day without 10 deep breaths, or four yoga poses? There's no difference, it is just that some activities feel more legitimate, or are better entrenched than others. This is just a mindset that we can change, just as we may hope that people change their mindsets around flying or eating meat, so we can change ours about our own care. Design a simple activity that you can do every day, and trial it for 60 days.

## Build connection with everything

We have spoken with so many who have felt alone, or who have doubted themselves and their impact. Many struggled with difficult emotions, burned out and needed to stop doing important work because the cost to their health and well-being was too great. Isolation is a real and present threat to anyone who seeks to change the status quo, so finding ways to connect, not just with like-minded others but with any humans, can reduce this. It may be that when you are feeling depleted your work colleagues (nice as they may be) are the last people you want to speak to. Cultivating a group of friends who have nothing to do with the climate crisis can create an escape hatch. Connection can also come from the natural world, and we explore this in more detail in the final chapter. We can also feel a sense of connection with institutions in which we believe and where we feel belonging, such as our children's school. Volunteering at the school fete might be an opportunity to switch off among the community. More formally, you might hire a coach or set up a mastermind group with others who are tackling a similar challenge to yours. If you are a coach, you might find a coach, mentor, or supervisor, to support your work on topics that can be more taxing than others.

### Manage your personal energy

We can all too easily assume that our energy reserves are like a bucket, from which we can take energy and then put it back. Perhaps we will have a nice spa day after that big report is in. In reality, the bucket has holes in the bottom and is constantly leaking energy, so instead of seeing this as a process of drawing down and filling up, we need to ensure that we are doing things *all the time* to manage our energy. One way we can keep an eye on this is to do a simple energy audit.

OVER TO YOU: Make a list of all the significant things that you have done in the last week or month. Take a piece of paper and draw four concentric rings, like a bullseye on a dart board. Put the things that have taken your focus and most energy in the centre ring, and the things that have had little of your time and energy in the outer rings. Stand back and look at it. Is it a good representation of what makes you fulfilled and happy? If not, what has happened to put the activities in the wrong circles? Perhaps you love to go running, but it is in the outer circle because you told yourself that work comes first. Perhaps your colleague's event is in the centre circle because you offered to help and it ballooned out of control? Don't beat yourself up, just notice and resolve to be more conscious. Draw the circles again and look at the coming week or month. What will you put in each ring this time, and what will that translate to in terms of how much time and attention they each get?

So by now, in the pursuit of creating behaviours that will support a balanced life you should have in your hand:

- a more helpful mindset around work and rest
- a set of boundaries to make work more comfortable and create space for other things
- a guide for how you best manage scarcity and overwhelm
- an understanding of your neglected values and what you can do to activate them
- some techniques for boosting your positivity and self-reflection
- ideas for how to manage isolation
- a sense of where your energy 'leaks' and how to be more conscious about using it.

There is now one final piece of the self-care jigsaw that relates to purpose, and that can free you from feeling that you have to say yes to or be good at everything, and that is to get really clear on what your contribution is.

## In practice

### Cultivating a solid inner ground

*Jenny Ekelund, director of engagement and green transition, The Partnering Initiative*

*Early in November 2021, I took a lengthy, frustrating train journey up to Glasgow to attend COP26. It was my first in-person work gathering since the start of the pandemic, and I felt a cocktail of emotions: hope, fear, excitement and uncertainty. Alongside this, I felt a deep need to connect and create with fellow climate practitioners and activists, something we'd largely been deprived of over the last 18 months. I usually love train journeys. They allow me to switch off, my brain to freewheel and for me to tune into my inner voice. Not this time. I spent much of the journey in a frenzy of preparation, checking and rechecking event times and notes, terrified that I would not make the most of the privilege of being there.*

*Every day, headlines came out of the UN 'Blue Zone' where negotiators were working feverishly on international commitments, and the streets rang with the sound of protestors' drums. It all felt so momentous and my own contribution so minimal in comparison. Was I being vocal enough? I did my best work that week when I sought out quiet. I travelled back to the little waterfront town where I was staying half an hour outside Glasgow, and I went for a walk around the sand dunes to process and reflect. I felt my sense of self recover and with it my energy and ideas.*

*Usually, in my professional life, allowing myself time to reflect helps me to approach conversations with compassion, to bring about change in a quiet and gentle way. I take time to understand what matters to people and then look for ways to link what is important to them with climate action. However, when I am met with defensiveness, or even aggression, my sense of self is rocked disproportionately. I once put a lot of work into a back-to-office report highlighting the opportunities for the organization I was working with to tackle climate change through its core business. There had been some big shifts in the industry, technologically and philosophically on climate, and I was excited to share them. I made some specific suggestions for areas that I felt we could explore further, and I received positive feedback and offers to discuss it. One colleague, however, sent a curt note chastising me for raising the issue and warning of what they considered the smoke and mirrors surrounding climate change. Our subsequent face-to-face discussion was tense, frank and emotional. This hit me hard. Who am I if I am not a kind person, who can listen and collaborate and find consensus on climate? I turned the incident over in my mind and that quiet reflection time was again what I needed. I realized that I was tensing up in response to my colleague's fear. They felt threatened and were reacting accordingly. The charged, uncomfortable conversation was not a failure on either of our parts, but a necessary, honest step, and staying true to my compassionate self was even more important. Instead of feeling discouraged, I began to feel grateful my colleague had been so unflinchingly honest. The next time we met, I opened by asking what I could do to help address some of the challenges they were facing. This shifted the nature of our interaction into something more collaborative and constructive.*

*My work is inextricably bound up with my sense of self. I have learned that if I do not allow myself quiet reflection time, even, and especially, when I am busy; I falter. In Glasgow, I needed that walk along the dunes to remind myself that I was making a valid contribution. In the last year, I have started blocking out one morning a week for reflection, writing and thinking time – and it has hugely helped me to safeguard a stable sense of self, and how I can be of best use to the climate movement. It is especially important to my identity to be kind, to under-stand others' perspectives and to find compromise. I still have a lot of work to do on keeping a stable sense of self in the face of conflict. I am, by nature, somebody that dislikes giving a strong opinion before others have spoken. I feel most con-fident when I have had time to listen, reflect, process and prepare before speak-ing. I no longer beat myself up for daydreaming or looking out the window when 'I should be working' – it is those moments that ground me.*

## Driving principles and your personal theory of change

One of the mental traps that we can fall into when working on any big, systemic problem, is to convert our deep sense of responsibility into a crushing feeling of it all being 'on us'. Logically, we know that there are many others working on the problem, but emotionally we may feel driven to meet the scale of the need. A practical way to manage this is to define your personal theory of how you make an impact. Organizations call this a 'theory of change' and we have helped many individuals to develop them too. Drawing on the Japanese idea of Ikigai, our the-ory of change brings together your sense of purpose and the big 'why' that you are working towards, with your unique skills and experience, and the needs of the world in relation to both.

### What you feel drawn to do

Revisit Chapter 11 and use the purpose tools on yourself. Once you have that, move beyond an intangible purpose and apply it in a more grounded way. For example, an intangible purpose of 'leaving the world a better place' could trans-late into knowing that what consistently feels purposeful for you is to 'help people to feel competent, not ill-equipped'. You might then decide to apply that to the climate crisis by becoming someone who shares information about the climate in a really accessible way. It is useful to keep this piece broad so that it can apply to lots of activities.

### What you have the skills, talents and experience to do

Here it is useful to list all of the things that you are good at or care about doing, from the hard to the soft skills. Where the purpose piece helps you understand *why* you do the work you do, this piece will give you a sense of *how* you yourself uniquely make change happen. We meet many people for whom recognizing their existing knowledge and skills is a tonic because they have instead been criticizing themselves for the skills they lack. This is not to say that unless you have experience you cannot bring your ideas and dreams to life; you absolutely can, it just may take a little longer. We champion anyone to radically change direction with little experience of the terrain; we would not be coaches if we did not believe in change.

### What the world needs, who particularly needs it and when

Finally, map your desire for change and your ability to effect it onto the world at large, to find the best people to enact it with. People often fall into the trap of assuming that 'everyone needs this', and while there may be some truth to that need (especially when we are looking at social change), it is not always the case that what someone needs aligns with what they want. We cannot make someone do something that they are not motivated to do. Think about the precise problem you want to solve and who most experiences that problem. Then consider *when in particular* they experience that problem. It may not be that 'all young people experience climate anxiety' but rather that 'a specific group of young people experience climate anxiety at key moments'. If you want to work with those young people, understanding the exact moment at which your support will be most useful can liberate you from feeling compelled to talk to all young people, all of the time. This is how you start to release the pressure to do everyone's job, and get really clear on the change you specifically contribute to making.

A theory of change can help us to 'stay in lane' in the healthiest sense, and can help us to articulate our choices. When we ourselves wanted to work on the climate crisis we assumed that we would have to stop coaching and retrain as climate experts in one form or another. Yet our guiding purpose was to help people to move away from powerlessness and towards belief, and our skills and experience suited that purpose. All that was left was to work out who particularly needed that approach and when. If we have got that right, hopefully, that was you. Understanding your driving purpose, the way that you bring it to life, who you do it with and when you do it, can set you free from trying to change all of the systems, all of the time. While we know that we are not responsible for changing the world, the gravity of the situation and the pain we feel about it can hijack us, and we beat ourselves up for not doing more. Sometimes, of course, we succumb to those internal voices of 'one more late night' or 'climate work should be all-consuming'. When we do so, we set ourselves up for joyless work, stress and burnout. So that is where we are going to visit next.

# 20 When the going gets tough

## How to recover from stress and burnout

Our well-being and mindset play a powerful role in our relationships with others. Their 'stuff' is one side of the conversation, but ours is on the other. If we don't acknowledge how tough this work can be, it can be easy to push ourselves too hard and damage our health. In the process, our relationships can also suffer. Tragically, so too can our ability to have an impact, which is the very thing in service of which we have risked our health in the first place. In this chapter we will explore what happens to us physiologically when we are dealing with challenge and stress, and what happens when this goes beyond 'manageable' and tips into burnout. We take a look at the different factors – both internal and external – that can affect our stress levels. Most crucially, we offer some tools to support you to manage stress and recover from burnout.

## What is stress and how does it affect us?

Stress is a natural by-product of modern life and should not be considered altogether bad. We have spoken to many involved in climate action for whom stress is a helpful companion to achievement. Not all stress is damaging, rather it is the amount of stress, how long we experience it for and how we deal with it that matters. Our stress response is a physiological reaction to a real or perceived threat. When we feel threatened our bodies go into what is known as our fight or flight response, producing a torrent of stress hormones that help us respond quickly to keep us safe. Our heartbeat quickens, our breathing becomes shallow and fast, our muscles tense up; all the things we physically need to make a quick escape from danger. Emily and Amelia Nagoski, in their book *Burnout*, describe how our bodies do not differentiate between a physical threat and a psychological one: 'stress is a biological event that really happens inside your body', and explain how stress has a cycle that has a beginning, a middle and an end, which we need to complete in order to recognize that our bodies are once again safe.[156] The problem is that most of us get stuck somewhere in the stress cycle, with on-going micro-stressors layering one stress upon another, so that our bodies never return to homeostasis.

Luckily, the key is not to remove the stressor, since many of our stresses, particularly those that are climate related, are beyond our control, but rather it is about enabling the conditions for us to return to a feeling of safety within our own bodies. The ideas we share here are designed to help with this process. This kind of cumulative stress is familiar to many of us. Many of us have experienced times at which we think we are managing OK, and then some small thing happens and we fall apart. That could be because that outwardly 'small' thing holds big meaning

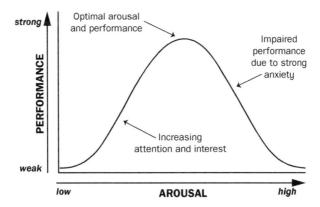

**Figure 20.1** The Yerkes Dodson Law

for us, but it can also be because we have been functioning on the edge of a tipping point, and been tipped into a chronic or acute stress response. These do not need to be big life stressors either; relentless negative news about climate change can be an ongoing low-grade stressor that over time wears down our nervous system. This brings to mind the famous analogy of the frog being boiled slowly. We would notice if overnight all of these pressures landed on us at once, making it quite plain the amount that we are dealing with, but when pressures build cumulatively we may not notice our rising stress levels or feel able to tell ourselves or others that we need help.

The Yerkes Dodson Law shows this relationship between our arousal (or stress) levels and our ability to function at our best (Fig. 20.1). As the model shows, we underperform when there is no pressure at all, but as pressure increases (and along with it our stress levels) we reach a zone of 'optimal arousal'. This is when our prefrontal cortex (which governs our executive function and our ability to plan and reason) is sharp and responsive, and where stress helps us to perform at our best. Beyond this point, however, we begin to experience a chronic heightened stress response, pushing us out of our optimal performance and into a potential danger zone. This is a highly adaptive survival system that was designed to keep us safe in the short term. However, if we get stuck in a chronic stress response, unable to complete the cycle, our bodies cannot return to a normal state of safety, and our physical, mental and emotional health begins to suffer as a consequence. If we stay there long enough, we may burn out.

The point at which we reach the zone or 'window' of optimal arousal is different for everyone. What makes you stressed may not bother a colleague. Baseline arousal levels and the nervous system's sensitivity to stimulation are different for everyone. Introverts, for example, can have a higher baseline of arousal, and so need less stimulation to reach their optimal window of arousal. Extroverts, on the other hand, have a lower baseline and need more stimulation to reach their peak performance window. Do not discount your needs because someone else experiences ease, while you feel stressed. Work by Elaine Aron[157] also shows that approximately 20 per cent of people in any population are more sensitive to sensory stimuli than others, and therefore are more affected by their environment, such as sights, sounds, smells, social cues and social dynamics. Greater sensitivity is therefore likely to mean that a person reaches a point of over arousal and overwhelms more quickly than someone with a less sensitive nervous system.

In societies and organizations that reward people who are able to 'push through' relentlessly, sensitivity can be stigmatized as a weakness, which can mean that many people spend much of their time working at the upper limit of their performance zones and risk tipping into unhealthy stress as well as low self-esteem. Aron suggests, however, that people with higher sensory sensitivity have served an important function throughout history, as they are often the philosophers, sages, prophets, counsellors, coaches and so on that serve a reflective function in society.[158] In a coaching session or trusted conversation, introducing the idea that some people are more sensitive than others, normalizes this characteristic, and creates a new perspective in which their sensitivity can be seen as a gift. When we work *with* this sensitivity and honour it, rather than trying to override, it also supports our self-care. This is more important when we are working on climate action because someone who is more sensitive is more likely to be adversely affected by the harsh responses of the people they are trying to influence. Our familiarity with a task can also affect our stress levels. While we are in the midst of change or growth, we are building new neural pathways. Until a habit has been firmly established, this pathway building is an activity that requires more of our internal resources. Therefore when we are learning we are more likely to go beyond our optimal window and have less capacity to manage other forms of pressure.

The context of the pressure also matters in how stress accumulates. We may handle stress at work provided we can return home to a calmer environment. But if our home becomes stressful in addition to our work, that new stressful context can mean we have no respite. This context-dependent nature of stress is often missed because few of us are able to see all of each other's lives. A colleague may not appreciate all of the pressures we are under. Another important element not to overlook is the stress that comes from feeling out of control of the decisions that govern our lives. Choosing to undertake a stressful activity can be much less stressful than having a decision to do that activity imposed on us. The Covid-19 pandemic put strain on many people who were otherwise managing well, creating a background anxiety of uncertainty and powerlessness.[159] This is an example of both a cumulative stressor as well as a contextual one. When working with climate change we are dealing not just with the stress of the work itself, but with the global context to which that work contributes. As a context in which we are out of control, the climate crisis has the potential to create a low level, continual cause of anxiety in our lives, making everything harder to do.

OVER TO YOU: What amount of 'stress' is enough to place you in optimal performance? What tips you over the edge into unmanageable stress? What can you do to dial these back when you get close to the tipping point?

## Triggers of stress

Climate change can bring up emotions that we may find particularly hard to deal with because of previous experiences in our lives. If we have a negative association with a particular experience- or if we have experienced threat or trauma before- we may have an unconscious, automatic stress response from which our arousal levels suddenly increase in response to similar stimuli. You could call these triggers because they create a strong, negative emotional response in us. In psychology, a trigger relates to a past trauma, which is ignited by a current event. Here we want to expand that out beyond *past* trauma, to include attitudes that threaten our sense of present or future safety. When we feel triggered, we can respond instinctively and rapidly, from the primal part of our brain (amygdala), rather than allowing time for our prefrontal cortex to assess the situation more rationally. Someone may tell us something fairly small scale and trivial, but if it triggers us then we can easily connect it to the extinction of human life and respond as if they have the codes to the nuclear button.

Scarcity is a key driver of a threat response, so it is worth noticing how your triggers around the climate crisis change depending on your perception of scarcity. A specific hope that we will decarbonize in 10 years may feel spacious and non-threatening now, but may feel very triggering in 9 years time when not enough has been done. This can reduce our ability to contextualize and manage the severity of our responses. Learning to separate the facts from the thoughts we have about those facts is a powerful way of managing our triggered responses. As psychologist Tara Brach said:

> We do not have to keep living in it [threat] but in order to wake up from it… We need to begin to call on mindfulness… The point is not to get rid of thoughts, but to just know that they're there so you do not mistake them for reality…[160]

Waking up from our thoughts can keep us (and others) safe from our triggers. We will never be able to prevent emotional hijack – this is an important response to danger – but being able to slow down how we *react* when hijacked, can enable us to respond more logically. On the other hand, if we are not able to manage our triggers, we may constantly exist in a state of threat, which can put us in a chronic condition of stress.

Finally, a strong inner critic can also increase our stress levels, continually heaping additional anxieties on us by catastrophizing. Stanford lecturer and coach Shirzad Charmine[161] describes a hyper vigilant critic that can push us into fight–flight–freeze responses by telling us we have committed wrongdoing, putting us in a state of high alert from which we are more likely to experience scarcity and overwhelm. Our inner critics can lead us to compare ourselves to others or to an ideal that we can never attain. They undermine and deplete our resilience by telling us to push through when our bodies are asking for recuperation. Inner critics can also be particularly damaging when they act on our values and strengths, pushing us to overplay these until they become weaknesses. For example, you may have a value of commitment, which allows you to stay the course when you suffer a setback, but overplayed and egged on by an inner critic that says 'you should', you might start signing up for lots of things when you are at capacity. Similarly, you may be known for being a quick thinker, but if you expect yourself to do that kind of thinking all day, all week, you'll find that your brain slows down. Our inner critics are as clever as we are because they are a part of us, so their voices can be very persuasive. Our fears are closely related to the things we care most deeply about, but it is important to scrutinize our thoughts to tease this apart from doubt.

OVER TO YOU: What do you know triggers you, or has triggered you from the stories in this book? What does your inner critic say to make you stressed? List them and write how your triggered-self wants to respond and then how your rational-self does.

## Support to manage stress

There is a myth that resilience is our ability to keep going when we are exhausted and to push through stress. In fact, resilience is our ability to support ourselves to 'come back' from stress and challenge by recognizing our needs. Resilience also does not mean expecting ourselves never to feel stressed at all, working relentlessly like automatons with no impact on our performance. We can only sustain intense pressure for so long before it begins to affect our health. Elite athletes know that their rest time is as important to their competitive edge as their training. Once we all accept this then the emphasis can shift to how we build a range of modes into our daily lives, to increase our resilience. A coaching approach emphasizes listening and accepting, but how often do we listen to and accept ourselves that well? One of the biggest factors in supporting our personal resilience is our ability to recognize, listen and respond to our own needs. If you are experiencing stress, tune into yourself and ask what you need to regain equilibrium. To bounce back from stress requires us to both maintain a healthy balance as much as we can and to recognize when it is time for us to step back and recuperate before we reach a state of burnout. Here are some ways in which you can do this:

### Notice and name it

Noticing the first signs of stress, naming the pressure or lack of control you feel and acknowledging it (either to yourself or to others) can make a big difference to your experience of stress. If we are able to acknowledge our stress to sympathetic others, we feel validated and understood, which decreases our threat response, and makes us more open to practical support. Contracting can help when you want to begin a conversation about how stressed you feel, as you can explain what you do and do not want the other person to do. Many of us just want to be listened to and empathized with. If you know that what you need is to pour it all out, rage, cry or just sit in silence for a while, then tell the other person that. You might say, 'I just want you to listen and not tell me what to do.'

### Complete the stress cycle

Even when the stressor itself has passed we need to make sure that we are constantly resolving the stress within our own bodies so that we can reset our nervous system to its baseline. Activities that physically signal to your body that it is once again safe include energetic exercise (high-intensity cardio exercise can also help to metabolize stress hormones secreted in your body), getting out into nature, connecting with someone you really care about or having a really good laugh.

### Stop reading the fire reports

We know the house is on fire, so why do we keep reading the fire reports? It may be necessary to keep up with news and developments in the climate crisis for your work but know the line between information gathering and doom scrolling. Researchers found that watching the news can have negative after effects and went so far as to suggest that relaxation techniques could be used after we consume the news, to help us recover.[162] Notice what happens in your body as you scroll the news or social media. If your stress response is triggered, take time away from the news cycle or limit your exposure. You could also compensate with alternative solution-focused media such as Positive News to see stories of beneficial systems change and proactive individuals who are acting on their values.

### Eat plenty of colourful, in-season fruits and vegetables

Not only is this a planet-friendly choice, it will help to supercharge your system and support your immune and adrenal function and energy production among other positive benefits. Stress commonly depletes a number of vitamins and minerals in the body that fruits and vegetables can help replace. They are also full of antioxidants that help to manage damage caused by stress in the body.[163]

### Develop a mindfulness practice

Many people give up on meditation because they are not able to stop or control their thoughts and conclude that they are not 'good at it'. This is a common misconception and is not the true purpose of meditation. Meditation's aim is to practise bringing ourselves back into the present moment and create awareness of our thoughts. It is also something that we have to work at and practise, like building a muscle, not magically be able to do. You can choose a daily seated practice, but if that is not for you, there are many other ways to bring mindfulness into your day, by paying close attention to the whole physical experience of washing your dishes, eating a meal or walking in nature.

### Breathe deeply

The state of our breath can tell us a lot about how stressed we are. Often our breath becomes shallow and we can even find ourselves holding our breath in times of great stress. Setting aside just a few minutes a day to consciously take deep breaths can help to signal to our bodies to switch from the sympathetic (fight or flight) nervous system to our parasympathetic (rest and digest) nervous system. Alternatively, you might take deep breaths every time you finish a task or notice stress rising in you. Simply take long, deep breaths right into the belly, making the exhalation twice as long as the inhalation, such as four counts in and eight counts out. Do this for 1 to 3 minutes and notice how you feel afterwards.

### Find something you love

Yes, there is urgency in our climate crisis work, but it is important also to step outside it to enjoy other pursuits that can bring us fun and fulfilment. It might be a creative pursuit, yoga, hiking, crafting or some other hobby. Being in flow with something we really love can be a powerful anti-stress remedy.

## In practice

### The urgency paradox: coaching climate activists

*Jessica Serrante, Radical Support Collective*

*My team and I started Radical Support Collective, a US-based organization that coaches climate and social justice leaders, because we believe that the global climate justice movement will be successful if its leaders are thriving, creative, and well-resourced to stay in it for the long haul. We each became coaches for climate justice activists after more than a decade of organizing for climate justice ourselves. We have led campaign teams to retire coal-fired power plants, transform the destructive palm oil industry and empower young people to take bold climate action. We have run ourselves ragged in an attempt to race the ticking climate clock and learned the hard way that always working from a place of urgency is not sustainable.*

*There is a central paradox of effective climate activism that we support our clients to navigate. This is that, while addressing the climate crisis requires bold, immediate action, when we do everything with a quality of urgency, we become less creative, flexible and compassionate, especially to ourselves. Paradoxically, then, we need to slow down to do quality work, sustain ourselves and have the impact we need in the time we have.*

*The hard truth of our climate reality is that the situation we are in is urgent, frightening and necessitates swift action. As climate activists, we have to acknowledge this truth to do our work, but it becomes trouble for us when we are being, or embodying, this urgency. When we are living in a 'crisis mode' response to the urgency of this moment, we buy into worries of scarcity and stories of martyrdom. We work beyond our body's capacity, skip rest and meals, and justify sidelining the things that matter most to us, like our families, bodies and creative practices. Our organizations often perpetuate this culture of urgency, encouraging staff and volunteers alike to work too-long hours and reinforcing the story that there's no time for rest until 'the work is done'. This way of working is a guaranteed recipe for burnout. If our movements continue to work in this way it will cost us the ability to build ecological and social justice at the scale that we need.*

*Ultimately, this frayed, exhausting way of working is frustrating because it contradicts our vision for the just and loving world that we are working so hard to create. When new clients come to me, they are typically ready to discover how to do things differently. Even if they don't know how, they know that if something does not change they will not be able to continue to act on behalf of what they love. For these leaders, their burnout can become an opportunity to learn how to live and work in a way that's satisfying, sustainable, and where each action they take is intentional, rather than reactive.*

*When coaching climate activists, it is important to fully honour and acknowledge the urgency and uncertainty they are feeling. To be effective climate coaches, we should not try to take the discomfort of this reality away from our clients, but we must be brave enough ourselves to be their companions amid the pain and discomfort of what is happening in our world. I often say to clients, 'Yes, this is urgent and scary. Yes, there is much for us to do, you are right.' Then I pause. I want to give them a moment to feel that I believe them entirely and I am not trying to take their experience away from them. If burnout is part of what we are working on together, I might ask, 'Are you showing up to your life and work with urgency? When you are being driven by urgency, what is your experience like?' The answer I often hear is: 'I'm exhausted but I don't know how else to do it. I just know that I cannot keep doing it this way.' This is typically when the magic can begin to happen. Telling the truth that this way of working is not actually working means that some new way of being can emerge. I'll often say:*

> *What a privilege it is that you have your eyes and heart open to what's happening on Earth right now! Our world needs you, and you are hearing the call. A new way of doing your activism is knocking at your door. You see that the way you've been doing it isn't working, so how do you want to show up now?'*

*When my clients shift their attention from the overwhelming nature of the climate crisis to their longing to contribute, they light up, their breathing steadies and they start to listen to themselves for the first time in a while. They suddenly focus where it really matters – on the compassionate, loving, brilliant being that they really are, and how they would truly love to show up to the courageous work of facing the climate crisis. When they slow down for a moment and see themselves in this way, they begin to see new possibilities for doing their work in a way that lights them up. From there, they can begin to take surprisingly simple next steps that turn that possibility into reality. These next steps can be as small as committing to take a lunch break or asking for help, but they can be a game changer. If we can learn to slow down enough to observe what we are doing and shift in the direction of our vision, over time, we will move steadily and consistently toward the just and life-sustaining world that we work so hard for.*

## Coaching yourself back from burnout

Just as with all things in nature, we are not a limitless resource and if we push ourselves too hard for too long, ignoring our natural rhythms, we can eventually burn out. Burnout is a state of mental, emotional and physical exhaustion that we reach when we have lived with prolonged, excessive stress. Not only do we become mentally exhausted and less productive, but our immune system becomes impaired and we become more vulnerable to illness such as colds and flu. Most sadly perhaps, burnout can result in a 'deadening' or hardening of our emotional responses, and we can experience hopelessness, cynicism, resentment and a feeling that nothing we do makes a difference. You can be vulnerable to burnout despite being deeply committed and engaged in your work, just as you can burn out from a job you loathe. You can burn out from unpaid work just as easily as you can from a salaried job. Workaholics Anonymous, for example, exists to help sufferers of compulsive working to recover and makes no distinction between what type of work that is, including housework, volunteering and paid work. For some of us, work can become a compulsion, and sometimes we only begin to take notice (and action) once we reach burnout or when our external performance is suffering, ignoring many of our internal signals along the way leading to this point. If you are feeling burnt out, here are some methods of recovery.

### Recognize where you are and who you are

The first step is to recognize and acknowledge that we are burning out. This can be hard to do, but consider how if you saw a loved one struggling as you are, you would want them to seek help. Notice if you have an unconscious belief that finding things difficult is a sign of weakness. Sometimes a very active inner judge can get between us and our self-compassion. Often we undermine our own internal psychological safety by judging and criticizing ourselves for our needs and limits rather than working with them. We can find it hard to treat ourselves, the one relationship that we have through our entire lives, with care and compassion and instead compare ourselves to others who appear more able to manage. A burnout does not define us, it teaches us something important about ourselves that we can use to manage our energies better in the future. None of us is immune to stress and challenge.

### Restore your health

If we are burnt out then lovingly nurture and restore yourself back to health. This is about calming the nervous system and returning to optimal brain function. Saundra Daulton-Smith[164] lists seven different kinds of rest, depending upon what kind of pressure has led us to feel depleted:

1. **Mental rest**: Give your brain a rest from taxing thinking.
2. **Sensory rest**: Reduce external stimuli; from dimming lights to turning off social media.
3. **Creative rest**: Do something inspiring and take a break from being productive.
4. **Emotional rest**: Allow room to process emotions rather than holding them back.
5. **Social rest**: Spend time alone.
6. **Spiritual rest**: Connect with something beyond the physical and mental.
7. **Physical rest**: Sleep and allow your body to slow down and relax.

Ways to do this include:

- Spending time in nature, either moving around or sitting, has been proven[165] to raise the level of NK (natural killer) cells in our bodies, which form part of our immune system.
- Develop a regular breathing practice, such as pranayama breathing, which can have a calming effect on the parasympathetic nervous system. As anxiety rises physically in our bodies, we can calm this by taking deep breaths down into our diaphragm.
- Meditate and explore mindfulness to train your mind to become present and develop the ability to recognize thoughts as distinct from facts or truths.

Mindfulness need not take a long time. You might simply take three deep breaths to ground and centre yourself.
- Practise self-compassion. There are many online resources for this, or groups that you can attend in person or virtually to become more familiar with it. Kristin Neff is a leading thinker.
- Notice when your inner dialogue is being dominated by your inner critic and if this is too potent on its own, consider seeking the support of a coach, or a professional trained in cognitive behavioural therapy or acceptance and commitment therapy.
- Build a self-management manual for yourself that you might even share with others, to remind you how to manage your stress. Include tips for yourself about the types of language you use when stressed and the way your demeanour changes.
- Create some simple mantras to support your recovery. These could be as simple as 'one day at a time', 'I will recover' or 'I can experience joy'.
- Develop a gratitude practice specifically around burnout, such as expressing gratitude for the lesson this experience has taught you and the support of those who have assisted you.
- Learn to listen to your body's needs as much as to your mind's. Our inner worlds (our emotions, physical sensations, energy levels) can sometimes feel like they get in the way of us coping, but when we learn to make room for them, they often hold the key to us thriving.
  - What do you notice here? What is that like for you? What do you need?
  - My heart is …
  - My body is …
  - My mind is …
  - My spirit is …
- Accept that life has cycles; like the apple tree it cannot bear fruit all year around. What might it mean to see your life through the prism of wintering, germinating, blossoming, fruiting and shedding? Katherine May writes: 'Wintering brings about some of the most profound and insightful moments of our human experience, and wisdom resides in those who have wintered.'[166]

OVER TO YOU: What does it mean to you to 'surrender' to the need to rest and winter?

## In practice

### Back from burnout

*Gillian Benjamin, founder of corporate climate training company, Make Tomorrow*

> *You aren't doing enough. You aren't moving fast enough or having enough impact. Do you really think you can be a social entrepreneur tackling the climate crisis? You want to affect the lives of millions, but you are struggling. Are you really cut out for this?*

*Welcome to the voice of my inner critic, having a go at me at 3:00 a.m. in the morning sometime in April 2020. Real and self-created work stress and anxiety had started affecting my sleep about a year prior. It began with an innocent half-hour toss and turn. Fast forward to peak burnout and I was barely sleeping 4 hours a night, before getting up early in the morning to try and push through a day of work. The best way I can describe burnout is that the passion and drive that usually animated my work was snuffed out. And I was terrified it would not return. My self-confidence was shot. I was questioning my line of work, my skills and talents, and my value. I struggled to focus, and even the smallest, simplest tasks took me twice as long to complete. I felt rudderless, scared and confused.*

*Otto Scharmer said that the future we create depends on the place that we act from inside ourselves. I knew that my inner world was completely off kilter, but I did not know what to do about it, except to push myself to continue to work. After realizing that what I was experiencing was more than just a 'low period', I sought out therapy and happened upon Workaholics Anonymous, which surprisingly really is a thing, and has been transformational. Cue taking some time out to rest, slowing down and spending time with friends and family.*

*Slowly I began to tease apart my critical voice from the simple facts. I had chastised myself for not being productive enough, not getting through my daily to-do list or failing to prioritize. This was turbocharged by the urgency of the climate emergency plus unhelpful entrepreneurship narratives about getting ahead by working all hours. Rather than acknowledging what I had achieved, or making my to-do lists more realistic, I would instead work into the evening and over weekends. I had a huge vision that needed a small army of collaborators, and yet every day I criticized myself for not getting further with it on my own. When I stood back, I realized that I was expecting the impossible from just one person. I had zero implementation capability besides a few freelancers who occasionally helped me out. I was criticizing myself for not being enough when the fact was I lacked implementers and a community.*

*I now know I cannot build the vision without a team of capable people making it happen. We are now a growing team of passionate people. I work happily in the knowledge that there are others to shoulder some of the responsibilities, and most importantly, to have fun with. I have also come to terms with the fact that I'll end most days feeling like I could have done more, but that this is not an excuse to try to squeeze in more in the hours where I should be resting or connecting with others. Importantly, I'm slowly changing the way that I speak to myself. I'm taming my inner critic and turning her into a more empathetic companion. Through time my inner voice is growing more accepting, kind and loving.*

*Pre-burnout, I lived a mono-culture life where work was the most important thing and everything else was secondary. Trying to work every available hour meant that I was putting very little energy building a life and identity outside of work. Which meant I had zero resilience when I had a work-related meltdown. In the months after my burnout I focused on some personal rewilding. I took a ceramics course and am now a member of a clay co-op; I took up tennis again and I put more time into seeing friends and family. Work still sometimes displaces weekend downtime – this is by no means a perfect process – but I have a better appreciation of the replenishing value of time out. Most importantly though, living a rich and varied life helps me take work a little less seriously, which means I can sleep better at night and be more effective in the long run.*

## A climate change coach's way of looking at productivity in the new paradigm

Most of us would agree that what is required is a paradigm shift (or shifts) in the way that we manage global resources, our economic philosophies and the expectations of what a life well lived looks like. Yet many of us are unintentionally aping the current paradigm in the way that we behave, by working long hours, neglecting our health and failing to attend to our emotional pain. Paradigms change as much because of changes in laws as because of changes in consciousness and behaviour, and all of us can model the latter now, not just because it is good for us, but because it shows others a different way to live and work, and nudges the new paradigm into existence.

Consider the world that you are working to build. How would the citizens of that world behave? Would they be less materialistic, consume less, travel less? What might they do more of, with all that extra time? Perhaps they would rush less, be kinder or more in community with others? Would we value with dignity those who achieved nothing 'more' than happiness? Every day we model the behaviours of the world we seek to build, right here and right now.

We tell you this to put power in your arm to prioritize your relationships and well-being. Serving the planet can be a two-way process, and we can also allow nature to cradle and nurture us. If you are working so hard that you experience high stress or even burnout, be kind to yourself; it very likely comes from a deep well of love and care. We hope that in quiet moments you might stand back and remember that love and the bigger paradigm that you want to shift over the years, not just today's to-do list, and allow yourself to proudly (and loudly) model leaving work on time, relishing lazy days with (or without!) your family if you have one, and holding the future lightly enough that you do not miss the beauty of the present moment. Just as the planet is not a limitless resource, so neither are we. We can treat ourselves with the same level of love and responsibility that we feel for the planet and in the process become a more sustainable resource ourselves. We may also feel more joy, happiness and contentment.

# 21 Loving, grieving and hoping

We wanted to end our book on love, grief and hope, because we know they sit below so many of the reasons that we take action on climate change. Underneath our grief is lodged a deep love for our home planet, our communities and our loved ones, and from that springs hope. This might be love and hope for the world we wish to build as much as protect. Every time we witness an act of climate destruction, whether because humankind is destroying nature or because our emissions have turned nature upon us, we can feel a range of emotions, from grief at the loss, to guilt that we have not done more, empathy, dread and even relief that it has happened elsewhere. Often we push these feelings down because we fear they will overrun us and prevent us from doing our work. Here we invite you to open the box and come into a deeper relationship with your grief as a pathway to hope and love.

## Staying with despair, becoming friends with grief

For many of those involved in the climate space, grief and despair are becoming familiar faces. Despair can be triggered by bad news, such as forest fires, glacial collapse or another species hitting the watch list or by seeing examples of society moving in a damaging direction. We may feel conflicted in places of high carbon consumption like airports or shopping centres. One client told us that going to a children's birthday party and seeing the torrent of single-use plastic reawakened her sense of hopelessness. If we have the courage to stay with those feelings of despair long enough, we can arrive into grief. This is helpful for several reasons.

The first is that if we attempt to outrun despair it will not only exhaust us, but it will eventually catch up with us as the climate crisis worsens. If, on the other hand, we can face this feeling fully we are able to understand why our despair becomes grief and to feel *into* those deep emotions. NASA climate scientist Peter Kalmus describes how:

> Sometimes a wave of climate grief breaks over me … without warning, I'll feel my throat clench, my eyes sting, and my stomach drop as though the Earth below me is falling away … I feel with excruciating clarity everything that we are losing – but also connection and love for those things.[167]

This connection does not diminish our grief's potency; we may still grieve deeply and long. But we can recognize that inside our grief, are love and hope. Climate grief can be a door into a wider range of emotions that may have been suppressed by our professional contexts. It can help us to live more fully and love more deeply.

Our emotions are our internal signalling systems; they tell us what we need. Are we able to listen to, acknowledge and allow our emotions, or do we block them? Our emotions can sometimes feel like the things that get in the way of us moving forward, but in fact they hold the key to our well-being and resilience. We feel despair because we feel love and are committed to our relationship with our planet. Yet many people working in this space don't feel able to openly share their despair with others. Those in a position of influence often feel a responsibility to magic away their own despair into positivity, lest they put others off climate action. This can be a huge burden to carry and create a sense of incongruence if those individuals don't find a safe place to discharge those feelings, or a means of sharing their hopelessness in a way that speaks of their love and commitment. Being able to accept that our feelings of despair are a normal human response to the crisis, and not feel ashamed of them, also protects us from turning that shame into contempt for the wider system, which is only demotivating to ourselves and others.

While climate grief can be triggered by outside events, often those events just push us towards our edge, not over it. We are most likely to cross the threshold into grief not because of outside forces but because we are internally in a state to accept it. This does not mean when we are well rested with a clear diary; often it means the exact opposite. It is more likely that when we are already burning out and exhausted or when our creativity is at a low ebb and we feel out of solutions, that we are more susceptible to our despair, and from there our grief. When we feel a strong sense of self-efficacy, meaning and impact, we can keep our feelings of despair at bay with our efforts to address the problem. We can try to 'solve' our way out of it. But at times when we feel small, doubt our capability or impact or have become physically or emotionally rundown, the combination of overwhelming bad news and our sense of personal insignificance can open the door to despair. If we can pause at those moments, we can move down into grief, which can bring us to a deep sense of love.

Climate grief differs from other forms of grief in that rather than one loss, we experience multiple losses, both past, current and anticipated. Where the pain of a more traditional form of loss may lessen over time, climate grief may keep returning, ever sharp and differently textured, with each new loss. If we can learn to grieve well – making space for it, and knowing that it will not destroy us – we can accept grief as a familiar friend, and that acceptance might allow us to experience a deeper connection to our world. If we can do that, then there is a gift in this time of sadness and anxiety; in surrendering to grief, we can connect to love.

OVER TO YOU: Write 100 or more words about your experience of climate grief. What is the love that your grief is expressing? Write another 100 words about that love.

## In practice

### On grief and love

*Sophie Tait, climate campaigner, activist and founder of Trash Plastic*

*For me it happened slowly then all at once. First there was a drip, drip, drip of news that started to find its way in. An uneasy, disquieting shifting of sands. Then came the flood. Ten minutes into a YouTube video and I felt the ground disappear – metaphorically and physically. I crumpled on the kitchen floor and howled. Everything changed in an instant.*

*I sometimes wonder if we have run out of words. We are borrowing familiar language from familiar feelings and applying it to the most unfamiliar of circumstances. When we have no common reference, when uncertainty blurs the edges of even our most solid foundations, it is no wonder it can feel like the bottom has fallen out of the world. The big feelings associated with climate change are often compared to grief, and I get that. But unlike the grief we feel at the loss of a loved one, what I feel cannot so easily be named. I'm mourning a sense of all the lives, creatures and habitats that we have lost already and of those that will come. But I am also mourning the death of my future hopes, dreams and expectations. I am having to relearn the world and accept that the only certainty is uncertainty.*

*The writer Jamie Anderson wrote:*

> *Grief, I've learned, is really just love. It's all the love you want to give, but cannot. All that unspent love gathers up in the corners of your eyes, the lump in your throat ... in that hollow part of your chest. Grief is just love with no place to go.[168]*

*I too feel the love gather in the corner of my eyes, the same lump in my throat. But I do not feel a hollow in my chest; quite the opposite. When my eyes prick with tears, I feel my chest expand and my heart get bigger. For me, climate grief is love with EVERYWHERE to go.*

*The knowing – the love – has propelled me to a life with far greater meaning. It has woken me up to the beauty and wonder of the earth like never before. I am glad that I have let the pain in enough to fall in love with this world, again and again and again. Every day some small miracle in nature stops me in my tracks and fills me with newfound wonder and awe. That's not to say I do not feel great sadness. There are moments of deep sorrow too. I cycle through anger, frustration, guilt and disbelief that this can even be happening. This grief, this love, is both heart-making and breaking. The cycles are a part of me now. I know better than to avoid what is bubbling beneath. I know when I'm hiding, when maybe I need to needle out a truer response. I know when to spot that a big feeling is coming. Usually I can move through it, knowing that on the other side there will be renewed determination to 'do'. Because within the ebb and flow of emotions, the constant that keeps me in balance is doing.*

*At the beginning, simply aligning my feelings with the choices I made at home felt enough. But 'notenoughism' is a thing in a crisis, so I constantly think and rethink how best to show up at this moment. As I have personally gone deeper, so has my work. I have let my heart lead my head, trusting that the 'knowing' is the North Star I need. My grief – 'my love with everywhere to go' – is a never-ending journey. It is constantly looking for new spaces and places to inhabit. A home where my thoughts, my feelings, my skills and experience can usefully and peacefully reside. And when those stars align, there is simply nothing that makes me feel more alive.*

## Love and loss in the natural world

Many societies today see death as an end not a rebirth. This may be part of why we feel ashamed to publicly express grief. Yet if we think back to Chapter 1 and the idea of returning to older narratives of existence, we can find more circular stories available to us than these linear ones. Unsurprisingly, these originate in a time when we understood our place in the cycles of the natural world and our role as stewards of it. Nature offers us a healthier perspective on birth, decay, death and rebirth, and helps us to see ourselves as part of a complex web of organisms, rather than a hero on a solitary journey. In nature there is a regenerative cycle in which life and death coexist, one complementing the other. As the plants rot down into compost so the forest floor is fed with vital nutrients. As some animals die so carrion animals eat. Notice that even after the major fires in New South Wales, Australia, in 2019, there was regrowth of plants and flowers several weeks after the devastation.

Coming to peace with the concept of our own decay and death can help us connect to our place in the ecosystem. For many this contemplation is relatively easy when we are young and healthy – when it feels more theoretical – and becomes harder as we age. The concept of mortality salience describes the conflict we experience as we instinctively try to outrun death while intellectually knowing it must come to us eventually. If that sounds far-fetched, consider how much advertising is aimed at *not* decaying, by staying either eternally youthful or relevant. Poet Ocean Vuong goes further than meditating on this in nature, by doing it in the presence of the dead:

> I walk around the cemetery … and I do a death meditation … then all the pettiness – the little angers that you have with those you love … falls away. It's so small, when the ultimate, lasting reality is death … how do we live a life worthwhile of our breath? … thinking about death [is] part of my own nurturing of my own mental health.[169]

If you would prefer a less extreme method, you might invent a simple visualization of yourself at the end of your life, looking back and reckoning with all that you have been and done. You may at the same time connect to your sense of purpose or of legacy.

Reflecting on our legacy can help us to shift from feeling great sorrow for what we are losing, to finding great love in what we might build together. As much as our love can come from wishing to protect and conserve, there is a palpable dream available to us all if we can embrace lives that consume less resources and therefore require less effort and time on the economic treadmill. In that world we may have more time for each other, for joy and community, and less stress and polarization. We can locate love and longing in this vision of the future, just as much as we can yearn lovingly for our world to be safe and for climate change to be brought under control.

Nature can also help us to find peace and calm. By demonstrating the impermanence of everything, nature has a wonderful way of helping us to recognize our insignificance and dispensability. When we feel we are carrying the burden of the climate crisis for the world, time in the natural landscape can serve to literally put us in our place and remind us that our toil is no more important or different to that of the grasshopper or woodlouse. In more dramatic scenery we can also experience the sublime, that concept of 'an agreeable kind of horror'[170] in which we at once awed at the beauty and humbled by the power that the natural world holds over us. The nineteenth-century Romantic painter Caspar David Friedrich famously depicted this as the lone man, standing among cloud-like mist in rocky mountains, gazing at the breathtaking yet life-threatening terrain below him.[171] In these large landscapes, whether deserts, mountains or coast lines, we are confronted by our smallness and our vulnerability, and this can be surprisingly soothing.

Just as nature can be a place to contemplate vulnerability and death, so it can calm and rejuvenate us. The health benefits of spending time in nature are well researched. 'Nature' can mean a variety of settings, not only acres of wilderness but also our gardens and public parks. Physician and writer Oliver Sacks noted the effect of gardens on his patients:

> [We are] simultaneously calmed and reinvigorated, engaged in mind, refreshed in body and spirit. The importance of these physiological states on individual and community health is fundamental and wide-ranging. In forty years of medical practice, I have found only two types of non-pharmaceutical 'therapy' to be vitally important ... music and gardens.[172]

What music and gardens have in common of course is that they invite us to be more in our bodies than our minds, and gardens particularly appeal to all of our senses. It can be no coincidence that many visualization exercises invite us into a place in nature and not a multi-storey car park. We feel a sense of peace and connection, and come into a relationship with the other life forms (the plants and animals) around us.

Nature can inspire us to act and remind us of why we love and care for it. Its intricacies and small miracles of life lived against all the odds can revive our failing hope and help us to keep going. Many coaches, such as Ruma Biswas in Chapter 8, hold sessions in fields or woods, and even public parks, and ask nature to be a partner in the session. It is even possible to coach someone on the phone this way, provided that both of you are in a natural environment, not just the coachee. Going further, as individuals we can make time for 'solos' into nature, in which we leave phones and notepads aside and find a place to sit and contemplate the immediate environment around us without any distraction. Here we are looking for stillness as much as connection, but what we are not seeking is productivity or a laundry list of the things we saw. Rather this is an exercise in noticing our desire to get up and out of relationship with this patch of earth and to stay with rather than move past those feelings. You may do this for 10 minutes, an hour, a day or even more than that. Recognized as an early founder of these ideas and the link between time spent in nature and our stewardship of it, John P. Milton's Way of Nature programme offers solos and vision quests of up to 44 days on his land in Colorado, as Hamish Mackay-Lewis experienced.

## In practice

### Connection to nature and ourselves

*Hamish Mackay-Lewis, leadership and life coach and nature-based facilitator*

*I arrived in my spot in the Coloradan wilderness laden with all my camping gear, some troubled thoughts and increasing intolerance to the beating sun and incessant mosquitos. I had committed to a month-long vision quest with John P. Milton of Way of Nature. This involved finding an isolated place on his land and pacing out a circle approximately one hundred metres across, within which I would set up camp and stay for the duration of a full lunar cycle. In the absence of any of the distractions of modern life, I wanted to confront my fears, find peace and build a more intimate relationship with myself and nature. This was as much to me about honouring as it was about discovering.*

*John told me that nature connection is just like any relationship, if you are to build trust you must give as much as you receive. But how could I possibly give back to whom I felt so indebted? Our gift to nature in these moments, I was told, can simply be our gratitude, appreciation and love. I also took tobacco, deliciously sweet to smell and sacred to the Native American culture whose love of the land I so admired. I offered it to the 11 directions as part of a daily morning and evening ceremony taught to me by John. Not accustomed to ceremony like this in my life, it felt contrived at first, but soon served as a meaningful and restorative act of kinship, giving me both a sense of participation with and belonging to the life around me.*

*During the first few days I was very much an observer, watching all the different beings going about their business. Then, a little more relaxed, my long-forgotten wildness tentatively came out to play. Hummingbirds, dragonflies and chipmunks inched closer and I found myself gently called into a different way of being. I wandered barefoot, and learned to move silently through the landscape. One day, as I sat in stillness at the base of a tree marvelling at the ants crawling over my hands, a squirrel ventured down the tree and almost bumped into me before shrieking back up the tree in shock. The pulsing fabric of the land was consuming me and the only sane response was to surrender to it. Day by day the boundaries of myself melted away, my inner nature and outer nature merged until I felt no perceivable difference between me and the trees, the river or the clouds racing overhead. The illusion of myself as separate and fixed was shattered.*

*I came away from this experience humbled, revived and somehow healed. The greatest gift, however, has been the unbreakable bond I now have, not just with the nature at my solo site but with the Earth itself. Everywhere I go on the face of this Earth I have a wise and compassionate friend and teacher whom I completely trust. I have more compassion, knowing that I am fundamentally connected to every living thing; more curiosity knowing that I am part of an unfolding mystery; and more meaning, knowing that I have a responsibility to serve the ecological integrity of this planet.*

*At the core of John's philosophy is the cultivation of relaxation and presence, in solitude and with nature. It is a process in which the heart naturally opens, joy arises, and the beginnings of a profound trust in Life is founded. I now offer this to my clients and they find meaning, wisdom and inspiration from nature, even if just from a short walk in the park. I encourage them to go back to their favourite place over and over again, to slow right down, to come to their senses and to witness the Earth as it is – a living and sentient being to whom they belong.*

OVER TO YOU: Plan a time in your week when you will spend some time in nature, however that is possible for you. Leave your devices and notepad at home and simply be with nature, the way you would with a friend. What did you notice from the experience?

## The audacity of hope

Against the backdrop of the destruction of a habitable planet for humankind, it would be easy to only focus on the difficulties of working on the climate crisis. This context, combined with the lack of resources traditionally put into this area have led many to burnout. Yet to focus on the harsh realities alone would be to miss the inspiration and motivation that draw so many to engage with this issue. Why otherwise do we work on problems that are so intractable when we could choose a quiet, easier life? As humans we have boundless creativity and resilience, and our attraction to such difficulty only serves to demonstrate this. Very many of us seek to either improve things, to protect them or both. We feel hopeful that our efforts will contribute to changing our own ways or to protecting our planet from runaway climate collapse. Yet the idea of hope can be hard to hold onto when we read the news, and (as we said in Chapter 3) it feels more acceptable to be cynical rather than hopeful.

In her podcast on this theme Dr Jennifer Atkinson detailed the many criticisms of hope, including Lauren Berlant's concept of 'cruel optimism' – that we are being tricked by the system into hoping that improvements in technology will save us from having to change the way we live at all.[173] She explained feeling conflicted about instilling hope in her students until she received letters from former students telling her that they had pursued environmental careers precisely because they had been inspired by her and her colleagues to believe that they could make a difference. She says:

> I had felt like a fraud, feeling one thing privately but projecting another publicly … [but] I started to understand hope in a … [new] way. When there does not seem to be a way forward but the hope you practice brings others on board in the effort, a path starts to emerge.[174]

Atkinson herself cited the Chinese writer Lin Yutang who wrote the beautifully simple sentence: 'Hope is like a road in the countryside; there was never a road, but when many people walk on it, the road comes into existence.'

Hope is key to our ability to collectively shift and change our systems, and as climate academic Elin Kelsey[175] has noted, hopelessness springs from not recognizing what has been done alongside what still needs to be done. Her research has shown that hope can create a ripple effect, bringing others on board who then also take action. Though it can sometimes feel like it, we are not at the starting line; many things are working, and attitudes and organizations are rapidly changing. There is so much more to be done, and so many more people wanting to know how to come on the journey with us. Perhaps we can reach not for blind or cruel optimism, or indeed for fear-mongering and polarization, but rather, propelled by our love for our world, our children and each other, we can seek to inspire a sense of what Joanna Macy calls active hope, in ourselves and others:

> Active Hope is not wishful thinking.
> Active Hope is not waiting to be rescued …
> by some savior.
> Active Hope is waking up to the beauty of life
> on whose behalf we can act.
> We belong to this world.
> The web of life is calling us forth at this time.
> We've come a long way and are here to play our part.
> With Active Hope we realize that there are adventures in store,
> strengths to discover, and comrades to link arms with.
> Active Hope is a readiness to discover the strengths
> in ourselves and in others;
> a readiness to discover the reasons for hope
> and the occasions for love.
> A readiness to discover the size and strength of our hearts,
> our quickness of mind, our steadiness of purpose,
> our own authority, our love for life,
> the liveliness of our curiosity,
> the unsuspected deep well of patience and diligence,
> the keenness of our senses, and our capacity to lead.[176]

## What gives us hope

We end with the thoughts of some of the people whose stories you have read here: 'What gives me hope is …

… the chance to care about something deeply enough to help bring it about in the world. To be confident in our journey. To relish the path rather than the destination. To cultivate hope by doing rather than by having.

(Adam Lerner)

… love and work. Work, because every incremental step toward understanding, engagement, and connection sustains my energy and builds momentum, and dissipates the paralysis of despair. And love, because love persists and nourishes in all circumstances; because in the eyes of my ancestors, looking at me from a few surviving photographs across two genocides, three continents, and a tumultuous century of insurmountable challenges, I see integrity and courage, and a call to have faith that love prevails.

(Irina Feygina)

... that none of us is alone, either in our fear for the future or our determination to do what we can to tackle the climate crisis. In an era when individualism often appears to have the upper hand, I'm inspired by the knowledge that so many people are joined in this collective endeavour.

(Alison Maitland)

... seeing how much of the climate movement has shifted from being head-centred to heart-centred in just a few short years. I do not believe that more science or technology or data will save us (though all of those things are helpful), but I do believe that our relationships might. A climate movement that prioritizes the strength and quality of our relationships just might succeed, against all odds.

(Elizabeth Bechard)

... the huge change I have seen in public awareness of climate change over the last 6 years. TV, radio and newspapers frequently carry reports on what is happening, with reality TV shows explaining the climate action we can all take. I hear friends exchanging far more ideas now on how to live sustainable lives. Many politicians understand the need for urgent action, we are heading the right way, we can encourage everyone to move further and faster.

(Jill Bruce)

... to frame this as Joanna Macy's Great Turning, and to keep asking myself: 'How might the Great Turning happen through me?'

(Megan Fraser)

... where the public and media lead, politicians follow. Citizens have life-long values, politicians are ephemeral. People want action on climate change, they understand some of the wider benefits it brings for clean air, water, and better energy. Engaging the public with climate change is key to keeping these sense-making conversations flowing and engaging people is what gives me hope.

(Asher Minns)

... what I witness and learn from working closely with people who dare to go through change. Humans are far more malleable, willing, and eager to shift than my mind might have imagined. In this I find deep hope and the motivation and wherewithal to keep going.

(Kelly Isabelle De Marco)

... every time I witness someone connecting to, and speaking from their heart, I see the best of humanity. This gives me hope. From here I know that humans are capable of doing the right thing and living in balance with the world around them.

(Sarah Flynn)

... the goodness, vulnerability and capability that I see in people when they feel safe enough to say how they truly feel. I hold great hope that, if we can enlarge that sphere of safety, those things can powerfully shift what we value and how we bring everyone with us in change.

(Charly Cox)

OVER TO YOU: What gives me hope is ...

# Acknowledgements

Charly would like to thank …

A great deal of people form the supporting cast around an author, and I have been very lucky to have had the love, patience and insight of many people as we wrote this book.

The unsung hero, deserving the greatest thanks of all is my husband, Dimitri, who alongside a stressful job as a sleep medicine specialist, was a sounding board for the ideas here and a lone parent to two children under six during the final months of editing. As parents with busy day jobs, much of this book was written after 8 p.m. I would stumble into bed at 2 a.m., and Dimitri frequently took on the bleary-eyed early morning shifts when our girls decided 5 a.m. was a suitable rising time. He carried the strain of running our household solo for months. This book would not have been written without you, D. Thanks is not a big enough word.

I have already thanked Sarah and Emily in my preface, but I want to say again a heartfelt thank you to them, and to Laura Elliott, our administrator. Sarah, thank you for being such a supportive, insightful co-author. Your enthusiasm and curiosity kept me going. Emily, you stepped into the breach and went far beyond your brief, becoming an invaluable thinking partner, editor and cheerleader. Laura, you have been my right hand in all of this, as a moral support as well as practically managing so many technical aspects of this book and keeping the train firmly on the tracks. We are truly lucky to have you in our team, Laura; you are one in a million. This book would not only be poorer without you all, but it would have been a very lonely process to write.

Our publishers at McGraw Hill Open University Press have stayed close to us throughout. A special thanks to Zoë Osman for her tireless support and to Eleanor Christie and Ali Davis for their expertise. My personal thanks to Laura Pacey, Head of Open University Press Publishing, whose crazy idea this was in the first place and who believed in my ability to write this long before I did.

It is always important to thank your parents but mine really did play a big role in this book, not only in supporting me to write since I was a child but also because my dad came out of retirement as a cartographer to create our fantastic cover design, based on the threatened Flores coastline and Himalayas, and to lovingly redraw all of our figures. Dad, I know I gave you an excuse not to do mum's 'little jobs' for a few weeks but thank you both for your inspired joint idea for the cover and for taking a ragtag assortment of diagrams and making this book so elegant. Mum, he's all yours again now.

While I was writing, Zoe Greenwood, my friend and co-founder of the Climate Change Coaches, was steering our company's ship. It is testament to Zoe and all of our team that none of our clients noticed that I was not always there. Thanks for your understanding and comradeship, Zoe. I am so proud of what we are building together. I promise I will not write another book (for at least a year).

Sarah and I are both tremendously grateful to all of the people who wrote case studies for this book, many of whom did so in a very short timeframe and during COP26. A lot of these were written on the train to Glasgow. A big thank you also to my colleague and friend Elizabeth Bechard, who was an early reader of an enormous manuscript, offering invaluable feedback and cheering us on. A special thank you to Kim Nicholas, for writing our foreword, and for your comradeship, valuable time and immense knowledge. We are grateful to all of you for your generosity and insight, and humbled by the lack of ego you brought to collaborating. It was heartening to hear how much you all thought this book was worth doing.

I am told that every author has a favourite café in which they write, and as a parent needing space from marauding children, this was no less the case for me. A special thank you to Alex and the team at Gails, Oxford, who kept my brain functioning with coffee, cake and Wi-Fi, come rain, shine and Covid. I'm sure a lot of PhDs are written there. But I will be sending you my gym bill.

Finally, and most of all, I want to thank my children, one of whom was born during the writing of this book, and both of whom had to contend with the Covid-19 pandemic as well as Mummy being squirrelled away in our home office. Nina, you were my original reason for doggedly pursuing a link between coaching and climate change when people told me it was not my job. And you and Phoebe are the reason that I have dedicated the rest of my career to this. While I feel palpable grief for the future, you have given me so many opportunities to breathe in the beautiful present. Our endless love, tea parties on the carpet and fortifying cuddles make this world even more precious. This book is for you.

Sarah would like to thank …

We have always said that this book is about relationships, and writing it has been a journey of evolving relationships that I feel so very grateful for. Without them this book could not have been written.

First, I want to thank Charly and the brilliant team at Climate Change Coaches. Charly, when you first put a shout out for coaches who cared about the environment, you not only brought together a group of amazing people, but you also brought together two passions in my life: to live sustainably myself and to build resilience or 'human sustainability' through my work. You sparked something that I, like so many others, was hungry for. Thank you for asking me to be involved with your vision and the team. You have inspired me throughout by what you do and by how you do it. When I needed to take a step back you took on the truly gargantuan task of bringing this book to fruition with such grace, friendship and kindness. I can never thank you enough.

To my colleagues and friends at Climate Change Coaches, I have learned so much from each one of you. Thank you for your wisdom, encouragement and support, and for being such fun to work with. Zoe, thank you for your energy, clarity and commitment as we developed and delivered workshops together, building many of the ideas and practices at the heart of this book. Emily, we had always hoped that you would bring your editing expertise to the book, but I had no idea just how fundamental this would be. Your skill, care, patience and insight have been essential to making this book happen. Thank you so very much. Laura, I know just how much you have supported Charly and the rest of the team, coordinating contributors and meetings. Your work is, for me, inscribed here also.

To my wonderful friends. You have been sounding boards, reality checkers, support and comfort throughout the long process of writing this book; thank you for putting up with me, and not giving up on me when I missed so many of our coffees and get-togethers, kindly overlooking my being ever so slightly obsessed.

I would like to acknowledge our coaching clients, professionals in the sustainability field from whom we have learned so much. We wanted to write this book because we saw so many of you giving your all to your crucial work and yet doubting yourselves and feeling alone. Every time I work with you, I feel awed and inspired anew. You show me that courageous, vulnerable conversations connect us and make change possible.

Andrew, thank you for being my companion in writing this book as well as my companion in life. You have believed when I have felt stuck. Thank you for your intelligence, your hugs and your unerringly compassionate sense of humour. Thank you for the porridge.

To Herbie, the four-legged member of our family, thank you for reminding me we are all part of nature, and for helping me to do my best thinking when walking behind you in the woods and swimming beside you in the sea.

We talk in the book about relationships with the generations that came before and that follow. This book is for our wonderful boys, Ben and Charlie. Thank you for encouraging your old mum and for helping me to see that pursuing what we love shows us our path in life. To Mum and Dad, who would have been tickled pink that I have written a book. And, finally, to Maura, my mother-in-law, whom we lost as the writing of this book neared its end; thank you for teaching me how to be a mum, and for being as proud of me as if I were your own daughter. I miss telling you how things are going.

# Endnotes

1. Stott, P. (2021) *Hot Air: The Inside Story of the Battle Against Climate Change Denial.* London: Atlantic Books, pp. 2–4.
2. Anecdotal evidence from various coaching clients and participants of Climate Change Coaches workshops.
3. Clarke, J., Webster, R. and Corner, A. (2020) *Theory of Change: Creating a Social Mandate for Climate Action.* Oxford: Climate Outreach. Available at: https://climateoutreach.org/about-us/theory-of-change/ (accessed December 2021).
4. Ray, S.J. (2021) Climate anxiety is an overwhelmingly white phenomenon, Scientific America. Available at: https://www.scientificamerican.com/article/the-unbearable-whiteness-of-climate-anxiety/ (accessed December 2021).
5. Shields, F. (2019) Why we are rethinking the images we use for our climate journalism, *The Guardian.* Available at: https://www.theguardian.com/environment/2019/oct/18/guardian-climate-pledge-2019-images-pictures-guidelines (accessed December 2021).
6. Climate Visuals (n.d.) Available at: https://climatevisuals.org/ (accessed December 2021).
7. Moeller, S. (1999) *Compassion Fatigue: How the Media Sell Disease, Famine, War and Death.* London and New York: Routledge, pp. 35–37.
8. Lewis, S. and Maslin, M.A. (2018) *The Human Planet: How We Created the Anthropocene.* London: Pelican Books, pp. 1–16.
9. Buchleitner, J. (2021) Infinitely evolving, exploitable you: What's driving influencer coaching, L'Atelier BNP Paribas. Available at: https://atelier.net/insights/infinitely-evolving-exploitable-you-influencer-coaching (accessed December 2021).
10. Department of Economic and Social Affairs (2018) 68% of the world population projected to live in urban areas by 2050, says UN. Available at: https://www.un.org/development/desa/en/news/population/2018-revision-of-world-urbanization-prospects.html (accessed October 2021).
11. Mac Guill, A. (2017) 'I've been followed, attacked, spat on': women on feeling scared to walk alone. *The Guardian.* Available at: https://www.theguardian.com/inequality/2017/oct/20/ive-been-followed-attacked-spat-on-women-on-feeling-scared-to-walk-alone (accessed October 2021).
12. The Star, Newsroom (2017) Women have been warned not to walk alone following a terrifying assault by a masked attacker in the South Yorkshire countryside, *The Star.* Available at: https://www.thestar.co.uk/news/women-warned-over-attacker-loose-south-yorkshire-countryside-449344 (accessed October 2021).
13. TED (2010) Teach every child about food. Available at: https://www.ted.com/talks/jamie_oliver_teach_every_child_about_food#t-295297 (accessed October 2021)
14. Adams, T. (2020) Why a lack of squished bugs on the windscreen is a worrying sign, *The Guardian.* Available at: https://www.theguardian.com/commentisfree/2020/sep/06/why-a-lack-of-squished-bugs-on-the-windscreen-is-a-worrying-sign (accessed October 2021).
15. According to a study by the British Hedgehog Preservation Society and the People's Trust for Endangered Species, there were 36.5 million hedgehogs in the UK alone in the 1950s, and by the 1990s this had declined to just 1.5 million, with a further decline since then. Roos, S., Johnston, A. and Noble, D. (2012) UK hedgehog datasets and their potential for long-term monitoring, British Trust for Ornithology. Available at: https://www.bto.org/our-science/projects/gbw/publications/papers/monitoring/btorr598 (accessed December 2021).

16. Anecdotal evidence drawn from participants on Climate Change Coaches training.

17. Angela Duckworth's Grit, Carol Dweck's Mindset, Daniel Pink's Drive and Katy Milkman's How to Change, have all addressed how we set the conditions to sustain change in different ways.

18. Positions on the change curve, Kübler-Ross, E. and Kessler, D. (2014) *On Grief and Grieving: Finding the Meaning of Grief Through the Five Stages of Loss*. New York: Scribner.

19. Stott, P. (2021) *Hot Air: The Inside Story of the Battle Against Climate Change Denial*. London: Atlantic Books, pp. 15–16.

20. 'In today's advanced civilization, we face complex social realities … with a brain designed for surviving physical emergencies.' Goleman, D., Boyatzis, R. and McKee, A. (2006) *The New Leaders*. London: Sphere, p. 35.

21. Marsh, H. (2014) *Do No Harm: Stories of Life, Death and Brain Surgery*. London: W&N, p. 14.

22. O'Neil, S. and Day, S. (2009) Fear will not do it: Promoting positive engagement with climate change through visual and iconic representation, *Journal of Science Communication*, 30(3): 355–379.

23. Heglar, M.A. (2019) Home is always worth it, Medium. Available at: https://medium.com/@maryheglar/home-is-always-worth-it-d2821634dcd9 (accessed December 2021).

24. Atkinson, J. (2020) Podcast episode 5 'Is hope overrated?'. Available at: https://soundcloud.com/jenniferwren/episode-5-is-hope-overrated?utm_source=www.drjenniferatkinson.com&utm_campaign=wtshare&utm_medium=widget&utm_content=https%253A%252F%252Fsoundcloud.com%252Fjenniferwren%252Fepisode-5-is-hope-overrated (accessed December 2021).

25. Strong, C. and Ansons, T. (2020) The science of behaviour change, Ipsos Mori. Available at: https://www.ipsos.com/ipsos-mori/en-uk/science-behaviour-change (accessed December 2021).

26. Flynn, S. (2020) *The Flynn Model of Sustainable Change*, published in 'Resilience for your Sustainability Journey' webinar, Women in Sustainability Network. Online event, UK. Women in Sustainability Digital Community Hub.

27. The Climate Change Coaching training that Sarah Taylor refers to was provided to Natural England by the Climate Change Coaches, www.climatechangecoaches.com.

28. Nicholas, K. (2021) *Under the Sky We Make: How to Be Human in a Warming World*. New York: Putnam, p. 72.

29. Goleman, D., Boyatzis, R. and McKee, A. (2006) *The New Leaders*. London: Sphere, p. 54.

30. Woodrow Wilson said that friendship 'is the only cement that will ever hold the world together' in a speech to the Red Cross, New York, 18 May 1918.

31. Peter, L.J. (2021) Laurence J. Peter, quotes, Good Reads. Available at: https://www.goodreads.com/author/quotes/182617.Laurence_J_Peter (accessed December 2021).

32. Dweck, C. (2017) *Mindset: Changing the Way You Think to Fulfil Your Potential*. New York: Random House Inc., pp. 6–7.

33. The many people involved in agitating for change in the USSR could never have imagined that each individual component would one day, seemingly overnight, lead to the fall of the Berlin Wall, and with it the disintegration of the entire USSR.

34. Harris, R.(2008) *The Happiness Trap: Stop Struggling, Start Living*. Edinburgh: Robinson Publishing, p. 58.

35. Jimeoinofficial (2012) Jimeoin – Edinburgh Comedy Fest 2010. Available at: https://www.youtube.com/watch?v=8pGHKghKNG4&t=54s (accessed October 2021).

36. Our emphasis, Kaur, V. (2020) *See No Stranger: A Manifesto of Revolutionary Love*. New York: Penguin Random House LLC, p. 143.

37. Only about half of the US workforce (51 per cent) say they feel valued by their employer, more than a third (36 per cent) have not received any form of recognition in the last year and just 47 per cent say recognition is provided fairly. American Psychological Association

Center for Organisational Excellence (2014) Employee recognition survey. Available at: http://www.apaexcellence.org/resources/special-topics/employee-recognition/ (accessed December 2021).

38. Maister, D.H., Galford, R. and Green, C. (2002) *The Trusted Advisor*. London: Simon & Schuster UK, p. 14.

39. Pink, D. (2013) *To Sell Is Human*. Edinburgh and London: Canongate, p. 6.

40. Tolstoy, L. (1900) Three methods of reform, in L. Tolstoy, *Pamphlets: Translated from the Russian* (trans. Aylmer Maude). London: Free Age Press.

41. Gottman, J. and Silver, N. (2015) *The Seven Principles for Making Marriage Work*. New York: Harmony Books, p. 121.

42. Ibid., p. 125.

43. Solnit, R. (2021) Ten ways to confront the climate crisis without losing hope, *The Guardian*. Available at: https://www.theguardian.com/environment/2021/nov/18/ten-ways-confront -climate-crisis-without-losing-hope-rebecca-solnit-reconstruction-after-covid?CMP=Share_ AndroidApp_Other (accessed November 2021).

44. 'A common trait of all these trusted advisor relationships is that the advisor places a higher value on maintaining and preserving the relationship than on the outcomes of the current transaction, financial and otherwise.' Maister, D.H., Galford, R. and Green, C. (2002) *The Trusted Advisor*. London: Simon & Schuster UK, p. 14.

45. Ted Talk, TEDWomen 2018 (2018) Katharine Hayhoe the most important thing you can do to fight climate change: Talk about it. Available at: https://www.ted.com/talks/katharine_ hayhoe_the_most_important_thing_you_can_do_to_fight_climate_change_talk_about_it (accessed November 2021).

46. Wheatley, M. (2009) *Turning to One Another*. Oakland, CA: Berrett Koehler Publishers Inc., p. 30.

47. Gottman, J.M. and Silver, N. (1999) *The Seven Principles for Making Marriage Work*. New York: Three Rivers Press, pp. 26–27.

48. Brest, A.M. (2021) Seasonal contemplative walks for women in Portola Valley, Design your Life. Available at: https://www.designingyourlife.coach/events/2021/11/21/wise-women-walk-2021 (accessed December 2021).

49. As coaches we do not need to be expert in the client's work or life, but we do need to know how the information that they are telling us relates to the topic on which they want to work, and we do need to remain grounded and connected to our clients, and not get caught in our own minds, wondering how something fits.

50. Notice how these stages of experiencing scarcity also mirror the Kubler Ross stages of change and grief that we explained in Chapter 2.

51. Christie, A. (1920) *The Mysterious Affair at Styles*. London: Bodley Head.

52. Boxbee (n.d.) 10 interesting self-storage statistics, Boxbee.com. Available at: https://www.boxbee.com/self-storage-statistics/ (accessed December 2021).

53. Rhodes, M. (2013) Yes, FOMO is now a word in the dictionary, Fast Company. Available at: https://www.fastcompany.com/3016488/yes-fomo-is-now-a-word-in-the-dictionary (accessed December 2021).

54. Harris, R. (2019) *ACT Made Simple*, 2nd edn. Oakland, CA: New Harbinger Publications, p. 213.

55. Schwartz, S. H. (2012) An overview of the Schwartz Theory of Basic Values, Online Readings in Psychology and Culture, 2(1). DOI:10.9707/2307-0919.1116, pp. 3–15.

56. Enron famously had values of 'communication, respect, integrity, and excellence', yet spectacularly failed to live by them. It is easy to set lofty ambitions but fail to land them into meaningful action. Lencioni, P. (2002) Make your values mean something, *Harvard Business Review*. Available at: https://hbr.org/2002/07/make-your-values-mean-something (accessed December 2021).

57. Ibid.
58. Common Cause Foundation (2011) *Common Cause Handbook*. Available at: https://commoncausefoundation.org/_resources/the-common-cause-handbook/ (accessed December 2021), p. 24.
59. Williamson, M. (2009) *A Return to Love*. San Francisco, CA: HarperOne.
60. Macy, J. (2012) *Active Hope: How to Face the Mess We Are in Without Going Crazy*. Novato, CA: New World Library.
61. As recent research into forests has shown, rather than compete for natural light, trees in fact take it in turns to share the light, as if recognizing that their individual survival is determined by their collective thriving. Simard, S. (2021) Trees talk to each other. 'Mother Tree' ecologist hears lessons for people, too, NPR. Available at: https://www.npr.org/sections/health-shots/2021/05/04/993430007/trees-talk-to-each-[…]st-hears-lessons-for-people-too?t=1639127459019&t=1639486533863 (accessed December 2021).
62. Thoreau, H.D. (2017) *Walden*. London: Vintage, p. 286.
63. Jack, A., Boyatzis, R., Khawaja, M. et al. (2013) Visioning in the brain: An fMRI study of inspirational coaching and mentoring, *Journal of Social Neuroscience*, 8(4): 369–384.
64. BBC (2014) Maya Angelou: In her own words, BBC.com. Available at: https://www.bbc.co.uk/news/world-us-canada-27610770 (accessed December 2021).
65. Harris, R. (2021) ACT made simple: A beginner's workshop with Russ Harris, Contextual Consulting. Originally available at: https://contextualconsulting.co.uk/workshop/act-made-simple-a-beginners-workshop-with-russ-harris (accessed 4 February 2021 – no longer available online).
66. Ibid.
67. As quoted in Kornfield, J. (2000) *After the Ecstasy, The Laundry*. London: Rider, p. 32.
68. Cohen, G. and Sherman, D. (2014) The psychology of change: Self-affirmation and social psychological intervention, *Annual Review of Psychology*, 65: 333–371.
69. Dickinson, E. (1975) *Emily Dickinson, Complete Poems*. London and Boston: Faber and Faber, p. 561.
70. The coach to whom Antonia Godber refers in her piece is Charly Cox.
71. Nicholas, K. (2021) *Under the Sky We Make: How to Be Human in a Warming World*. New York: Putnam, pp. 69–71.
72. Yona, L. (2018) [Twitter] 8 October. Available at: https://twitter.com/LeehiYona/status/1049394285047558144 (accessed December 2021).
73. As more large-scale organizations seek to address their 'Scope 3' emissions related to their supply chain, they start to ask questions of the organizations that supply them. If yours is one such organization then in order to keep your contract, your bigger corporate client may require you to supply plans for how you intend to change in order to reduce your emissions.
74. Some organizational changes *may* happen without a clear decision being made, but the outcome is less certain and the pace much slower than when a clear line in the sand has been drawn.
75. Prochaska, J.O. and Velicer, W.F. (1997) The transtheoretical model of health behavior change, *American Journal of Health Promotion*, 12(1): 38–48.
76. ClimateX Team (2018) Series 3, Episode 1, The psychology of learning to change, a conversation with Renee Lertzman, 3 October. Available at: https://climate.mit.edu/podcasts/climate-conversations-s3e1-psychology-learning-change-conversation-renee-lertzman (accessed October 2021).
77. Anonymous personal interview between employee of a waste management company and Charly Cox, 30 April 2019.
78. MSF Federation green group. Example given to Charly Cox and Emily Buchanan during an in-person Climate Change Coaches workshop, Brussels, Belgium, 13 June 2019.

79. Examples of 'new broom' CEOs that set ambitious climate goals include Alison Rose at NatWest, appointed in 2019, and Amanda Blanc at Aviva, appointed in 2020, both of whom very shortly afterwards announced ambitious climate-focused strategies. It is also worth noting that Amanda Blanc previously led Zurich Insurance (whose sustainability story we feature here) and so would have been known for pioneering a sustainability agenda during the recruitment process.

80. NLM in Focus (2018) Dr Alan Kay talks about the future at annual Lindberg-Kinglecture. Available at: https://infocus.nlm.nih.gov/2018/10/04/dr-alan-kay-talks-about-the-future-at-annual-lindberg-king-lecture/ (accessed 30 October 2021).

81. The B-corporation movement is one such example of this shift from profit-only purpose to a wider social and environmental mission.

82. Toast Ale is an obvious example of a for-profit business that has a strong environmental purpose, using waste bread in its production, while Coutts Bank, also a B-Corporation, developed its wider purpose more recently.

83. Spence, Jr. R.M. and Rushing, H. (2009) *It's Not What You Sell, It's What You Stand For: Why Every Extraordinary Business Is Driven by Purpose*. New York: Portfolio/Penguin Group.

84. Glavas, A. and Piderit, S.K. (2009) How does doing good matter? Effects of corporate citizenship on employees, *Journal of Corporate Citizenship*, 36 (Winter): 51–70.

85. Deloitte (2015) Mind the gaps: The 2015 Deloitte millennial survey executive summary. Available at: https://www2.deloitte.com/content/dam/Deloitte/global/Documents/About-Deloitte/gx-wef-2015-millennial-survey-executivesummary.pdf29th (accessed October 2021).

86. Certified B-Corporation Who Gives A Crap sells toilet paper that donates 50 per cent of its profits to water and sanitation initiatives and attracts customers for whom this mission is attractive alongside their need for toilet paper and kitchen roll.

87. De Smet, A., Gagnon, C. and Mygatt, E. (2021) Organizing for the future: Nine keys to becoming a future-ready company, McKinsey and Company. Available at: https://www.mckinsey.com/business-functions/people-and-organizational-performance/our-insights/organizing-for-the-future-nine-keys-to-becoming-a-future-ready-company (accessed 30 October 2021).

88. When Gerry Anderson first became the president of DTE Energy, he did not believe in the power of higher organizational purpose, but the 2008 recession taught him how crucial it was in times of organizational hardship. Quinn, R.E. and Thakor, A.V. (2018) Creating a purpose-driven organization, how to get employees to bring their smarts and energy to work, *Harvard Business Review*. Available at: https://hbr.org/2018/07/creating-a-purpose-driven-organization (accessed December 2021).

89. Nearly two-thirds of US-based employees we surveyed said that Covid-19 has caused them to reflect on their purpose in life. And nearly half said that they are reconsidering the kind of work they do because of the pandemic. Millennials were three times more likely than others to say that they were re-evaluating work. Dhingra, N., Samo, A., Schaninger, B. and Schrimper, M. (2021) Help your employees find purpose – or watch them leave, McKinsey. Available at: https://www.mckinsey.com/business-functions/people-and-organizational-performance/our-insights/help-your-employees-find-purpose-or-watch-them-leave (accessed December 2021).

90. There are, of course, places in which these two things conflict, such as charity governance rules, in which charities may find it difficult to practically 'do no harm' beyond the defined remit of their activities. An example of this may be that a charity is mandated to invest its reserves in such a way as to generate the biggest return on the investment. This may mitigate against ethical investing, which produces a lower return but is part of a bigger role that the charity wishes to play in the world.

91. Anecdotal evidence for this is drawn from a county-wide working group on which Charly Cox sits in Oxfordshire, UK, which is seeking to support local businesses to better understand their own decarbonizing journey.

92. '… it's the activation of the left prefrontal cortex that gives us a motivating hope … That's what spurs us on, despite obstacles … Conversely, if we fixate on what's in our way … [we] activate the right prefrontal area and are plunged into a pessimistic view.' Goleman, D., Boyatzis, R. and McKee, A. (2006) *The New Leaders*. London: Sphere, p. 149.

93. American journalist Richard Louv is the author of *Last Child in the Woods: Saving Our Children From Nature-Deficit Disorder*. Louv, R. (2019) What is nature–deficit disorder? Available at: http://richardlouv.com/blog/what-is-nature-deficit-disorder/ (accessed December 2021).

94. Tree.fm offers the sounds of forests from around the world, free of advertising or disruptions.

95. Information on different ways that businesses canvas staff before declaring a climate strategy drawn from a workshop held by Business Declares at Meaning Conference, UK, 14 November 2019.

96. SME stands for 'small and medium sized' organizations.

97. An example of this is law firm Bates Wells, who took the decision to declare a climate emergency to all of their staff, as explained verbally by Abbie Rumbold, Bates Wells, during a workshop held by Business Declares at Meaning Conference, UK, 14 November 2019.

98. One example of this is Aviva, with 31,000 employees, where a new climate-related strategy was announced in April 2021 following the work of a relatively small number of senior colleagues. Source, personal conversation with senior Aviva staff member, April 2021.

99. An example of this from our own experience is Médecins Sans Frontières, whose green group was formed from across federal members of the MSF federation, who coalesced around the issue organically, led by a relatively junior member of staff.

100. Sherman and Cohen speak about the importance of affirming the sense of self to counterbalance the effects of being in a minority. Cohen, G. and Sherman, D. (2014) The psychology of change: Self-affirmation and social psychological intervention, *Annual Review of Psychology*, 65: 333–371.

101. Fuller, T. and Yumiao, T. (2006) Small and medium-sized enterprises and corporate social responsibility: Identifying the knowledge gaps, *Journal of Business Ethics*, 67(3): 287–304.

102. 'Sustainability concerns are now influencing consumer behavior among more than half the population: 53% of consumers overall and 57% in the 18–24 age group have switched to lesser known brands because they were sustainable. More than half of consumers (52%) say that they share an emotional connection with products or organizations that they perceive as sustainable. 64% say that buying sustainable products makes them feel happy about their purchases (this reaches 72% in the 25–35 age group).' Jacobs, K., Robey, J., Lago, C., et al. (2020) Consumer products and retail: How sustainability is fundamentally changing consumer preferences, CapGemini Research Institute. Available at: https://www.capgemini.com/gb-en/research/how-sustainability-is-fundamentally-changing-consumer-preferences/ (accessed December 2021).

103. Ramachandran, N. (2021) BBC, ITV reveal climate change goals ahead of COP26 conference, *Variety*. Available at: https://variety.com/2021/tv/global/bbc-itv-climate-change-goals-cop26-conference-1235096690/ (accessed December 2021).

104. Bechard, E. (2021) *Parenting in a Climate Crisis*. Brasstown NC: Citrine Publishing, p. 3.

105. Chandran, R. (2021). Can SE Asian workers take the heat? Researchers tackle rising temperatures, Reuters. Available at: https://www.reuters.com/article/us-climate-change-workers-trfn-idUSKBN29I0T5 (accessed January 2021).

106. Sigal, B. and O'Neill, O.A. (2016) Manage your emotional culture, *Harvard Business Review*. Available at: https://hbr.org/2016/01/manage-your-emotional-culture (accessed December 2021).

107. Receiving training by the Climate Change Coaches, between April and December 2021. Available at: www.climatechangecoaches.com/organisations

108. As expressed to the Climate Change Coaches team by internal coaches during training.

109. 'Dunbar and Dunbar (1998) suggested that when individuals experience social pain in the workplace from feeling isolated, for instance, the region of the brain that is activated is the same as if physical pain had been experienced.' Houston, E. (2021) The importance of positive relationships in the workplace, PositivePsychology.com. Available at: https://positive-psychology.com/positive-relationships-workplace/ (accessed December 2021).

110. NatWest took the Climate Change Coaches 12-week ICF-accredited coach training programme.

111. Kimberly Nicholas, our foreword author, writes of cathedral building. Nicholas, K. (2021) *Under the Sky We Make: How to Be Human in a Warming World*. New York: Putnam, p. 240.

112. 'For those at the top, the entire middle management will occasionally seem like a barrier. They're the "rock in the middle." Senior management wants to get on with change … but the rock is in the way … the answers so often given [are]: "they're tied to the past." "They cannot learn a new style." "They're protecting their jobs." Well yes, but these answers are pessimistic and condescending.' Kotter, J. and Cohen, D.S. (2002) *The Heart of Change*. Boston, MA: Harvard Business Review Press, pp. 104–105.

113. Edmondson, A.C. (2019) *The Fearless Organization; Creating Psychological Safety in the Workplace for Learning, Innovation, and Growth*. Hoboken, NJ: John Wiley & Sons, pp. 692–724.

114. Clark, T.R. (2020) *The 4 Stages of Psychological Safety: Defining the Path to Inclusion and Innovation*. Oakland, CA: Berrett-Koehler.

115. Clark, T.R. (2020) Is it expensive to be yourself? How this golden question reveals psychological safety, Forbes. Available at: https:www.forbes.com/sites/timothyclark/2020/09/04/is-it-expensive-to-be-yourself-how-this-golden-question-reveals-psychological-safety/ (accessed 29 October 2021).

116. John Wittington has written on constellations, including Wittington, J. (2020) *Systemic Coaching and Constellations: The Principles, Practices and Application for Individuals, Teams and Groups*. London: Kogan Page.

117. Gottman, J.M. and Silver N. (1999) *The Seven Principles for Making Marriage Work*. New York: Three Rivers Press, pp. 22–23 and 51–52.

118. The Cynefin framework as drawn by artist Sue Borchardt, with typed text added by David Cox.

119. In a personal communication between Nick Ceasar and Charly Cox, October 2021.

120. The Cynefin framework, as explained by Dave Snowden. CognitiveEdge (2010) The Cynefin framework. Available at: https://www.youtube.com/watch?v=N7oz366X0-8 (accessed December 2021).

121. In a personal interview between Professor Dave Snowden and Charly Cox, October 2021.

122. IBM Institute for Business Value (2021) Sustainability at a turning point. Available at: https://www.ibm.com/thought-leadership/institute-business-value/report/sustainability-consumer-research (accessed December 2021).

123. Climate Change Coaches co-founder, Zoe Greenwood, previously worked for EarthWatch taking executives to the charity's project sites to learn about biodiversity with scientists. While these were educational excursions, organizations can use such trips as team building and rewards for furthering the change agenda.

124. Gottman, J.M. and Silver, N. (1999) *The Seven Principles for Making Marriage Work*. New York: Three Rivers Press, pp. 22–23.

125. Alvero, A.M., Bucklin, B.R. and Austin, J. (2001) The importance of measurement and feedback towards reaching a goal: An objective review of the effectiveness and essential characteristics of performance feedback in organizational settings (1985–1998), *Journal of*

*Organizational Behavior Management*, 21(1): 3–29. Also, Aguinis, H., Gottfredson, R. and Joo, H. (2012*)* Delivering effective performance feedback: The strengths-based approach, *Business Horizons*, 55(2): 105–111

126. Tagliabue, M., Sigurjonsdottir, S.S. and Sandaker, I. (2020) The effects of performance feed-back on organizational citizenship behaviour: A systematic review and meta-analysis. *European Journal of Work and Organizational Psychology*, 29(6): 841–861.

127. Charlotte Sewell spoke during the 'Fireside chat with Cook and B-Corporation', Meaning Conference, 17 November 2016.

128. Stavros, J., Torres, C. and Cooperrider, DL. (2018) *Conversations Worth Having: Using Appreciative Inquiry to Fuel Productive and Meaningful Engagement*. Oakland, CA: Berrett-Koehler Publishers, Inc; McArthur-Blair, J. and Cockell, J. (2018) *Building Resilience with Appreciative Inquiry: A Leadership Journey Through Hope, Despair, and Forgiveness*. Oakland, CA: Berrett-Koehler.

129. This American Life (2010) Nummi, episode 403. 26 March. Available at: https://www.thisamericanlife.org/radio-archives/episode/403/nummi (accessed December 2021).

130. Kurzgesagt – In a Nutshell (2021) Can YOU fix climate change? Available at: https://www.youtube.com/watch?v=yiw6_JakZFc (accessed December 2021).

131. Wheatley, M. and Frieze, D. (2006) Using emergence to take social innovations to scale. Margaretwheatley.com. Available at: https://margaretwheatley.com/articles/emergence.html (accessed December 2021).

132. Ibid.

133. Ibid.

134. Discover John Muir website (2021) Quote from website landing page: 'When we try to pick out anything by itself, we find it hitched to everything else in the Universe' – John Muir (1838–1914), John Muir Trust. Available at: https://discoverjohnmuir.com/ (accessed December 2021).

135. TEDGlobal>NYC (2017) Per Espen Stoknes: How to transform apocalypse fatigue into action on global warming. Available at: https://www.ted.com/talks/per_espen_stoknes_how_to_transform_apocalypse_fatigue_into_action_on_global_warming (accessed 11 December 2021).

136. Rogers, E. (2003) *Diffusion of Innovations*, 5th edn. New York: Simon and Schuster.

137. While often we think of innovation in terms of technology and within that hardware, Everett defined it as any idea, practice or object that is perceived as new by an individual or other entity that could adopt it, like an organization. Ibid.

138. Berger, M.W. and Soane, J. (2018) Tipping point for large-scale social change? Just 25 percent, *Penn Today*. Available at: https://penntoday.upenn.edu/news/damon-centola-tipping-point-large-scale-social-change (accessed December 2021).

139. We also see this lookalike marketing in book publishing and film production, where one original idea spawns a range of similar titles, for example *Thes Kite Runner* and similar books such as *The Bookseller of Kabul*.

140. Electric car subscription services are a direct attempt to convert the early majority to electric vehicles by lowering the risks of doing so. Customers can subscribe for a month and then cancel if they do not like it, and full repairs and servicing are included in the cost.

141. Sources for ideas contained in the two loops model include: Frieze, D. and Wheatley, M. (2011) *Walk Out, Walk On*. San Francisco: Berrett-Koehler Publishers Inc.; Wheatley, M. and Frieze, D. (2006) Using emergence to take social innovations to scale, Margaretwheatley.com. Available at: https://margaretwheatley.com/articles/emergence.html (accessed December 2021); TEDxJamaicaPlain (2015) How I became a localist. Available at: https://www.youtube.com/watch?v=2jTdZSPBRRE (accessed December 2021).

142. Larger Us homepage. Available at: https://larger.us/ (accessed December 2021).

143. University of Exeter (2021) GSI Seminar Series: Dr Kimberly Nicholas – *Under the Sky We Make* book discussion. Available at: https://www.youtube.com/watch?v=3VcfPA6eChI (accessed December 2021).

144. Ibid.

145. King, M.L. (1963) Letter from a Birmingham Jail, April 16, 1963, The Martin Luther King, Jr. Research and Education Institute. Available at: https://kinginstitute.stanford.edu/building-world-house (accessed December 2021).

146. Evans, A. (2021) Let's make climate a culture war, Larger Us. Available at: https://larger.us/ideas/lets-make-climate-a-culture-war/ (accessed December 2021).

147. Wang, S., Corner, A. and Nicholls, J. (2020) *Britain Talks Climate: A Toolkit for Engaging the British Public on Climate Change.* Oxford: Climate Outreach

148. Kuhn, T.S. (1970) *The Structure of Scientific Revolutions.* Chicago: University of Chicago.

149. Bourdieu, P. (1977) *Outline of a Theory of Practice (Vol. 16).* Cambridge: Cambridge University Press.

150. Macy, J. (2012) *Active Hope: How to Face the Mess We Are in Without Going Crazy.* Novato, CA: New World Library, p. 33.

151. Solnit, R. (2021) Ten ways to confront the climate crisis without losing hope, *The Guardian.* Available at: https://www.theguardian.com/environment/2021/nov/18/ten-ways-confront -climate-crisis-without-losing-hope-rebecca-solnit-reconstruction-after-covid?CMP=Share_ AndroidApp_Other (accessed November 2021).

152. Wheatley, M. (2017) Islands of Sanity, Meaning Conference. Available at: https://meaning-conference.co.uk/videos/margaret-wheatley (accessed October 2021).

153. Chomsky, N. (2020) The world is at the most dangerous moment in human history, *New Statesman.* Available at: https://www.newstatesman.com/politics/2020/09/noam-chomsky-the-world-is-at-the-most-dangerous-moment-in-human-history (accessed December 2021).

154. Dweck, C. (2017) *Mindset: Changing the Way You Think to Fulfil Your Potential.* New York: Random House Inc., pp. 6–7.

155. Hammond, C. (2019) *The Art of Rest: How to Find Respite in the Modern Age.* Edinburgh: Canon Gate, p. 13.

156. Nagoski, E. and Nagoski, A. (2019) *Burnout.* London: Vermillion, p. 25.

157. Aron, E.N. (1999) *The Highly Sensitive Person: How to Survive and Thrive When the World Overwhelms You.* London: Thorsons, p. 18.

158. Ibid.

159. Davis, N. (2021) Crisis dramatically worsened global mental health, *The Guardian.* Available at: https://www.theguardian.com/world/2021/oct/08/covid-crisis-dramatically-worsened-global-mental-health-study-finds (accessed December 2021).

160. Brach, T. (2015) Learning to respond not react. Available at: https://www.youtube.com/watch?v=ymPF0q7U5oM (accessed December 2021).

161. Chamine, S. (2012) *Positive Intelligence: Why Only 20% of Teams and Individuals Achieve Their True Potential and How You Can Achieve Yours.* Austin, TX: Greenleaf Book Group Press, p. 45.

162. Szabo, A. and Hopkinson, K.L. (2007) Negative psychological effects of watching the news on the television: Relaxation or another intervention may be needed to buffer them!, *International Journal of Behavioural Medicine*, 14(2): 57–62.

163. Rink, S.M., Mendola, P., Mumford, S.L. et al. (2013) Self-report of fruit and vegetable intake that meets the 5 a day recommendation is associated with reduced levels of oxidative stress biomarkers and increased levels of antioxidant defense in premenopausal women, *Journal of the Academy of Nutrition and Dietetics*, 113(6): 776–785.

164. Daulton-Smith, S. (2019) *Sacred Rest: Recover Your Life, Renew Your Energy, Restore Your Sanity.* Brentwood, TN: FaithWords.

165. Li, Q., Morimoto, K., Nakadai, A. et al. (2007) Forest bathing enhances human natural killer activity and expression of anti-cancer proteins, *International Journal of Immunopathology and Pharmacology*, 20(2): 3–8.

166. May, K. (2020) *Wintering: The Power of Rest and Retreat in Difficult Times*. London: Rider, p.12.

167. Kalmus, P. (2020) Climate anxiety, what is it and how can we cope, the big Q. Available at: https://www.thebigq.org/2020/09/03/climate-anxiety-what-is-it-and-how-can-we-cope/ (accessed December 2021).

168. Maharaj, S. (n.d.) Grief – an act of love, Therapy Route. Available at: https://www.therapyroute.com/article/grief-an-act-of-love-by-s-maharaj (accessed December 2021).

169. Tippett, K. and On Being Podcast (2020) Ocean Vuong: A life worthy of our breath. Available at: https://onbeing.org/programs/ocean-vuong-a-life-worthy-of-our-breath/ (accessed December 2021).

170. Addison, J. (1773) *Remarks on Several Parts of Italy etc. in the years 1701, 1702, 1703.* 1773 edn, London: T. Walker, p. 261.

171. Friedrich, C.D. (1818) *Wanderer Above the Mist, Der Wanderer über dem Nebelmeer.* [Oil on canvas]. Hamburger Kunsthalle, Hamburg.

172. Sacks, O. (2020) *Everything in Its Place: First Loves and Last Tales*. London: Pan Macmillan, p. 243.

173. Atkinson, J. (2020) Podcast episode 5, Is hope overrated?. Available at: https://soundcloud.com/jenniferwren/episode-5-is-hope-overrated?utm_source=www.drjenniferatkinson.com&utm_campaign=wtshare&utm_medium=widget&utm_content=https%253A%252F%252Fsoundcloud.com%252Fjenniferwren%252Fepisode-5-is-hope-overrated (accessed December 2021).

174. Ibid.

175. Rachel Carson Center (2020) Climate change: A crisis of hope by Elin Kelsey. Available at: https://www.youtube.com/watch?v=5nOGVXvryrA&list=UUd9pm9iQsU8mna_hRpYJ-SuQ&index=25 (accessed December 2021).

176. Macy, J. (2012) *Active Hope: How to Face the Mess We are In Without Going Crazy*. Novato, CA: New World Library, p. 35, reproduced by kind permission of the authors.

# Bibliography

Adams, T. (2020) Why a lack of squished bugs on the windscreen is a worrying sign, *The Guardian*. Available at: https://www.theguardian.com/commentisfree/2020/sep/06/why-a-lack-of-squished-bugs-on-the-windscreen-is-a-worrying-sign (accessed October 2021).

Addison, J. (1773) *Remarks on Several Parts of Italy etc. in the years 1701, 1702, 1703*. 1773 edn, London: T. Walker.

Aguinis, H., Gottfredson, R. and Joo, H. (2012) Delivering effective performance feedback: The strengths-based approach, *Business Horizons*, 55(2): 105–111.

Alvero, A.M., Bucklin, B.R. and Austin, J. (2001) The importance of measurement and feedback towards reaching a goal: An objective review of the effectiveness and essential characteristics of performance feedback in organizational settings (1985–1998), *Journal of Organizational Behavior Management*, 21(1): 3–29.

American Psychological Association Center for Organizational Excellence (2014) Employee recognition survey. Available at: http://www.apaexcellence.org/resources/special-topics/employee-recognition/ (accessed December 2021.

Aron, E.N. (1999) *The Highly Sensitive Person: How to Survive and Thrive When the World Overwhelms You*. London: Thorsons.

Atkinson, J. (2020) Podcast episode 5, Is hope overrated?. Available at: https://soundcloud.com/jenniferwren/episode-5-is-hope-overrated?utm_source=www.drjenniferatkinson.com&utm_campaign=wtshare&utm_medium=widget&utm_content=https%253A%252F%252Fsoundcloud.com%252Fjenniferwren%252Fepisode-5-is-hope-overrated (accessed December 2021).

BBC (2014) Maya Angelou: In her own words, BBC.com. Available at: https://www.bbc.co.uk/news/world-us-canada-27610770 (accessed December 2021).

Bechard, E. (2021) *Parenting in a Climate Crisis*. Brasstown NC: Citrine Publishing.

Berger, M.W. and Soane, J. (2018) Tipping point for large-scale social change? Just 25 percent, *Penn Today*. Available at: https://penntoday.upenn.edu/news/damon-centola-tipping-point-large-scale-social-change (accessed December 2021).

Bourdieu, P. (1977) *Outline of a Theory of Practice (Vol. 16)*. Cambridge: Cambridge University Press.

Boxbee (n.d.) 10 interesting self-storage statistics, Boxbee.com. Available at: https://www.boxbee.com/self-storage-statistics/ (accessed December 2021).

Boyatzis, R., Melvin L., Smith Jr., and Van Oosten, E. (2019) *Helping People Change: Coaching with Compassion for Lifelong Learning and Growth*. Boston, MA: Harvard Business Review Press.

Brach, T. (2015) Learning to respond not react. Available at: https://www.youtube.com/watch?v=ymPF0q7U5oM (accessed December 2021).

Brest, A.M. (2021) Seasonal contemplative walks for women in Portola Valley, Design your Life. Available at: https://www.designingyourlife.coach/events/2021/11/21/wise-women-walk-2021 (accessed December 2021).

Bridges, W. (2019) *Transitions: Making Sense of Life's Transitions*. New York: Hachette Book Group Inc.

Bridges, W. and Bridges, S. (2017) *Managing Transitions: Making the Most of Change*, revised 4th edn. London: Nicholas Brealey Publishing.

Brown, B. (2010) *The Gifts of Imperfection: Let Go of Who You Think You Are Supposed to Be and Embrace Who You Are*. Minnesota: Hazelden Publishing.

Brown, B. (2013) *Daring Greatly: How the Courage to be Vulnerable Transforms the Way We Live, Love, Parent and Lead.* London: Portfolio Penguin

Buchleitner, J. (2021) Infinitely evolving, exploitable you: What's driving influencer coaching, L'Atelier BNP Paribas. Available at: https://atelier.net/insights/infinitely-evolving-exploitable-you-influencer-coaching (accessed December 2021).

Campbell, J. (2019) *The Resilience Dynamic.* UK: Practical Inspirations Publishing.

Chamine, S. (2012) *Positive Intelligence: Why Only 20% of Teams and Individuals Achieve Their True Potential and How You Can Achieve Yours.* Austin, TX: Greenleaf Book Group Press.

Chandran, R. (2021) Can SE Asian workers take the heat? Researchers tackle rising temperatures, Reuters. Available at: https://www.reuters.com/article/us-climate-change-workers-trfn-idUSKBN29I0T5 (accessed January 2021).

Chomsky, N. (2020) The world is at the most dangerous moment in human history, *New Statesman.* Available at: https://www.newstatesman.com/politics/2020/09/noam-chomsky-the-world-is-at-the-most-dangerous-moment-in-human-history (accessed December 2021).

Christie, A. (1920) *The Mysterious Affair at Styles.* London: Bodley Head.

Clark, T.R. (2020) Is it expensive to be yourself? How this golden question reveals psychological safety, Forbes. Available at: https:www.forbes.com/sites/timothyclark/2020/09/04/is-it-expensive-to-be-yourself-how-this-golden-question-reveals-psychological-safety/ (accessed 29 October 2021).

Clark, T.R. (2020) *The 4 Stages of Psychological Safety: Defining the Path to Inclusion and Innovation.* Oakland, CA: Berrett-Koehler.

Clarke, J., Webster, R. and Corner, A. (2020) *Theory of Change: Creating a Social Mandate for Climate Action.* Oxford: Climate Outreach. Available at: https://climateoutreach.org/about-us/theory-of-change/ (accessed December 2021).

Climate Change Coaches (2021) Organisations. Available at: www.climatechangecoaches.com/organisations (accessed December 2021).

Climate Visuals (n.d.) Available at: https://climatevisuals.org (accessed December 2021).

ClimateX Team (2018) Series 3, Episode 1, The psychology of learning to change, a conversation with Renee Lertzman, 3 October. Available at: https://climate.mit.edu/podcasts/climate-conversations-s3e1-psychology-learning-change-conversation-renee-lertzman (accessed October 2021).

CognitiveEdge (2010) The Cynefin framework. Available at: https://www.youtube.com/watch?v=N7oz366X0-8 (accessed December 2021).

Cohen, G. and Sherman, D. (2014) The psychology of change: Self-affirmation and social psychological intervention, *Annual Review of Psychology,* 65: 333–371.

Common Cause Foundation (2011) *Common Cause Handbook.* Available at: https://commoncausefoundation.org/_resources/the-common-cause-handbook/ (accessed December 2021).

Common Cause Foundation (2016) Perceptions matter: The Common Cause UK values survey. Available at: https://commoncausefoundation.org/wp-content/uploads/2021/10/CCF_survey_perceptions_matter_full_report.pdf (accessed December 2021).

Common Cause Foundation (2017) Discover and share: Ways to promote positive values in arts and cultural settings. Available at: https://commoncausefoundation.org/wp-content/uploads/2021/10/CCF_report_discover_and_share_promoting_positive_values_arts_cultural.pdf (accessed December 2021).

Daulton-Smith, S. (2019) *Sacred Rest: Recover Your Life, Renew Your Energy, Restore Your Sanity.* Brentwood, TN: FaithWords.

Davis, N. (2021) Crisis dramatically worsened global mental health, *The Guardian.* Available at: https://www.theguardian.com/world/2021/oct/08/covid-crisis-dramatically-worsened-global-mental-health-study-finds (accessed December 2021).

De Bono, E. (2009) *Six Thinking Hats: Run Better Meetings, Make Faster Decisions.* London: Penguin Life.

De Smet, A., Gagnon, C. and Mygatt, E. (2021) Organizing for the future: Nine keys to becoming a future-ready company, McKinsey and Company. Available at: https://www.mckinsey.com/business-functions/people-and-organizational-performance/our-insights/organizing-for-the-future-nine-keys-to-becoming-a-future-ready-company (accessed 30 October 2021).

Deloitte (2014) Culture of purpose: Building business confidence; driving growth. Available at: https://www2.deloitte.com//us/en/pages/about-deloitte/articles/culture-of-purpose.html (accessed 29 October 2021).

Deloitte (2015) Mind the gaps: The 2015 Deloitte millennial survey executive summary. Available at: https://www2.deloitte.com/content/dam/Deloitte/global/Documents/About-Deloitte/gx-wef-2015-millennial-survey-executivesummary.pdf (accessed 29 October 2021).

Department of Economic and Social Affairs (2018) 68% of the world population projected to live in urban areas by 2050, says UN. Available at: https://www.un.org/development/desa/en/news/population/2018-revision-of-world-urbanization-prospects.html (accessed October 2021).

Dhingra, N., Samo, A., Schaninger, B. and Schrimper, M. (2021) Help your employees find purpose – or watch them leave, McKinsey. Available at: https://www.mckinsey.com/business-functions/people-and-organizational-performance/our-insights/help-your-employees-find-purpose-or-watch-them-leave (accessed December 2021).

Dickinson, E. (1975) *Emily Dickinson, Complete Poems*. London and Boston: Faber and Faber.

Discover John Muir website (2021) Quote from website landing page: 'When we try to pick out anything by itself, we find it hitched to everything else in the Universe' – John Muir (1838–1914), John Muir Trust. Available at: https://discoverjohnmuir.com/ (accessed December 2021).

Dweck, C. (2017) *Mindset: Changing the Way You Think to Fulfil Your Potential*. New York: Random House Inc.

Eaton, G. (2020) Noam Chomsky: The world is at the most dangerous moment in human history, *The New Statesman*. Available at: https://www.newstatesman.com/politics/2020/09/noam-chomsky-the-world-is-at-the-most-dangerous-moment-in-human-history (accessed November 2021).

Edmondson, A.C. (2019) *The Fearless Organization; Creating Psychological Safety in the Workplace for Learning, Innovation, and Growth*. Hoboken, NJ: John Wiley & Sons.

Einzig, H. (2017) *The Future of Coaching: Vision, Leadership and Responsibility in a Transforming World*. Oxford and New York: Routledge.

Evans, A. (2021) Let's make climate a culture war, Larger Us. Available at: https://larger.us/ideas/lets-make-climate-a-culture-war/ (accessed December 2021).

Ferriss, T. (2020) *The Four Hour Work Week*. London: Ebury.

Figueres, C. and Rivett-Carnac, T. (2020) *The Future We Choose*. London: Manilla Press.

Flynn, S. (2020) *The Flynn Model of Sustainable Change*, published in 'Resilience for your Sustainability Journey' webinar, Women in Sustainability Network. Online event, UK. Women in Sustainability Digital Community Hub.

Friedrich, C.D. (1818) *Wanderer Above the Mist, Der Wanderer über dem Nebelmeer*. [Oil on canvas]. Hamburger Kunsthalle, Hamburg.

Frieze, D. and Wheatley, M. (2011) *Walk Out, Walk On*. San Francisco: Berrett-Koehler Publishers Inc.

Fuller, T. and Yumiao, T. (2006) Small and medium-sized enterprises & corporate social responsibility: Identifying the knowledge gaps, *Journal of Business Ethics*, 67(3): 287–304.

Glavas, A. and Piderit, S.K. (2009) How does doing good matter? Effects of corporate citizenship on employees, *Journal of Corporate Citizenship*, 36 (Winter): 51–70.

Godwin, S. (2008) *Tribes*, 1st edn. London: Piaktus.

Goleman, D., Boyatzis, R. and McKee, A. (2006) *The New Leaders*. London: Sphere.

Gottman, J.M. and Silver, N. (1999) *The Seven Principles for Making Marriage Work*. New York: Three Rivers Press.

Gottman, J. and Silver, N. (2015) *The Seven Principles for Making Marriage Work.* New York: Harmony Books.

Hammond, C. (2019) *The Art of Rest: How to Find Respite in the Modern Age.* Edinburgh: Canon Gate.

Harris, R.(2008) *The Happiness Trap: Stop Struggling, Start Living.* Edinburgh: Robinson Publishing.

Harris, R. (2019) *ACT Made Simple,* 2nd edn. Oakland, CA: New Harbinger Publications.

Harris, R. (2021) ACT made simple: A beginner's workshop with Russ Harris, Contextual Consulting. Originally available at: https://contextualconsulting.co.uk/workshop/act-made-simple-a-beginners-workshop-with-russ-harris (accessed 4 February 2021 – no longer available online).

Heglar, M.A. (2019) Home is always worth it, Medium. Available at: https://medium.com/@mary-heglar/home-is-always-worth-it-d2821634dcd9 (accessed December 2021).

Holborn, V. (2020) *How to Be an Activist.* London. Robinson.

Holmes, T., Blackmore, E., Hawkins, R. and Wakeford, T. (2011) *Common Cause Handbook.* UK: Public Interest Research Centre.

Houston, E. (2021) The importance of positive relationships in the workplace, PositivePsychology.com. Available at: https://positivepsychology.com/positive-relationships-workplace/ (accessed December 2021).

Huston, T. (2021) *Let's Talk: Make Effective Feedback Your Superpower.* London: Penguin Random House

IBM Institute for Business Value (2021) Sustainability at a turning point. Available at: https://www.ibm.com/thought-leadership/institute-business-value/report/sustainability-consumer-research (accessed December 2021).

Jack, A., Boyatzis, R., Khawaja, M. et al. (2013) Visioning in the brain: An fMRI study of inspirational coaching and mentoring, *Journal of Social Neuroscience,* 8(4): 369–384.

Jacobs, K., Robey, J., Lago, C. et al. (2020) Consumer products and retail: How sustainability is fundamentally changing consumer preferences, CapGemini Research Institute. Available at: https://www.capgemini.com/gb-en/research/how-sustainability-is-fundamentally-changing-consumer-preferences/ (accessed December 2021)

Jay, J. and Grant, G. (2017) *Breaking through Gridlock: The Power of Conversation in a Polarized World.* Oakland, CA: Berrett-Koehler Publishers, Inc.

Jimeoinofficial (2012) Jimeoin – Edinburgh Comedy Fest 2010. Available at: https://www.youtube.com/watch?v=8pGHKghKNG4&t=54s (accessed October 2021).

Kahn, W.A. (1990) Psychological conditions of personal engagement and disengagement at work, *Academy of Management Journal,* 33(4): 692–724.

Kalmus, P. (2017) *Being the Change: Live Well and Spark a Climate Revolution.* Gabriola Island, Canada: New Society Publishers.

Kalmus, P. (2020) Climate anxiety, what is it and how can we cope, the big Q. Available at: https://www.thebigq.org/2020/09/03/climate-anxiety-what-is-it-and-how-can-we-cope/ (accessed December 2021).

Kaur, V. (2020) *See No Stranger: A Manifesto of Revolutionary Love.* New York: Penguin Random House LLC,

King, M.L. (1963) Letter from a Birmingham Jail, April 16, 1963, The Martin Luther King, Jr. Research and Education Institute. Available at: https://kinginstitute.stanford.edu/building-world-house (accessed December 2021)

Kornfield, J. (2000) *After the Ecstasy, The Laundry.* London: Rider.

Kotter, J. and Cohen, D.S. (2002) *The Heart of Change.* Boston, MA: Harvard Business Review Press.

Kübler-Ross, E. and Kessler, D. (2014) *On Grief and Grieving: Finding the Meaning of Grief Through the Five Stages of Loss.* New York: Scribner.

Kuhn, T.S. (1970). *The Structure of Scientific Revolutions.* Chicago: University of Chicago.

Kurzgesagt – In a Nutshell (2021) Can YOU fix climate change. Available at: https://www.youtube
.com/watch?v=yiw6_JakZFc (accessed December 2021).

Lahey, J. (2015) *The Gift of Failure: How to Step Back and Let Your Child Succeed.* New York:
HarperCollins.

Laloux, F. (2014) *Reinventing Organizations.* Brussels, Belgium: Belson Parker.

Louv, R. (2019) What is nature–deficit disorder? Available at: http://richardlouv.com/blog/
what-is-nature-deficit-disorder/ (accessed December 2021).

Lencioni, P. (2002) Make your values mean something, *Harvard Business Review.* Available at:
https://hbr.org/2002/07/make-your-values-mean-something (accessed December 2021).

Lewis, S. and Maslin, M.A. (2018) *The Human Planet: How We Created the Anthropocene.* London:
Pelican Books.

Loy, D.R. (2018) *Ecodharma: Buddhist Teachings for the Ecological Crisis.* Somerville, MA:
Wisdom Publications.

Li, Q., Morimoto, K., Nakadai, A. et al. (2007) Forest bathing enhances human natural killer activity and expression of anti-cancer proteins, *International Journal of Immunopathology and
Pharmacology,* 20(2): 3–8.

Mac Guill, A. (2017) 'I've been followed, attacked, spat on': women on feeling scared to walk
alone, *The Guardian.* Available at: https://www.theguardian.com/inequality/2017/oct/20/ive-
been-followed-attacked-spat-on-women-on-feeling-scared-to-walk-alone (accessed October
2021).

McArthur-Blair, J. and Cockell, J. (2018) *Building Resilience with Appreciative Inquiry: A
Leadership Journey Through Hope, Despair, and Forgiveness.* Oakland, CA: Berrett-Koehler.

Macy, J. (2007) *World as Lover, World as Self.* Berkeley, CA: Parrallax Press.

Macy, J. (2012) *Active Hope: How to Face the Mess We Are In Without Going Crazy.* Novato, CA:
New World Library.

Maharaj, S. (n.d.) Grief – an act of love, Therapy Route. Available at: https://www.therapyroute.
com/article/grief-an-act-of-love-by-s-maharaj (accessed December 2021).

Marchall, G. (2015) *Do not Even Think About It: Why Our Brains Are Wired to Ignore Climate
Change.* New York and London: Bloomsbury USA.

Marsh, H. (2014) *Do No Harm: Stories of Life, Death and Brain Surgery.* London: W&N.

May, K. (2020) *Wintering: The Power of Rest and Retreat in Difficult Times.* London: Rider.

Miller, L.M. (2013) Transformational change vs. continuous improvement, *Industry Week.* Available at: https://www.industryweek.com/leadership/change-management/article/21960254/
transformational-change-vs-continuous-improvement (accessed 29 October 2021).

Moeller, S. (1999) *Compassion Fatigue: How the Media Sell Disease, Famine, War and Death.*
London and New York: Routledge,

Nagoski, E. and Nagoski, A. (2019) *Burnout.* London: Vermillion.

Net Impact (2012) Talent report: What workers want in 2012. Available at: https://netimpact.org/
sites/default/files/documents/what-workers-want-2012.pdf (accessed 29 October 2021),

Nicholas, K. (2021) *Under the Sky We Make: How to Be Human in a Warming World.* New York:
Putnam.

NLM in Focus (2018) Dr Alan Kay talks about the future at annual Lindberg-King Lecture. Available at: https://infocus.nlm.nih.gov/2018/10/04/dr-alan-kay-talks-about-the-future-at-annual-
lindberg-king-lecture/ (accessed 30 October 2021).

O'Neil, S. and Day, S. (2009) Fear will not do it: Promoting positive engagement with climate
change through visual and iconic representation, *Journal of Science Communication,* 30(3):
355–379.

Pascarella, P. and Frohman, MA. (1989) *The Purpose-Driven Organization: Unleashing the
Power of Direction and Commitment.* San Francisco, CA: Jossey-Bass.

Peter, L.J. (2021) Laurence J. Peter, quotes, Good Reads. Available at: https://www.goodreads.
com/author/quotes/182617.Laurence_J_Peter (accessed December 2021).

Pink, D. (2013) *To Sell Is Human*. Edinburgh and London: Canongate.

Prochaska, J.O. and Velicer, W.F. (1997) The transtheoretical model of health behavior change, *American Journal of Health Promotion*, 12(1): 38–48.

Quinn, R.E. and Thakor, A.V. (2018) Creating a purpose-driven organization, how to get employees to bring their smarts and energy to work, *Harvard Business Review*. Available at: https://hbr.org/2018/07/creating-a-purpose-driven-organization (accessed December 2021).

Rachel Carson Center (2020) Climate change: A crisis of hope by Elin Kelsey. Available at: https://www.youtube.com/watch?v=5nOGVXvryrA&list=UUd9pm9iQsU8mna_hRpYJSuQ&index=25 (accessed December 2021).

Ramachandran, N. (2021) BBC, ITV reveal climate change goals ahead of COP26 conference, *Variety*. Available at: https://variety.com/2021/tv/global/bbc-itv-climate-change-goals-cop26-conference-1235096690/ (accessed December 2021).

Ramachandran, N. (2021) *Profit from Passion and Purpose*. Upper Saddle River: Wharton School Publishing.

Ray, S.J. (2021) Climate anxiety is an overwhelmingly white phenomenon, Scientific America. Available at: https://www.scientificamerican.com/article/the-unbearable-whiteness-of-climate-anxiety/ (accessed December 2021).

Rhodes, M. (2013) Yes, FOMO is now a word in the dictionary, Fast Company. Available at: https://www.fastcompany.com/3016488/yes-fomo-is-now-a-word-in-the-dictionary (accessed December 2021).

Rink, S.M., Mendola, P., Mumford, S.L. et al. (2013) Self-report of fruit and vegetable intake that meets the 5 a day recommendation is associated with reduced levels of oxidative stress biomarkers and increased levels of antioxidant defense in premenopausal women, *Journal of the Academy of Nutrition and Dietetics*, 113(6): 776–785.

Rogers, E. (2003) *Diffusion of Innovations*, 5th edn. New York: Simon and Schuster New York

Roos, S., Johnston, A. and Noble, D. (2012) UK hedgehog datasets and their potential for long-term monitoring, British Trust for Ornithology. Available at: https://www.bto.org/our-science/projects/gbw/publications/papers/monitoring/btorr598 (accessed December 2021).

Rosenburg, M. (2015) *Nonviolent Communication*. Encinitas, CA: PuddleDancer Press.

Sacks, O. (2020) *Everything in Its Place: First Loves and Last Tales*. London: Pan Macmillan,

Satir, V., Banmen, J., Gerber, J. and Gomori, M. (1991) *The Satir Model*. Palo Alto, CA: Science and Behavior Books.

Scharmer, C.O. (2009) *Theory U: Leading From the Future as It Emerges: The Social Technology of Presencing*. San Francisco, CA: Berrett-Koehler.

Schecter, T. and Gould, M. (2020) *Lead from Your Heart: The Art of Relationship Based Leadership*. n.p.: HTI Institute Inc.

Schwartz, S. H. (2012) An overview of the Schwartz Theory of Basic Values, Online Readings in Psychology and Culture, 2(1). DOI:10.9707/2307-0919.1116

Shields, F. (2019) Why we are rethinking the images we use for our climate journalism, *The Guardian*. Available at: https://www.theguardian.com/environment/2019/oct/18/guardian-climate-pledge-2019-images-pictures-guidelines (accessed December 2021).

Sigal, B. and O'Neill, O.A. (2016) Manage your emotional culture, *Harvard Business Review*. Available at: https://hbr.org/2016/01/manage-your-emotional-culture (accessed December 2021).

Simard, S. (2021) *Finding the Mother Tree: Discovering the Wisdom of the Forest*. London: Allen Lane.

Simard, S. (2021) Trees talk to each other. 'Mother Tree' ecologist hears lessons for people, too, NPR. Available at: https://www.npr.org/sections/health-shots/2021/05/04/993430007/trees-talk-to-each-[…]st-hears-lessons-for-people-too?t=1639127459019&t=1639486533863 (accessed December 2021).

Sisodia, R.S., Wolfe, D.B. and Sheth, J.N. (2007) *Firms of Endearment: How World-class Companies Profit from Passion and Purpose*. Upper Saddle River, PA: Wharton School Publishing.

Skinner, B.F. (1965) *Science and Human Behaviour*. New York: Simon and Schuster.

Solnit, R. (2021) Ten ways to confront the climate crisis without losing hope, *The Guardian*. Available at: https://www.theguardian.com/environment/2021/nov/18/ten-ways-confront-climate-crisis-without-losing-hope-rebecca-solnit-reconstruction-after-covid?CMP=Share_AndroidApp_Other (accessed November 2021).

Spence, Jr. R.M, and Rushing, H. (2009) *It's Not What you Sell, It's What you Stand For: Why Every Extraordinary Business Is Driven by Purpose*. New York: Portfolio/Penguin Group.

Staddon, J.E.R. and Cerutti, D.T. (2003) Operant conditioning, *Annual Review of Psychology, Annual Reviews*, 54: 115–144.

Stavros, J., Torres, C. and Cooperrider, D.L. (2018) *Conversations Worth Having: Using Appreciative Inquiry to Fuel Productive and Meaningful Engagement*. Oakland, CA: Berrett-Koehler Publishers, Inc.

Stott, P. (2021) *Hot Air: The Inside Story of the Battle Against Climate Change Denial*. London: Atlantic Books.

Strong, C. and Ansons, T. (2020) The science of behaviour change, Ipsos Mori. Available at: https://www.ipsos.com/ipsos-mori/en-uk/science-behaviour-change (accessed December 2021).

Syed, M. (2019) *Rebel Ideas*. London: John Murray.

Szabo, A. and Hopkinson, K.L. (2007) Negative psychological effects of watching the news on the television: Relaxation or another intervention may be needed to buffer them!, *International Journal of Behavioural Medicine*, 14(2): 57–62.

Tagliabue, M., Sigurjonsdottir, S.S. and Sandaker, I. (2020) The effects of performance feedback on organizational citizenship behaviour: A systematic review and meta-analysis. *European Journal of Work and Organizational Psychology*, 29(6): 841–861.

Taylor, A. (2020) *Soft Skills Hard Results*. Tadley: Practical Inspiration Publishing.

TED (2010) Teach every child about food. Available at: https://www.ted.com/talks/jamie_oliver_teach_every_child_about_food#t-295297 (accessed October 2021)

TEDGlobal>NYC (2017) Per Espen Stoknes: How to transform apocalypse fatigue into action on global warming. Available at: https://www.ted.com/talks/per_espen_stoknes_how_to_transform_apocalypse_fatigue_into_action_on_global_warming (accessed 11 December 2021).

TEDx Talks (2015) How I became a localist. Available at: https://www.youtube.com/watch?v=-2jTdZSPBRRE (accessed December 2021).

Ted Talk, TEDWomen 2018 (2018) Katharine Hayhoe the most important thing you can do to fight climate change: Talk about it. Available at: https://www.ted.com/talks/katharine_hayhoe_the_most_important_thing_you_can_do_to_fight_climate_change_talk_about_it (accessed November 2021).

Thoreau, H.D. (2017) *Walden*. London: Vintage.

The Star, Newsroom (2017) Women have been warned not to walk alone following a terrifying assault by a masked attacker in the South Yorkshire countryside, *The Star*. Available at: https://www.thestar.co.uk/news/women-warned-over-attacker-loose-south-yorkshire-countryside-449344 (accessed October 2021).

Maister, D.H., Galford, R. and Green, C. (2002) *The Trusted Advisor*. London: Simon & Schuster UK.

This American Life (2010) Nummi, episode 403. 26 March. Available at: https://www.thisamericanlife.org/radio-archives/episode/403/nummi (accessed December 2021).

Thomas, K.W. and Kilmann, R.H. (2009–2021) Thomas-Kilmann model adapted from the TKI Model that is Copyright © 2009–2021, Kilmann Diagnostics LLC. Original figure is available at: https://kilmanndiagnostics.com/overview-thomas-kilmann-conflict-mode-instrument-tki/ (accessed December 2021)

Tippett, K and On Being Podcast (2020) Ocean Vuong: A life worthy of our breath. Available at: https://onbeing.org/programs/ocean-vuong-a-life-worthy-of-our-breath/ (accessed December 2021).

Toleikyte, G. (2021) *Why The F\*ck Can't I Change? Insights from a Neuroscientist to Show You that You Can*. London: Sphere.

Tolstoy, L. (1900) Three methods of reform, in L. Tolstoy, *Pamphlets: Translated from the Russian* (trans. Aylmer Maude). London: Free Age Press.

Turner, T. (2017) *Belonging: Remembering Ourselves Home*. Salt Spring Island, BC: Her Own Room Press.

University of Exeter (2021) GSI Seminar Series: Dr Kimberly Nicholas – *Under the Sky We Make* book discussion. Available at: https://www.youtube.com/watch?v=3VcfPA6eChI (accessed December 2021).

Wang, S., Corner, A. and Nicholls, J. (2020). *Britain Talks Climate: A Toolkit for Engaging the British Public on Climate Change*. Oxford: Climate Outreach.

West, L. (2015) *Coaching With Values*. Bloomington, IN: AuthorHouse.

Wheatley, M. (2009) *Turning to One Another*. Oakland, CA: Berrett Koehler Publishers Inc.

Wheatley, M. (2017) Islands of Sanity, Meaning Conference. Available at: https://meaningconference.co.uk/videos/margaret-wheatley (accessed October 2021).

Wheatley, M. and Frieze, D. (2006) Using emergence to take social innovations to scale. Margaretwheatley.com. Available at: https://margaretwheatley.com/articles/emergence.html (accessed December 2021).

Williamson, M. (2009) *A Return to Love*. San Francisco, CA: HarperOne.

Wilson, W. (1918) Speech to the Red Cross, New York, 18 May 1918. Available at: https://www.presidency.ucsb.edu/documents/address-opening-the-campaign-for-the-second-red-cross-fund-new-york-city (accessed December 2021).

Wittington, J. (2020) *Systemic Coaching and Constellations: The Principles, Practices and Application for Individuals, Teams and Groups*. London: Kogan Page.

Yona, L. (2018) [Twitter] 8 October. Available at: https://twitter.com/LeehiYona/status/1049394285047558144 (accessed December 2021).

# Index